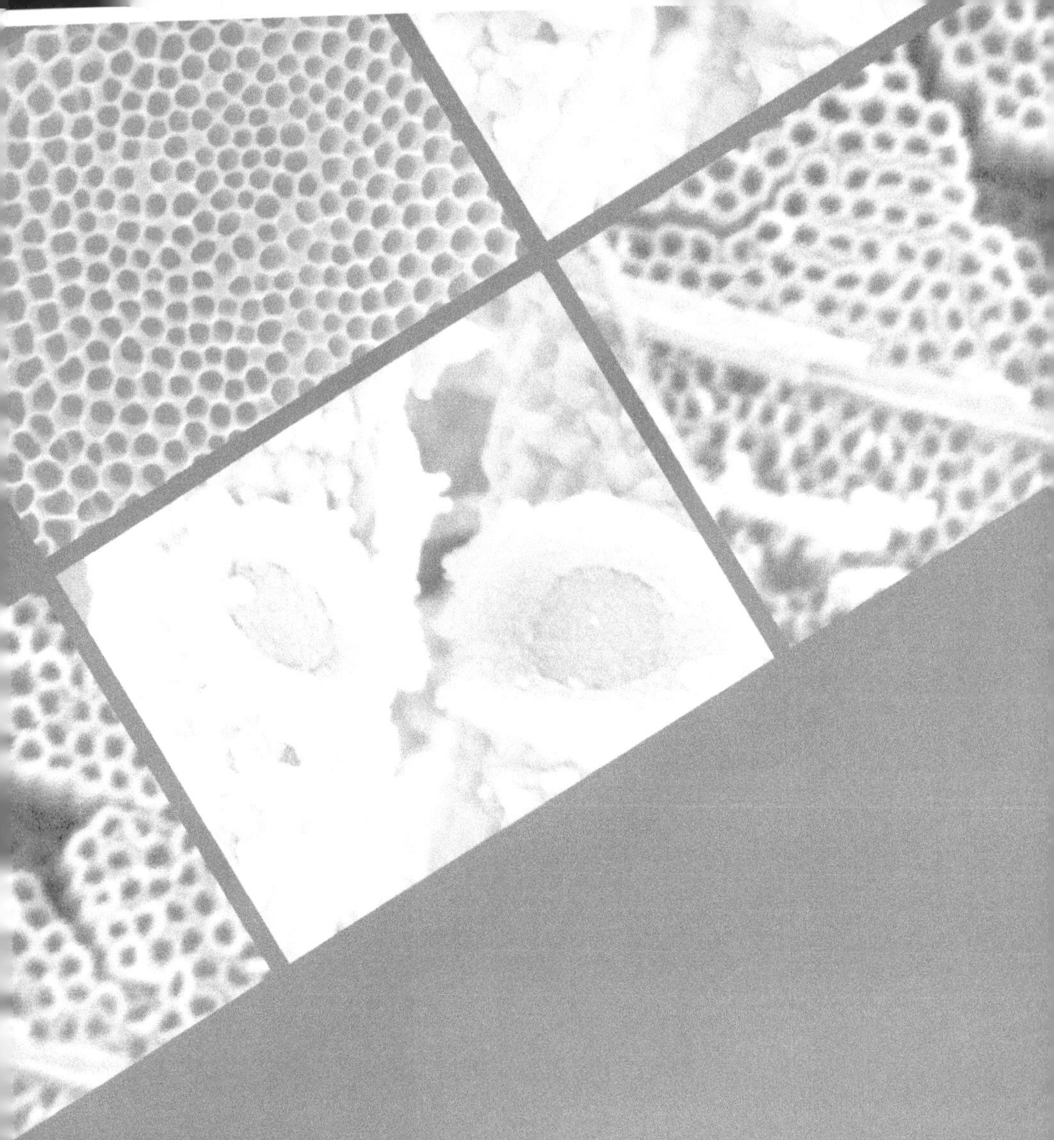

Nanostructured Titanium Dioxide in Photocatalysis

Nanostructured Titanium Dioxide in Photocatalysis

It-Meng Low
Hani Manssor Albetran
Victor Manuel de la Prida Pidal
Fong Kwong Yam

JENNY STANFORD
PUBLISHING

Published by

Jenny Stanford Publishing Pte. Ltd.
Level 34, Centennial Tower
3 Temasek Avenue
Singapore 039190

Email: editorial@jennystanford.com
Web: www.jennystanford.com

British Library Cataloguing-in-Publication Data
A catalogue record for this book is available from the British Library.

Nanostructured Titanium Dioxide in Photocatalysis

ISBN 978-981-4877-07-7 (Hardcover)
ISBN 978-1-003-14853-1 (eBook)

Contents

Preface

TiO_2 has attracted considerable attention as a promising inorganic material. It is an attractive raw material for various applications, such as water purification, gas sensing, storage and delivery, and energy generation. It is also a versatile material with distinct advantages such as low cost, nontoxic nature, stability, and excellent photocatalytic activity. Photocatalysis is one of the most promising fields for sustainable chemistry in the twenty-first century. It can contribute to solve environmental, energetic, and chemical problems, and to sustainably produce commodities in the near future. Nanotechnology has offered the possibility to carefully tune the properties of the semiconducting TiO_2 materials used in photocatalysis. This has resulted in a tremendous improvement in the efficiency of photocatalytic processes in many sectors. TiO_2-based materials today represent some of the major photocatalysts, and their reactivity is strictly related to their nanostructure, physical and chemical characteristics, as well as their phase relations.

Hitherto, most of the published articles on nanostructured TiO_2 have appeared in journals and the dissemination of the published work is quite incoherent. There is an impending need to present the work on these materials in a coherent manner. This book presents a comprehensive review of the latest advances in the synthesis, characterization and applications of nanostructured TiO_2 in several fields and discusses various parameters controlling its synthesis and crystallization.

It-Meng Low
Hani Manssor Albetran
Victor Manuel de la Prida Pidal
Fong Kwong Yam
November 2020

Part I

Introduction and Background

Chapter 1

Introduction and Literature Review

1.1 Introduction

Although several decades are yet to pass before fossil fuels deplete on Earth, climate has changed quickly since the Industrial Revolution a century ago. Contaminants, chemical pollutants, dust, gusty winds, and sandstorms are harmful to human beings and for the environment. In air, these harmful substances can exist in the form of liquid droplets, particulate or gas, whereas, in water, they are present as dissolved substances. The level of air pollution inside houses is often two to five times higher than the outdoor level [1]. Indoor air pollution has been confirmed by the Environmental Protection Agency because it is one of the top human health risks. Household products, including paints, wood preservation, aerosol sprays, disinfectants and cleaners, automotive products, and dry-cleaned clothes are some major or potential sources of indoor air pollution. Moreover, the rapid depletion of both fossil fuels and safe drinking water could result in severe political and economic conflicts among countries with rapidly growing population and increasing environmental pollution [1].

Green energy and other sustainable alternative sources have become a major issue. Solar energy is estimated to play a crucial role in the world's energy consumption in the future. It can be used for solving global warming or climate change–induced problems

Nanostructured Titanium Dioxide in Photocatalysis
It-Meng Low, Hani Manssor Albetran, Victor Manuel de la Prida Pidal, and Fong Kwong Yam

ISBN 978-981-4877-07-7 (Hardcover), 978-1-003-14853-1 (eBook)
www.jennystanford.com

since solar energy sources can generate electricity without emitting carbon dioxide and other greenhouse gases. The first practical solar cell was fabricated in 1954 using silicon, and then a number of efficient solar cells with varied materials and structures have been developed for high power conversion efficiency and low cost [2].

Titanium dioxide (TiO_2) has attracted considerable attention as a promising inorganic material and an attractive raw material for various applications such as water purification, gas sensing, storage and delivery, and energy generation, due to its versatile advantages such as inexpensiveness, nontoxic nature, stability, and excellent photocatalytic activity [3–12].

TiO_2 offers two important polymorphs in nature: anatase and rutile, which are considered mainly in this thesis. Anatase and rutile exhibit different properties and consequently different photocatalytic performances. Anatase is a metastable phase, which transforms to the stable rutile phase at an elevated temperature.

In recent years, research on heterogeneous photocatalysis has accelerated with the rapid emergence of new materials and devices, suggesting promising performance characteristics. To achieve further in a safe photocatalytic reaction, it is necessary to choose a suitable photocatalyst, fabricate a device with a proper nanostructure, and understand the physics and chemistry behind the photocatalyst. Thus, this study focuses on both synthesis and characterization of nanostructured TiO_2 for photocatalytic and other related applications.

The importance of titanium oxide has increased because it can lead to fascinating applications in the areas of environment and alternative energy sources. It is an attractive material due to its wide-ranging applications, for example, in ceramic interference coatings, paints, sunscreens, optical devices, gas sensing, solar cells, and biomedical products [3–5, 9, 11, 13, 14]. With the advent of nanotechnology, nanostructured semiconducting TiO_2 has also attracted considerable attention, particularly due to its high potential for being applied in diverse areas, including photocatalysis of pollutants, photo-splitting of water, and as transparent conducting electrodes for dye-sensitized solar cells. In particular, elongated TiO_2 nanostructures (nanofibers or nanotubes) and nanopowders have generated much interest because of their high photoactivity and photodurability [15, 16]. In addition, electrospun nanofibers have a wide range of applications, including water filtration, because of the ease and simplicity of the fabrication process together with the high porosity and high surface area of these fibers. Electrospun TiO_2 nanofibers rapidly and drastically degrade various contaminants in water and wastewater to remove organic pollutants [17–20].

This book focuses on both synthesis and characterization of nanostructured TiO_2 mainly for photocatalytic application, as well as other applications in general. TiO_2 nanofibers, nanopowders, and nanotubes were synthesized using electrospinning, sol–gel, and anodization methods, respectively.

The electrospun TiO_2 nanofiber, sol–gel TiO_2 nanopowders, and anodized TiO_2 nanotubes were synthesized and characterized with different analytical techniques, including scanning electron microscopy (SEM), energy dispersive X-ray spectroscopy (EDS), transmission electron microscopy (TEM), X-ray diffraction (XRD), in situ high-temperature synchrotron radiation diffraction (SRD), differential thermal analysis (DTA), thermogravimetric analysis (TGA), differential scanning calorimetry (DSC), X-ray photoelectron spectroscopy (XPS), and UV/visible spectrophotometer.

1.2 Literature Review

Titanium dioxide or titanium (IV) oxide, also known as titania, is the naturally occurring oxide of titanium, with the chemical formula TiO_2. It is a white pigment that can provide a basic white background color, capable of making paints, inks, plastics, paper, ceramics, candy coating, glazed fruit, toothpaste, sunscreen, and cosmetic products impenetrable by light as well as reflecting light across most of the visible spectrum. It is widely used in the food industry as a colorant and classified as E171, which has long been used in many different industrial fields as a white pigment in foodstuff. For instance, it can be used in Mozzarella cheese to give the whole appearance and coloring of skim milk and flour [4, 5, 9, 21].

Following the discovery of photocatalytic water splitting using TiO_2 in the late 1970s by Fujishima and Honda [22], a new era of TiO_2-based materials has begun. TiO_2 has emerged as a novel material of choice for a variety of applications, because it is inexpensive and both chemically and biologically stable. TiO_2 has drawn more and more attention from researchers in the past few decades, owing to its wide range of potential environmentally friendly applications. Since the synthesis of nano TiO_2, the amount of published material has been growing exponentially every year. Figure 1.1 shows that the number of published papers increased from 27 papers in 2004, and these numbers are likely to grow even faster in the coming days.

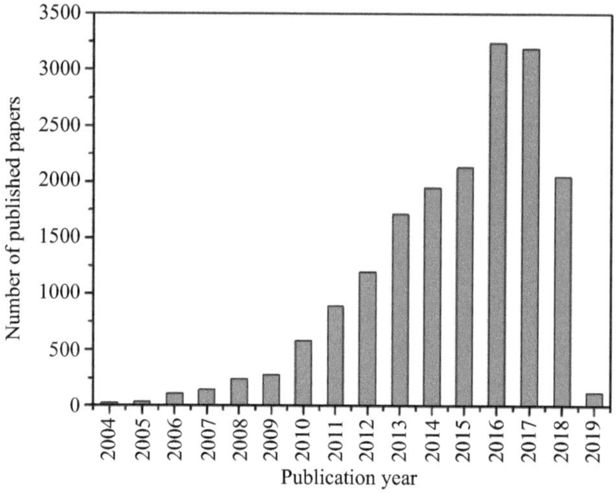

Figure 1.1 The number of published papers related to nano TiO_2 for 2004–19. (Data were collected from ProQuest Materials Science and Engineering Collection using "Nano TiO_2" as keywords.)

The introduction review section provides a brief review of titania fabrication, assembly, and application that focuses on the most common TiO_2 phases, namely anatase and rutile. Anatase and rutile are found in nature and particularly play a significant role in industrial applications [23, 24]. They are popular industrial or engineering materials with promising prospects for everyday human life because of their wide existing and potential applications along with their low-cost advantage [6, 25]. Figure 1.2 shows the average price of anatase to be ~\$3.6 per kg, whereas the average price of rutile is ~\$3.8 per kg, which is slightly higher for rutile than the price of anatase. This difference in the prices of the two suggests that TiO_2 phases could be related to the type of their applications. The anatase phase is utilized in the industrial sector as an efficient photocatalyst, whereas the rutile phase is used for photovoltage application.

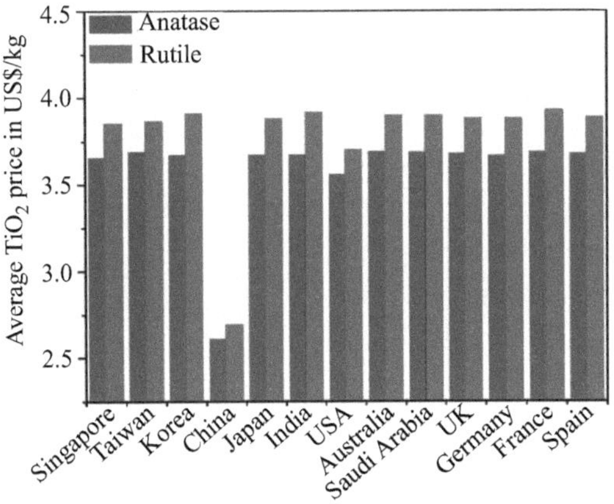

Figure 1.2 The prices of anatase and rutile in different countries. Data were taken from Petrosil (www.petrosil.com).

1.2.1 Crystal Structure of TiO_2

TiO_2 occurs in a variety of crystal structures and exhibits diverse electronic and optical properties. It has four polymorphs in nature, namely anatase, rutile, brookite, and bronze [TiO_2 (B)] [26]. Moreover, columbite (TiO_2 II) and hollandite [TiO_2 (H)] are two additional high-pressure rutile polymorphs, which have been synthesized in the laboratory [27, 28]. Baddeleyite (TiO_2 III) and ramsdellite [TiO_2 (R)] or any other TiO_2 modifications have been synthesized in the laboratory too [29, 30]. The composition is the same for TiO_2 phases, but each titania phase represents a different arrangement of the atoms, making up a different crystal structure.

The application of TiO_2 is usually based on one or a mixture of TiO_2 crystalline phases. Since then, the study of TiO_2 crystalline phase has also become important for TiO_2 applications. The three common polymorphs of titanium dioxide, namely anatase (tetragonal), rutile (tetragonal), and brookite (orthorhombic), are considered in this section; even bronze (monoclinic) is rare because it is mostly synthesized by high-pressure treatment and can be readily synthesized in the laboratory.

In 1916, a structural investigation of anatase (tetragonal) and rutile (tetragonal) was originally conducted by Vegard applying Brag's ionization method [31], whereas the brookite crystal structure (orthorhombic) was first determined by Pauling and Sturdivant using powder X-ray diffraction in 1928 [32]. In 1955, the acceptable crystal structures of anatase and rutile were revised by Cromer and Herrington using powder X-ray diffraction as well [33].

In the common titania forms, titanium (Ti^{4+}) atoms are coordinated to six oxygen (O^{2-}) atoms, forming TiO_6 octahedra. Crystalline anatase and rutile are tetragonal structures with which the octahedra share edges at (001) and (001) planes, respectively. On the other hand, both edges and corners of brookite are shared to give an orthorhombic structure. Anatase and rutile appear more often than brookite in lab-synthesized materials and in many other device applications involving titania [34].

1.2.1.1 Anatase

Anatase is one of the most widely studied TiO_2 polymorphs. It is widely employed as a photocatalyst because it is the most photoactive among the three common TiO_2 polymorphs [35]. It has the highest density of localized state and consequent surface-adsorbed hydroxyl radicals and the slowest charge carrier recombination compared to rutile and brookite. This is due to the differences in crystal structures and associated exposed planes among the titania phases, where anatase has a lower surface enthalpy and lower surface free energy than rutile [36].

Anatase and rutile have a tetragonal structure as shown in Figs. 1.3 and 1.4, respectively, but the distortion of the TiO_6 octahedron is slightly larger in anatase than in rutile; and they have different space groups and atoms per unit cell (Z) [37–39]. Table 1.1 shows the structural parameters, as well as physical and mechanical properties of anatase. The lattice anatase parameters are $\boldsymbol{a} = \boldsymbol{b} = 0.3784$ nm, and $\boldsymbol{c} = 0.9515$ nm at room temperature. Anatase has higher unit cell symmetry when compared with brookite, since there are only five different bond lengths in its representative octahedron. There are four Ti–O distances of 0.1937(3) nm and two Ti–O distances of 0.1964(9) nm.

Table 1.1 Structural parameters, together with physical and mechanical properties of anatase

Property		Value	Reference
Crystal system		Tetragonal	
Space group		$I4_1/amd$	
Atoms per cell (Z)		4	
MD basic cell		32 ($4\boldsymbol{a}\times4\boldsymbol{a}\times2\boldsymbol{c}$)	
Lattice parameters (25°C) (nm)	***a***	0.3784(5)	
	c	0.9515(5)	
Volume of cell (25°C) (nm^3)		0.1363	
Density (25°C) (kg/m^3)		3894	
O–Ti–O bond angle (°)		77.7 92.6	
Mean interatomic distances (nm)	Ti–Ti	0.304	
	Ti–O	0.195	
Thermal expansion coefficients (TECs) $\times10^{-5}$ ($°C^{-1}$) at 900°C	Linear (α_a)	0.53	
	Linear (α_c)	1.04	[33, 38–47]
	Volumetric (β)	2.17	
Molecular weight (g/mol)		79.88	
Molar volume ($\times10^{-6}$ m^3/mol)		20.51	
Optical band gap (eV)		3.20(5)	
Light absorption (nm)		~388	
Refractive index		2.52(3)	
Mohs hardness		5.5–6.0	
Resistivity (25°C) (Ωcm)		10^5	
Conductivity (25°C) (1/Ωcm)		10^4–10^7	
Bulk modulus, K (TPa)		0.21	
Melting point (°C)		Transformation to rutile	
Solubility in water		Insoluble	

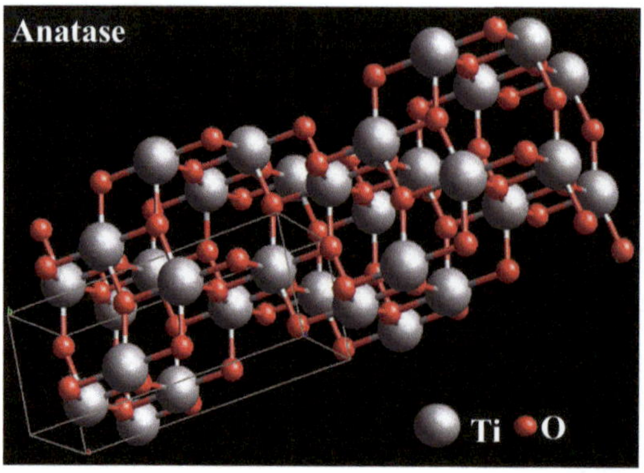

Figure 1.3 The basic unit cell crystal structure of anatase (ICSD 202242).

1.2.1.2 Rutile

Rutile has been studied extensively because it has the simplest TiO_2 structure and the most common and stable TiO_2 form. The tetragonal rutile structure is defined by three crystallographic parameters: two lattice parameters ***a***, ***c*** and the oxygen fractional coordinate. Anatase and rutile have the same tetragonal crystal structure, but they have a different space group and atoms per cell (Z), *P*42/mnm sharing two edges for rutile, and, *I*41/amd sharing four for anatase (see Tables 1.1 and 1.2). Figure 1.4 shows the basic unit cell crystal structure of rutile. The cell parameters are ***a*** = 0.45936(5) nm and ***c*** = 0.29587(3) nm, and the oxygen parameter is 0.03056(6) nm at room temperature. There are four Ti–O distances of 0.1946(3) nm and two Ti–O distances of 0.1984(4) nm (see Table 1.2).

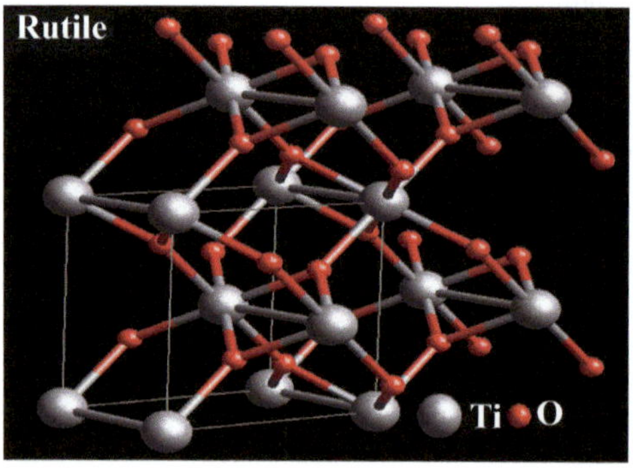

Figure 1.4 The basic unit cell crystal structure of rutile (ICSD 64987).

Table 1.2 Structural parameters, together with physical and mechanical properties of rutile

Property	Value		Reference
Crystal system	Tetragonal		
Space group	P42/mnm		
Atoms per cell (Z)	2		
MD basic cell	96 (4***a***×4***a***×6***c***)		
Lattice parameters (25°C) (nm)	***a***	0.45936(5)	
	c	0.29587(3)	
Volume of cell (25°C) (nm^3)	0.0624		
Density (25°C) (kg/m^3)	4130		
O–Ti–O bond angle (°)	81.2 90.0		
Mean interatomic distances (nm)	Ti–Ti	0.296	
	Ti–O	0.196	
Thermal expansion coefficients (TECs) $\times 10^{-5}$ (°C^{-1}) at 900°C	Linear (α_a)	0.87	
	Linear (α_c)	1.01	[33, 38, 40–50]
	Volumetric (β)	2.83	
Molecular weight (g/mol)	79.9		
Molar volume ($\times 10^{-6}$ m^3/mol)	18.80		
Optical band gap (eV)	3.00(5)		
Light absorption (nm)	~413		
Refractive index	2.9 or 2.75		
Mohs hardness	7.0–7.5		
Resistivity (25°C) (Ωcm)	10^{12}		
Conductivity (25°C) (1/Ωcm)	10^4–10^7		
Bulk modulus, K (TPa)	0.23		
Melting point (°C)	1825		
Boiling point (°C)	2500–3000		
Solubility in water	Insoluble		

The rutile phase is most stable at elevated temperature and pressure up to 50 kbar. Both anatase and brookite structures transform to the rutile phase with a certain condition and parameter (see Section 1.2.3, Kinetics of TiO_2 Phase Transformation).

Among the three common polymorphs of TiO_2, the highest symmetric structure is possessed by rutile. It contains six atoms per unit cell with the TiO_6 octahedron slightly distorted. The rutile crystal structure is defined as sharing of two edges and two vertices of the octahedron, which leads to a higher symmetry of the crystal structure.

1.2.1.3 Brookite

Brookite belongs to the orthorhombic crystal structure (see Fig. 1.5). The titanium atom is located near the center, with oxygen atoms at the vertices. The brookite unit cell is constituted by octahedra, and the crystal structure is formed when the octahedra share three edges. The Ti–O and O–O bond lengths are different, leading to a distorted octahedron. The brookite cell volume is larger than either anatase or rutile cell volume, with eight atoms per cell (Z), compared with four for anatase and two for rutile. The brookite cell volume is 0.32172 nm^3, whereas it is 0.1363 nm^3 and 0.0624 nm^3 for anatase and rutile, respectively (see Table 1.3).

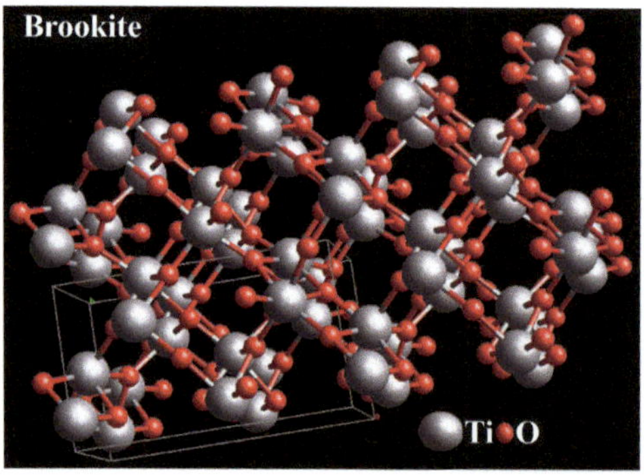

Figure 1.5 The basic unit cell crystal structure of brookite (ICSD 015409).

Although many reports showed interest in the brookite structure, only a few of them presented the synthesis of pure brookite [40]. The metastable anatase may transform into brookite thermodynamically under standard condition, and then the metastable brookite transforms to the stable rutile upon calcination at an elevated temperature, but this is rare due to the fact that it is difficult for the brookite crystal structure to form (see Section 1.2.3, Kinetics of TiO_2 Phase Transformation).

Table 1.3 Structural parameters, together with physical and mechanical properties of brookite

Property		Value	Reference
Crystal system		Orthorhombic	
Space group		P b c a	
Atoms per cell (Z)		8	
MD basic cell		18 (2***a***×3***b***×3***c***)	
Lattice parameters (25°C) (nm)	*a*	0.9174(5)	
	b	0.5439(5)	
	c	0.5163(5)	
Volume of cell (25°C) (nm^3)		0.32172	
Density (25°C) (kg/m^3)		4250	
O–Ti–O bond angle (°)		77.0~105	
Mean interatomic distances (nm)	Ti–Ti	0.302	
	Ti–O	0.196	
Thermal expansion coefficients (TECs) $\times 10^{-5}$ (°C^{-1}) at 900°C	Linear (α_a)	0.68	[38, 40, 42–44, 47, 49, 51]
	Linear (α_b)	0.69	
	Linear (α_c)	1.05	
	Volumetric (β)	2.40	
Molecular weight (g/mol)		79.866	
Molar volume ($\times 10^{-6}$ m^3/mol)		19.35	
Optical band gap (eV)		3.35(5)	
Light absorption (nm)		~370	
Refractive index		2.63	
Mohs hardness		5.5–6	
Resistivity (25°C) (Ωcm)		10^5	
Bulk modulus K (GPa)		0.22	
Melting point (°C)		Transformation to rutile	
Solubility in water		Insoluble	

1.2.2 TiO_2 Band Gap, Doping, and Modifying

Figure 1.6 presents the schematic illustration of the energy band gap. An electron (e^-) from the valence band (VB) is excited to the conduction band (CB) with the formation of a hole (h^+) by a light radiation of energy equal to or greater than the TiO_2 band gap.

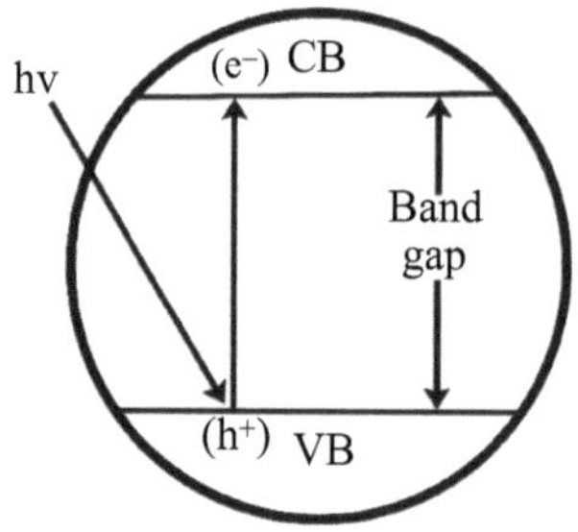

Figure 1.6 Schematic illustration of energy bands.

The conductance band of the semiconductor must be more negative than the reduction potential of the chemical species, whereas the valence band of the semiconductor must be more positive than the oxidation potential of the chemical species [52]. Among these semiconductors, titania phases have received the largest attention as a photocatalytic material because of titania's superior photocatalytic activity, chemical stability, low cost, and nontoxicity. Titania (TiO_2) is an n-type semiconductor because of oxygen deficiency. TiO_2 valence band is O_{2p} in one character, while the CB is Ti_{3d} and Ti_{4s} in the other character. It is a large band gap semiconductor, and the optical band gap of amorphous TiO_2 was found to be in the range of 3.30–3.50 eV [53]. This value decreased slightly for single-crystalline anatase (3.20 eV) and crystalline rutile (3.00 eV) [54].

The rutile band gap is slightly lower than the anatase band gap at absorbing the solar spectrum, but both can be excited only under ultraviolet irradiation, which occupies only 3–5% of the total solar irradiation at the Earth's surface. For anatase, a wavelength (λ) shorter than 387 nm is required to excite an electron to the CB (e_{CB-}) in order to generate a positive hole in the valence band (h_{VB+}) according to:

$$TiO_2 + h\nu \rightarrow h_{VB+} + e_{CB-} \quad (1.1)$$

Controlling the electrical conductivity of TiO_2 is an important step toward the application of electronics and photoelectronics. If the TiO_2 semiconductor with any forms is irradiated with photons of energy more than their band gap, the electron is promoted from the valence band to the CB (see Fig. 1.6).

Many of the TiO_2 uses are closely related and dependent on its optical properties, and one drawback appears because of its wide band gap. Several approaches for TiO_2 modification have been proposed to realize an efficient TiO_2 by changing the absorption edge from UV to visible region, including metal loading, metal-ion doping, anion doping, metal-ion implantation, organic dyes, polymer attachment, and hydrogen plasma reduction [55].

Coating the TiO_2 surface with a thin layer of dye is one strategy for TiO_2 modification and for reducing the band gap. This method is called dye sensitization and has been successfully applied to solar cell devices. Photodegradation of organic pollutants in water has been achieved on the TiO_2 surface modified with thionine and eosin Y dyes using visible light. The dyes (sensitizer) adsorbed on the titania surface are excited by absorbing visible light and affect charge injection into the CB at sub-band gap excitation and the catalytic processes follow through an interfacial electron transfer [56].

Doping of TiO_2 with metal (Cr, Cu, Ni, Mn, Mo, Nb, Ru, Au, Ag, Pt, Co, V, and Fe) or nonmetal elements (N, B, P, I, F, C, and S) is a common method and currently attracting considerable interest as a promising route to extend the absorption of TiO_2 from UV light to visible light range [24, 57–69].

These five electronic mechanisms can explain different optical absorptions of metal- and nonmetal-doped TiO_2 [70]: (a) narrowing of band gap by the newly formed VB by mixing O_{2p} of TiO_2 with doping states; (b) and (c) doped TiO_2 with a localized and new dopant level near the VB and the CB, respectively; (d) formation of color centers in the band gap as the origin of the visible light absorption band; and (e) sensitization mechanisms in which melamine condensation products as visible light sensitizer.

The next section reviews methods of reducing TiO_2 band gap with metallic and nonmetallic species, including ion implantation, sol–gel, and heterostructure methods.

1.2.2.1 Ion-implantation method

In 2000, Anpo applied the ion-implantation method to improve the electronic properties of TiO_2 and present the second titania generation photoactivity under visible light [57]. In the ion-implantation method, metal ions, including V^+, Cr^+, Fe^+, V^+, Mn^+, among others, are accelerated in the electronic field and injected into the deep bulk of TiO_2 sample target as the ion beam. The interaction of implanted ions with the sample surface depends on the kinetic energy. High kinetic energies from 150 to 200 keV are accelerated enough to implant metal ions (Fe^+, Mn^+, V^+, etc.) into TiO_2 bulk samples by ion implantation. Depending on the type and amount of implanted metal ions, the UV-VIS absorption spectra of TiO_2 photocatalysts were found to shift toward visible light regions [63]. The UV light to visible light (UV-Vis) absorption spectra of some metal-ion-implanted TiO_2 photocatalysts such as V, Cr, Mn, Co, Ni, and Cu were shifted to near the visible light regions [67, 71]. However, ion implantation with Ti, Ar, or Na was not effective at all to modify the TiO_2 properties to make a shift in the absorption band to the longer wavelength region [63].

1.2.2.2 Sol–gel doping and other methods

Doping with metal and nonmetal using the sol–gel method is a powerful strategy for modifying the electronic structure of TiO_2 for two reasons. First, depending on the type and amount of metal used, TiO_2 can be activated at a range of wavelengths from ultraviolet to visible light. Second, the sol–gel process is one of the most versatile methods to prepare titania nanopowder materials since it has a high ratio of surface area to volume, which increases the decomposition rate of water and air pollutants and allows for photocatalytic reactions to occur more rapidly on the titania photocatalyst surface. It has been found that some doped elements and types play a significant role in the TiO_2 band gap.

Nonmetal doping was investigated more intensively than nitrogen-doped TiO_2 films by Asahi et al. in 2001 [58]. The nitrogen-doped TiO_2, which narrowed the TiO_2 band gap, contributed to the visible light absorption, because the N_{2p} state in nitrogen forms hybrids with the O_{2p} state in the anatase TiO_2, in which their energies are very close.

Following N-doped TiO_2, a series of nonmetal-doped TiO_2 were successfully synthesized, and they were photoactive under visible light such as iodine, flour, phosphorus, boron, sulfur, and carbon [59–62, 64, 65].

Doping TiO_2 with anion elements such as nitrogen, carbon, sulfur can reduce the TiO_2 band gap and exhibit visible photoresponse, as the atomic orbitals of the nonmetal elements (N_{2p}, C_{2p}, and S_{3p}, respectively) have higher potential energy than titania O_{2p}. This will form a new valence band instead of a pure O_{2p} atomic orbital that minimizes the band-gap energy without changing the conduction band level.

Among nonmetal elements used for doping TiO_2, carbon did not receive high attention like nitrogen, even though C-doped TiO_2 powder was determined to be five times more active than N-doped TiO_2 [62]. Another study reports preparation of C-doped TiO_2 nanotube arrays with high aspect ratios for maximizing the photo-cleavage of water under white-light irradiation. The synthesized C-doped titania nanotube arrays showed much higher photocurrent densities and more efficient water splitting under visible light illumination (>420 nm) than pure titania nanotube arrays [72]. Additionally, a photocatalytic performance of a composite of titania nanotubes and carbon nanotubes (TNT-CNT) was used to study benzene degradation in the gas phase as a testing reaction. The results demonstrated that a certain amount of CNT doping into the matrix of TNT slightly affects the conversion ratio of benzene degradation [10]. Moreover, N-doped TiO_2, C-doped TiO_2, and (C, N) co-doped TiO_2 with different nitrogen and carbon contents were used to reduce the TiO_2 band gap and utilize visible light more efficiently in photocatalytic reactions [66]. The nitrogen atoms weaved into the titania lattice and induced new band states in the TiO_2 band gap, through replacing the sites of oxygen atoms, whereas the carbon atoms could form a layer composed of a complex carbonaceous mixture on the TiO_2 surface. It was found that co-doped TiO_2 exhibited the highest photocatalytic activity under visible light than C-doped or N-doped TiO_2 due to the assigned synergistic effect of doped C and N atoms.

Different metal elements such as chromium, indium, and vanadium were used to prepare doped titania, and the optical absorption edge could be shifted to the visible range of solar irradiance. Chromium-doped TiO_2 was synthesized using various morphologies and methods. Mishra et al. [73] synthesized low-content Cr-doped TiO_2 nanotubes by electrochemical anodization methods of Ti–Cr alloys in ethylene glycol electrolyte. Cr^{3+}-doped TiO_2 with crystalline anatase was prepared by a combination of the sol–gel process with hydrothermal treatment [74]. Cr^{3+}-doped anatase TiO_2 photocatalytic activity was investigated under UV and visible light irradiation. The investigation showed good ability for absorbing visible light because of the excitation of the 3d electron of Cr^{3+} to the TiO_2 CB. Doping of chromium ions effectively improved the photocatalytic activity under visible light irradiation with optimal doping concentrations of 0.15% and 0.2%, respectively. Moreover, a visible-light-active mesoporous Cr-doped TiO_2 photocatalyst was synthesized by applying an evaporation-induced self-assembly approach [75]. The effect of Cr^{3+} doping concentration on the photocatalytic activity of mesoporous TiO_2 was investigated from 0.1 to 1 mol percent. The Cr-doped TiO_2 extended the photo-absorption edge into the visible light region and exhibited higher photocatalytic activities when compared with pure mesoporous TiO_2 under visible light irradiation [76].

Doped TiO_2 nanofibers with 1.0 and 5.0 wt.% V were also successfully incorporated into the crystal lattice of TiO_2 nanofibers and could extend the visible light absorption of TiO_2 nanofibers. The obtained V-doped TiO_2 nanofibers possessed high activity for the photodegradation of organic pollutant methylene blue (MB) [68].

1.2.2.3 Mixed titania phases: Heterojunction (heterostructure)

Mixed-phase titania, which includes anatase/rutile, anatase/brookite, rutile/brookite, and anatase/TiO_2 (B), has been recently developed as a strategy to circumvent the wide band gap of titania [77–80].

The common mixed-phased titania photocatalyst is Degussa P25, which contains a combination of mainly crystalline anatase (~80%) and crystalline rutile (~20%). The photocatalytic activity of the mixed titania phases is better than one of them, because the rutile conduction band is more positive than that of anatase, which means that the crystalline rutile phase may act as an electron sink for photogenerated electrons from the conduction band of the anatase phase. The high photocatalytic activity of this mixed titania phase enhances the separation of photogenerated electrons and holes and results in reduced recombination.

Composites of two semiconductors with different band gaps can suppress photocatalytic performance. Many efforts have been made in the synthesis of different heterostructure semiconductors such as CdS/TiO_2, Bi_2S_3/TiO_2, Fe_2O_3/TiO_2, and ZnO/TiO_2, which can be present as a core-sell geometry or sandwich structure [81–84]. These couplings resulted in extended photo-absorption bands in the visible form and were reported to be more effective for the photocatalytic degradation of a dye under visible light irradiation, as compared to pure TiO_2. The coupling of a smaller band gap semiconductor such as Bi_2S_3 (1.42 eV) with the TiO_2 band gap is of great interest using solar radiation. In principle, the heterostructure of anatase and rutile with ZnO seems useful to achieve a more efficient electron–hole pair separation under visible light irradiations due to the increased lifetime of the photo-produced electron–hole pairs, which is transferred between the two semiconductors.

Li et al. reported that two semiconductors Al_2O_3/Degussa P25 TiO_2 composite exhibited visible-light absorption, which was attributed to nitrogen doping during high-temperature combustion, and to alterations in Ti species with the electronic structure induced by Al addition [85]. The optimal molar ratio of Al_2O_3 to Degussa P25 TiO_2 was 1:1.5, and the composite of semiconductors exhibited a large specific area of 152 m^2/g, with enhanced photocatalytic activity, and large surface positive charges. The degradation rate of anionic methylene orange was enhanced by the N-doped Al_2O_3/TiO_2, which was 43.6 times greater than pure Degussa P25 TiO_2. The high visible-light photocatalytic activity of anionic methylene orange was attributed to the synergetic effects between the two semiconductors, Al_2O_3 and TiO_2, N-doping a large specific surface area (see Section 1.2.4, Nanostructured TiO_2), and low recombination efficiency of photoexcited electrons and holes.

1.2.3 Kinetics of TiO_2 Phase Transformation

TiO_2 (titania) belongs to the transition metal oxides family, and it is known to exist in various polymorphs and crystalline phases. Amorphous TiO_2 transforms into crystalline anatase, and the crystalline anatase transforms to crystalline rutile at an elevated temperature. Rutile is the only stable phase at normal condition, whereas anatase and brookite are metastable and undergo irreversible transformations into rutile at an elevated temperature.

Some studies focused on the amorphous-to-anatase transformation [53]. However, many studies were conducted on the anatase-to-rutile transformations, because crystallinity was an important parameter that influenced the narrowing of titania band gap, and thus titania applications. The crystalline anatase-to-rutile transformation temperature, as reported in the literature, ranges from 400 to 1200°C and is affected by various conditions, including calcination time, synthesis method, level of impurities, doping, calcination time, the deviation of stoichiometry, surface area, grain and particle size, reaction atmosphere type and condition, method of synthesis, and so forth [25, 39, 50, 86–100].

These transformations do not have a unique temperature, and the processes that are involved in the titania transformation, as well as the methods to promote or inhibit this transformation, have not yet been reviewed and discussed comprehensively.

The fraction of anatase-to-rutile transformation was estimated using the Spurr–Myers equations [101]:

$$A(\%) = 1/[1 + I_R/1.26I_A] \times 100\% \tag{1.2}$$

$$R(\%) = 1/[1 + I_R/0.8I_R] \times 100\% \tag{1.3}$$

$$A(\%) = 100 - R(\%) \tag{1.4}$$

where A (%) and R (%) are the relative weight fraction of crystalline anatase and rutile, respectively. I_A and I_R are the X-ray integrated intensities of anatase (101) reflection and rutile (110) reflection, respectively.

It should be noted that the crystal structure of titania is not completely understood by XRD patterns. For example, TiO_2 nanotubes are not pure TiO_2, but crystal modifications of TiO_2 (anatase and rutile) with some polytitanic acids and organic sources, which make it difficult to identify the modified titania phases (TiO_{2-x}). The titania nanotubular structure is another difficulty because the sample is not flat, as the orientation will affect the XRD results, and hence the relative weight calculated from the Spurr and Myers method will not be identified accurately, where only the peak intensities of anatase (101) and rutile (110) are used. However, for in situ XRD/SRD high-temperature experiments, the phase identification will be more difficult due to the thermal expansion of the lattice parameter at an elevated temperature. Moreover, the structure is relatively unstable and can undergo further phase transformation during a thermal treatment.

This section reviews the titania phase transformation by discussing the thermodynamics of the titania phase transformation and the factors affecting its observations.

1.2.3.1 Temperature

Anodized TiO_2 nanotube arrays were heated at different temperatures from 300°C to 800°C with a step of 100°C in an air atmosphere for 2 h with the heating rate of 10°C/min [93]. The crystal structures and crystallinity of the samples were analyzed using the XRD patterns. Anodized TiO_2 nanotube arrays indicate that the diffraction peaks of the Ti foils were only observed before calcination, thus indicating the amorphous nature of TiO_2 nanotubes. The amorphous-to-anatase transformation can be seen clearly at 300°C, which includes the characteristic peak of crystalline anatase. The anatase-to-rutile transformation started when the TiO_2 nanotubes were calcinated at 600°C until the rutile phase began to appear in the XRD patterns.

The weight fraction of rutile increased dramatically as the temperature increased with the transformation of metastable anatase into rutile at elevated temperatures, which were 65%, 91.8%, and 100% for 600°C, 700°C, and 800°C, respectively.

1.2.3.2 Calcination time

The effect of calcination time on the anatase-to-rutile transformation was studied. Anatase was heated in air at 950°C for periods of 4, 24, and 48 h. After each treatment, the anatase and rutile were calculated from XRD measurements. The analysis shows that an increase in calcination time increased the rutile percentage and decreased the anatase percentage. The relative rutile percentages were 5% for 4 h, 65% for 24 h, and 92% for 48 h [102].

The calcination times of 1, 2, 4, 8, and 12 h at 500°C with a heating rate of 10°C/min were used to study the effect of calcination time for anatase-to-rutile transformation of TiO_2 nanotube arrays [93]. XRD patterns of the anodized TiO_2 nanotubes for different calcination times have only the (101) anatase peak of the crystallographic plane appear following 1 h calcination. For 2 h calcination, (101) and (200) anatase crystallographic planes began to appear. The formation of rutile phase (40.5%) was observed for 4 h calcination. The anatase phase transformed completely into the rutile phase when the calcination time was 12 h.

1.2.3.3 Heating rate

TiO_2 nanotube arrays were calcined for 2 h at 500°C in an air atmosphere with heating rates of 1, 5, 10, and 20°C/min to determine the crystalline phases [93]. Crystalline anatase was just observed for all different heating rates, which only affected the intensity of (101) and (200) crystallographic planes, and the amorphous-to-anatase transformation. The relative anatase crystallinity was the highest for the 10°C/min heating rate. It should be noted that the intensity of nanotubes was affected by the morphology of nanotubes. Moreover, the crystallite size varied from 50.9 to 90.4 nm with the increasing heating rate from 1 to 20°C/min, indicating that the higher heating rate promoted the anatase grain growth.

1.2.3.4 Atmospheres

The effects of different reactive atmospheres, including air, argon, nitrogen, hydrogen, vacuum, hydrogen–nitrogen, chlorine, argon–chlorine atmospheres, on TiO_2 transformation were reported by various researchers [39, 50, 86, 92, 102–106].

For instance, it was reported that the anatase-to-rutile transformation rate was lower in a vacuum than in a hydrogen atmosphere, whereas an increase in the oxygen partial pressure decreased the transformation rate [107]. It was also reported that the anatase-to-rutile transformation was lower in a vacuum than in air [86]. Various studies showed that the atmosphere enhanced the TiO_2 transformation due to the formation of oxygen vacancies and the diffusion necessary for the cryptographic rearrangement [86].

The oxygen vacancy defects in TiO_2 acted as color centers, changing from white to grey or dark blue, with increasing non-stoichiometry (TiO_{2-x}) [86, 108].

The oxygen vacancies, which lower the anatase-to-rutile transformation temperature, can be created by annealing TiO_2 under certain atmosphere. The effects of oxygen vacancies on the anatase–rutile transformation of TiO_2 thin films were studied and annealed at various temperatures of 400–800°C in air, vacuum, and H_2 atmospheres. In both H_2 and vacuum atmospheres, the oxygen vacancies concentration increased with increasing temperature, whereas it remained constant in air. The amorphous-to-anatase and the anatase–rutile transformations occur at lower temperatures in hydrogen than in air or vacuum atmosphere because of the higher oxygen vacancy concentration in the TiO_2 film [105].

The anatase-to-rutile phase transformation in argon–chlorine atmosphere is 300 times faster than in air atmosphere at 950°C. The transformed rutile percentages were 5%, 65%, and 92%, when anatase was heated in air for periods of 4, 24, and 48 h, respectively. In contrast, when the anatase was heated in argon–chlorine for just 10 min, the rutile fraction was 99% [102].

1.2.3.5 Impurities, presence of foreign elements, or doping

The presence of foreign ions affects the kinetics of the anatase-to-rutile transformation at elevated temperatures. The role of TiO_2 dopants is considered to facilitate an increase in surface-adsorbed species levels, improvement in charge carrier separation, introduction of mid-gap states, and reduction in the band gap [109].

The nature of the presence of foreign ions appears to control the stoichiometry of the TiO_2 by the formation of both titanium interstitial and oxygen vacancy concentration. The formation of oxygen vacancies was expected to accelerate the anatase-to-rutile transformation, whereas the titanium interstitial formation inhibited the transformation. There are some reports that show that doping TiO_2 with Nb, Al, Ni, Ga, Ta, and W delays the anatase-to-rutile transformation, whereas Mn, Fe, Cu, and Zn are generally believed to promote the phase transformation [95, 96, 98, 100, 110–112].

The presence of foreign ions affects the titania defect structure and, therefore, the transformation rate, which enters either substitutionally or interstitially. The presence of interstitial ions inhibits the anatase-to-rutile transformation. However, substitutional ions may accelerate or inhibit the transformation depending on the type and size of substitutional ions. Smaller ionic radius than Ti, e.g., Cu^{2+}, Li^{+}, or Co^{2+}, can enter the TiO_2 structure substitutionally and would be expected to increase the oxygen vacancy concentration, and thus the phase transformation. However, larger ions, e.g., S^{6+} and P^{5+} would be expected to decrease the oxygen concentration and the rate of anatase-to-rutile transformation, because it will be easy to make rearrangement of the TiO_2 structure [86].

The anatase-to-rutile transformation in the presence of Fe_2O_3 was investigated in air and argon atmospheres [90]. In the presence of Fe ions, the transformation rate in air is higher, and the transition temperature is lower than the corresponding one in undoped TiO_2. Moreover, the anatase-to-rutile transformation in the presence of Fe^{3+} in argon atmosphere is more rapid than in air because of the formation of oxygen vacancies. Thus, the transformation rate was strongly dependent on impurities and the weight percentage. The influence of doping on the anatase-to-rutile phase transformation of different forms of TiO_2 was studied. Table 1.4 shows some literature values for different titania forms and foreign elements.

The effects of silver on the phase transformation of electrospun TiO_2 nanofiber were also investigated. Pure and Ag-doped TiO_2 nanofibers were calcined for 4 h at different calcination temperatures of 0, 400, 450, 500, 600, 700, and 800°C. An analysis of the XRD patterns of the pure and Ag-doped electrospun TiO_2 nanofibers provided the influence of Ag on the anatase-to-rutile transformation. For both samples, they were initially amorphous, with crystalline anatase first appearing at 400°C. The presence of the rutile phase was observed after the calcination of pure TiO_2 nanofiber at 450°C and at 600°C for Ag-doped electrospun TiO_2 nanofibers [113].

Table 1.4 Compilation of literature values on the effect on impurity/doping on anatase-to-rutile transformation

Wt. % of foreign ions	Anatase (%)	Rutile (%)	Heating time, and temperature	Reference
Pure	80	20		
5% SrO	100	0		
5% CaO	90	10		
5% BaO	90	10	3 days, at 850°C	
5% K_2O	50	50		
5% Na_2O	30	70		
5% Li_2O	0	100		
Pure	5	95		
1% NaO	100	0		
1% MnO_2	100	0		
1% WO_3	100	0		
1% Fe_2O_3	98	2	3 h, at 800°C	[86, 107, 114]
1% NiO	97	3		
1% Cr_2O_3	88	12		
1% MoO_3	84	16		
1% CoO	79	21		
Pure	5	95		
5% Cl^-	51	49	1 h, at 708°C	
5% Zn^{++}	24	76		
5% PO^{4--}	100	0		
5% SO^{4--}	71	29	1 h, at 870°C	
5% Al^{3+}	28	72		

1.2.3.6 Synthesis method

The effect of the preparation method on titania transformation was studied. The study concluded that a major reason for the differing transformation temperature ranges was the degree of crystallinity of the precipitate [107]. It is also known that the method of anatase preparation caused wide variation in the anatase-to-rutile temperature, which can probably be due to the impurity of anatase. Anatase can be prepared by the hydrolysis of titanium sulfate, and this is possible due to the presence of sulfate ion impurity in the anatase. Thus, the transformation of impure anatase occurs at a much higher temperature than that of pure anatase [115].

Anodization time has an appreciable effect on anatase-to-rutile transformation [7]. Uniform and highly ordered TiO_2 nanofibers were fabricated by the electrochemical anodic oxidation on Ti foil (see Section 1.2.5.6, Anodization Method). The anodization process was performed in Nalco solution with a concentration of 0.5 M under an applied voltage of 20 V for 5 or 10 min. The crystallization and crystal growth of TiO_2 nanofibers were characterized using in situ synchrotron radiation diffraction (SRD) at the temperature range of 25–700°C. The anodized TiO_2 nanofibers were initially amorphous, but crystallized to anatase at ~400°C and to rutile at ~550–600°C. The rutile weight fractions for nanofibers anodized for 5 min were 54.2 wt.% at 550°C, 54.7 wt.% at 600°C, 54.8 wt.% at 650°C, and 56.7 wt.% at 700°C. The corresponding values for 10 min anodized nanofibers were 55.0 wt.% at 550°C, 55.2 wt.% at 600°C, 55.5 wt.% at 650°C, and 59.2 wt.% at 700°C. This implies that a longer anodization time with the Nalco solution favors the fraction of anatase-to-rutile transformation [7].

1.2.3.7 Particle/grain size

The kinetics of titania's structural transformation from anatase to rutile were studied extensively by various researchers using microcrystalline or nanocrystalline anatase. The nanocrystalline rutile is thermodynamically stable relative to nanocrystalline anatase and brookite under ambient condition. However, anatase is more stable than rutile at particle diameters below 14 nm [116]. The influence of particle size on TiO_2 phase transformation was studied using a mixed titania of anatase (5.1 nm, 46.7 wt.%) and brookite (8.1 nm, 53.3 wt.%), both isothermally and isochronally, at a reaction time of 2 h in the temperature range 325–750°C [117]. Results show that both anatase and brookite were transformed into rutile, but anatase was transformed before brookite. The anatase-to-rutile and brookite-to-rutile transformations depend on their initial particle size [91]. It was also reported by the same authors that the kinetics of nanocrystalline anatase phase transformation and the activation energy of nucleation are size dependent [91]. Anatase and brookite phases were transformed into the rutile phase after reaching a certain particle size [118].

1.2.3.8 Surface area

Anatase with the highest purity was used to study the effect of surface area, which seemed to favor the anatase-to-rutile transformation. The anatase surface area of ~54 m^2/g and ~25 m^2/g was heated at 700°C. The observation showed that the larger surface area slightly accelerated the anatase-to-rutile transformation [115].

1.2.4 Nanostructured TiO_2

Nanostructured materials have existed in this world for centuries. The field of nanoscience is relatively young, and several new terms have appeared, some of which are inconsistent. The modern boom in nanotechnology and nanoscience is, to some extent, a result of the recent advances in the methods of investigating and manipulating small objects. Based on the nanometer (nm), the nanoscale exists specifically from 1 to 100 nanometers, but definition can be used to some micrometers [3]. Developing tools for probing and manipulating atomic-scale objects was mostly driven by improvement in the resolution, reliability, and availability of electron microscopy, including scanning electron microscopy (SEM), electron backscatter diffraction (EBSD), transmission electron microscopy (TEM), X-ray spectroscopy (EDS), X-ray diffraction (XRD), synchrotron radiation diffraction (SRD), X-ray photoelectron spectroscopy (XPS), and atomic force microscopy (AFM).

In general, TiO_2 has a relatively low surface area compared to other semiconductors. This reduces its photocatalytic activity due to the adsorption of organic pollutants on the surface of TiO_2 [112]. The strategy of morphological modifications, such as increasing surface area and porosity, has been adopted for improving the photocatalytic efficiency of titania or other optical properties.

TiO_2 nanostructures have a reduced size and an increased surface area, which can lead to such interactions having a significant impact. A high surface area leads to a higher density of localized states when electrons are involved from the VB to the CB, and the higher localized states provide valuable charge separation in the form of trapping sites for photogenerated charge carriers, by keeping electrons present due to terminated and unsaturated bonds on the surfaces [15]. Reducing TiO_2 surface features down to the nanoscale can lead to unique electronic, physical, and chemical properties, which can offer the development of novel applications, owing to its quantum effect, regional confinement of matter, and ultra-high surface [16].

Nanostructured TiO_2 exhibits superior photocatalytic activity compared to conventional bulk TiO_2 because of its high surface area, and it is one of the most intensively researched substances of the past decade [120]. The specific surfaces of TiO_2 are in the range of 200–300 m^2/g for nanotubes produced by hydrothermal method, 20–50 m^2/g nanofiber and nanorods produced by hydrothermal method, and <20 m^2/g for nanotube arrays produced by the anodization method [3].

Nanostructured TiO_2 has been synthesized in a wide variety of shapes, including symmetrical spheres and polyhedrons, cylindrical tubes and fibers, or random and regular pores in solids. Unfortunately, the definition of nanostructured TiO_2 of various morphological forms has not been developed in a careful fashion, which has resulted in some confusion over their use. In the following sections, various shapes are discussed.

1.2.4.1 Zero-dimensional nanostructures

Nanoparticles/quantum dots: Nanoparticles are microscopic particles with size less than 100 nm, whereas quantum dots are nanoparticles scaled down to further lower dimensions between 2 and 10 nm. The most commonly used titania morphology is that of monodispersed nanoparticles/nanopowders such as Degussa P25, wherein the diameter is controlled to give benefits from the small crystallite size of ~25 nm (high surface area, reduced bulk recombination) without the detrimental effects associated with minuscule particles.

1.2.4.2 One-dimensional nanostructures

The applications of TiO_2 are mainly based on powdered TiO_2, or later evolved into colloid, particles, and thin films. Further, the maturation of synthesis techniques and the invention of modern characterization technologies allow the development of one-dimensional (1D) nanostructures with high aspect ratio, in the form of belts, fibers, rods, tubes, and wires, through numerous approaches such as hydrothermal, template-assisted, and electrochemical anodization (see Chapter 2, Material Synthesis and Methodologies).

One-dimensional TiO_2 nanostructures are extremely important because they can efficiently transport electron/photon and optical excitations. This enhancement of morphology has a larger specific surface and pore volume as compared to TiO_2 nanoparticles. The directional transport in 1D nanostructure is vital to functioning and device integration at the nanoscale.

In general, 1D TiO_2 nanostructures have recently received much attention because of their unique properties, exploitable for device applications, and structures that are ideal for studying the dependence of physical properties on directionality [121]. Elongated structures are of particular importance, as long, thin nanotubes, nanorods, nanowire, nanoneedles, nanobelts, and nanofibers can provide a high specific surface area during the photocatalytic reaction [3].

TiO_2 nanorods (TNR), nanowires (TNW), nanoneedles (TNN), nanobelts (TNB): Nanorods, nanowires, nanoneedles, and nanobelts are long and solid cylinders with a circular base. Major differences between them are length and thickness. TiO_2 nanowires (TNW) were obtained by ultrasonically dispersing TiO_2 nanofibers for 0.5 h in acetic acid, and a thick film was developed using the NW paste [121]. NWs are produced by a bottom-up approach of having self-assembly of the target material from vapor, liquid, or solid phases through nucleation and growth under optimized conditions of pressure, temperature, and concentration.

Electrospinning is one of the emerging bottom-up techniques for fabricating 1D nanostructures (see *Electrospinning Method, Section 1.2.5.7*), including nanowires (NWs), nanorods (NRs), nanofibers (NFs), and nanoneedles (NNs), which work on the principle of asymmetric binding of a charged liquid jet when it is enhanced by a longitudinal electric field.

TiO_2 nanobelts were prepared by a hydrothermal method [122] (see Section 1.2.5.2, Hydrothermal Method). TiO_2 nanobelts are 1D nanostructures with a length of several tens of micrometers, more than 100 nm width, and 20–25 nm thickness [122].

TiO_2 nanofibers (TNFs): Nanofibers are 1D long, solid, and parallel piped nanostructures like other elongated TiO_2 structures with a diameter typically in the nano-range, but the length can be several tens of micrometers. Figure 1.7 provides the SEM image of electrospun nanofibers. The nanofibers have smooth surfaces with a random distribution.

Figure 1.7 Electrospun TiO_2 nanofibers.

One-dimensional electrospun nanofibers composed of well-aligned nanofibrils in the fiber direction are beneficial for the enhancement of the surface area, crystallinity, and the photocatalytic activity for hydrogen evolution. This methodology of electrospun TiO_2 offers a straightforward guide to the immobilization of novel nanofiber photocatalysts for water splitting and hydrogen evolution. One-dimensional electrospun nanofibers with extremely aligned bundled nanofibrils are beneficial for the enhancement of the higher photocatalytic activity, large surface area, and crystallinity. The 1D electrospun polymer/TiO_2 nanofibers were prepared by combining electrospinning and sol–gel techniques. For example, electrospun PVP/TiO_2 nanofibers were formed with lengths up to several centimeters and diameter size in nanometers [123]. These amorphous PVP/TiO_2 nanofibers were converted into crystalline titania phases (anatase, and rutile) without changing their morphology via calcination at elevated temperatures. Pure TiO_2 nanofibers with smaller diameter size were obtained from calcination of the as-spun nanofiber at elevated temperatures. These nanofibers were utilized in the photocatalytic activity for hydrogen evolution, which showed the highest activity, when compared with a nanofiber prepared by other methods, e.g., hydrothermal method and also higher than reference commercial of TiO_2 nanoparticles [23].

TiO_2 nanotubes (TNT): Nanotubes are nanometer-scale tubes, with a hollow center, as compared to other 1D nanostructures. Nanotubes are long cylinders having a hollow cavity positioned at their center and lying along their length. The nanotube structure is extremely attractive because of its high surface area and thermal stability against high-temperature calcinations [124].

The first nanotubes were carbon nanotubes in 1991, which were discovered and developed into multi-walled carbon nanotubes by Sumio Iijima and many other scientists from the NEC Corporation, Fundamental Research Laboratories, Japan [125]. The preparation of finite carbon structure consists of needle-like tubes. The nanotubes grow at the negative end of an electrode used from an arc discharge. Electron microscopy revealed that each needle comprised coaxial tubes of graphitic sheets. On each tube, the C-atom hexagons were arranged in a helical fashion around the needle axis. The carbon nanotubes formed were in a few tens of nanometers in diameter.

Scientists have also synthesized nanotubes from various oxides such as aluminum, zinc oxide, and hafnium. The TiO_2 nanotubes (TNT) were synthesized by many approaches such as anodic oxidation of titanium, deposition with templates, and hydrothermal treatment with concentrated alkali solutions [126–133]. The TiO_2 nanotubes were synthesized by the hydrothermal method, where the tubes can have a multilayered onion shape with a number of 2–10 layers and are usually straight with relatively constant diameter and being open at both ends. Moreover, TiO_2 nanotubes are produced by the anodic oxidation of titanium with one open end and another close end. Through anodization methods, TiO_2 nanotubes bear similarities to alumina nanotubes. A template of aluminum or titanium can be used for the synthesis of aluminum nanotubes or TiO_2 nanotubes, respectively. Arrays of hexagonally ordered TiO_2 nanotubes can be properly synthesized in the amorphous state by using combined methods of laser interference lithography (LIL) and electrochemical anodization [134], as well as through atomic layer deposition (ALD) technique by depositing titanium (IV) tetraisopropoxide gaseous precursor inside the nanopores of anodic alumina templates [135].

TiO_2 nanotubes are highly exciting material systems, which can be employed in a wide range of present and potential applications. TiO_2 nanotube arrays are used as electrodes for solar cell devices and batteries, and for carbon in fuel cells due to the tube's highly effective supporting matrix. Gas sensors, water photoelectrolysis, and photocatalyst devices are other potential applications of TiO_2 nanotube arrays [136–142]. In particular, 1D semiconductor TiO_2 nanotube devices suggest improved energy conversion efficiencies in DSSCs [138, 143, 144].

1.2.4.3 Two-dimensional nanostructures

TiO_2 nanofilms and nanosheets: Nanosheets and nanofilms are very thin two-dimensional (2D) nanostructures (10 × 100 × 100 nm). They have two types of layers: single nanosheets and multilayer nanosheets. The two types can be folded to make single or multilayered nanotubes via the alkaline hydrothermal route [3].

Nanosheets can be converted to nanofibers instead of scrolling into nanotubes under alkaline hydrothermal conditions. An increase in Ti source concentration can result in a faster growth rate of nanosheets, with less effect on the rate of scrolling of nanosheets. Thus, the thicknesses of nanosheets have a particular value where they become too rigid to bend, before curving can occur to form nanofibers rather than nanotubes [3].

TiO_2 nanosquares (TNS): Titania nanosquares are another 2D nanostructures, which were prepared by a mild solvothermal method [145]. Titania is composed of square-like crystals having a length of 10–20 nm and an area of ~200 nm^2.

1.2.4.4 Three-dimensional nanostructures

The three-dimensional (3D) nanostructures consist of thin films ranging from nanometers to micrometers in dimensions with several nanometers in all dimensions, such as metals, semiconductors, and magnetic materials. The functionality of thin films is particularly enhanced when they are made of multiple layers combining together the functionalities of all the layers. This structure saves cost and allows manipulation of the refractive index to a higher degree [146].

Superlattices: The periodic structure of alternating thin films of a semiconductor such as TiO_2 forms an arrangement known as superlattices. This structure is made of different bandgap semiconductors and results in the formation of quantum well, which confines the direction of charge flow through the lattice [146].

1.2.5 Synthesis Methods of Nanostructured TiO_2

A wide range of bulk TiO_2 synthesis techniques were used, the major ones being sol–gel, hydrothermal treatment, solvothermal, microemulsion, vapor deposition, and mechanical milling. In the last decade, nanostructured TiO_2 was an important property and was developed by utilizing the same synthesis methods and creating new synthesis methods, such as the sol–gel method, hydrothermal, solvothermal, electrospinning, and anodization. Hitherto, various techniques were used to synthesize nonporous materials, nanoparticles, nanowires, nanorods, nanotubes, and nanofibers of TiO_2 [7, 23].

1.2.5.1 Sol–gel

The sol–gel method is a preferred technique as it is a low-temperature and cost-effective process. The sol–gel method was developed as one of the most promising ways of synthesizing porous TiO_2 materials. Substances with controlled pore TiO_2 structure such as pore size distribution, porosity, and pore shape are useful in photocatalytic applications. Compared to other synthesis methods of photocatalysts, the sol–gel method presents certain advantages, such as the possibility of depositing on substrates with complex shapes, highly pure and homogeneous products at low temperature, uniform aperture size, and easy control of the doping level [99, 147]. The most common way to impart crystallinity of the thus formed polymeric gel is calcination at elevated temperature, but the prepared crystalline TiO_2 has increased particle size and decreased surface area. Moreover, the sol–gel method is used to dope TiO_2 with various forms. It is also used as a combination method with electrospinning techniques for the preparation of doped electrospun TiO_2 nanofibers [113].

1.2.5.2 Hydrothermal method

In 1998, TiO_2 nanotubes were the first nanostructured TiO_2 forms to be synthesized by applying the hydrothermal method [148]. It is a simple method for the preparation of TiO_2 nanotubes, involving the treatment of amorphous TiO_2 with a concentrated solution of NaOH. In a typical process, several grams of nanotubes can be synthesized from TiO_2 raw material at between 110–150°C, followed by washing with water and HCl. The formation of TiO_2 nanotubes occurred spontaneously and was characterized by a wide distribution of morphological parameters, with a random nanotubular orientation. TiO_2 nanotubes prospered via the alkaline hydrothermal method were observed only in multilayered walls. The key factors affecting the titania nanotube structure, according to the hydrothermal method, include the washing process, calcination temperature and atmosphere, and the important role of residual sodium ions in the stabilization of the tubular framework structure [149]. This method can also be used to produce TiO_2 nanofibers (TNF) besides the TiO_2 nanotubes (TNT) under the hydrothermal condition through the treatment of anatase TiO_2 powder with NaOH for 1 week at 130°C and 110°C, respectively. The nanofiber has elongated morphology with a typical width being in the range of 20–200 nm and a length of several micrometers [150, 151].

Multilayered TiO_2 nanosheets can be obtained by the hydrothermal treatment method. Nanosheets have an important role in the mechanism of nanotube formation by this method. TiO_2 nanotubes can contain a few multilayered TiO_2 nanosheets, which can be scrolled or folded into a nanotubular morphology [152, 153].

1.2.5.3 Template method

This method is called the template method because the pores in the nanoporous membranes are used as templates. The template method is an old approach and a classic bottom-to-up method, which has become extremely popular since the last decade [154]. The template is particularly suitable for the generation of thin films and powders with a well-defined porosity. Moreover, the template synthesis uses a nanoporous membrane as a template to make nanofibers, nanowires, and nanotubes. The most important feature of this method is that tubes and fibers composed of electronically conductive polymers and semiconductors can be prepared in a nanostructure such as titania nanotubes and nanofibers [155]. Although the template is a directed method that provides a straightforward and reliable stratagem for processing 1D nanostructured TiO_2, it involves multiple (at least three) steps: template fabrication, filling or coating of the template surface with titania precursor, and the selective removal of templates [123].

The preparation of TiO_2 nanotubes involves the controlled sol–gel hydrolysis of titanium solutions, followed by the polymerization of TiO_2 in the self-assembled template molecules or deposition of TiO_2 onto the surface of template aggregates [3]. This method has some disadvantages in most cases; the template material is sacrificial and needs to be destroyed after synthesis, leading to increased cost of materials. Moreover, it is also important to maintain a high level of surface cleanliness [3]. The template synthesis has been used to synthesize nanofibers in recent years. Templated PAN nanofibers with different diameter and densities were synthesized using the PAN precursor solution into a solidifying solution under a necessary pressure [156]. The template can be an anodic aluminum oxide membrane with a thickness of 60 mm, a diameter of 13 mm, and a porosity consisting of an array of parallel, straight channels.

1.2.5.4 Chemical vapor deposition

Chemical vapor deposition (CVD) is a chemical method like the sol–gel process, but at relatively high temperatures, it can form a crystalline material of a solid film under certain crystal structures. The TiO_2 film can be obtained by chemical reaction between the heating TiO_2 substrate and the vapor, where increasing temperature is necessary to maintain the chemical reaction and crystalline TiO_2.

1.2.5.5 Layer-by-layer method

The layer-by-layer self-assembly method was developed to produce multilayer nanofilm structures containing semiconductor nanoparticles. In this method, polyelectrolytes are employed to provide electrostatic attractions with the inorganic material in the next layer and, therefore, reduce the formation of defects within the multilayered structures [157].

Nanoporous TiO_2/poly(sodium-4-styrenesulfonate) (PSS) multilayered films were synthesized by applying layer-by-layer self-assembly films [158]. A TiO_2/PSS solution was rinsed extensively with water. Then an ordered multilayered TiO_2/PSS film was obtained. The TiO_2/PSS films deposited on a quartz substrate were heated to elevated temperature for 2 h in the ambient atmosphere.

Moreover, TiO_2 nanofilms assembled with TiO_2 nanoparticles with similar polyelectrolyte component [titanium (IV) bis(ammonium lactato) dihydroxide (TALH)] were fabricated by employing the layer-by-layer self-assembly method at room temperature [159]. In the first step of the procedure to fabricate the $(TiO_2/TALH)_{30}$ nanofilms using the layer-by-layer self-assembly method, a negative substrate, glass, and Si/SiO_2 were obtained by KOH treatment in ultrasonication for 5 min and then rinsed in water. The substrates were immersed into TiO_2 colloidal solution and polyelectrolyte for 10 min and TALH solution from 10 s to 10 min in the second and third steps, respectively. The second and third steps were repeated 30 times in the fourth step. The synthesized $(TiO_2/TALH)_{30}$ multilayered nanofilms showed a high transmittance in the visible range for optical applications with refractive index (ca. $n = 1.75$), and high photocatalytic property by decomposing methyl orange molecules gradually, according to UV light irradiation [14].

The combination of the electrospinning method with layer-by-layer technique was used to fabricate hollow titania-containing multilayered nanofibers. In the electrospun process, 1.5 g of polystyrene (PS, MW =185,000 g/mol) was dissolved in a solvent mixture of tetrahydrofuran (THF) and N,N-dimethylformamide (DMF) with a volume ratio of 1:1. The solution was stirred for 2 h to get a homogeneous sol–gel. Then the sol–gel was loaded into a 5 gauge syringe with stainless needle, which was connected to a high-voltage supply capable of generating voltage up to 15 kV. An automatic syringe pump was used to control the feeding rate of the sol–gel, and the needle-to-collector was 10 cm (see Section 1.2.5.7, Electrospinning Method). To obtain the hollow multilayer TiO_2-containing nanofiber, the synthesized electrospun PS nanofibers were selected as the template and immersed into PEI/TiO_2 solution. Four multilayered polymers of $(PEI/TiO_2)_4$-coated electrospun PS nanofiber were immersed into THF to remove the PS core. To remove the PEI solution inside the nanofiber, the nanofiber was centrifuged at 1000 rmp for 1 min and washed with a solvent. The diameter of the obtained hollow multilayered nanofiber is ~700 nm, and its shell thickness is ~140 nm. These nanostructured titania nanofibers have the highest photocatalytic activity to degrade methylene blue (MB) solution compared with a TiO_2 film because of their unique hollow structures [14].

1.2.5.6 Anodization method

An alternative to the hydrothermal method, which facilitates a structured array of nanotubes with a narrower distribution of morphological parameters, is the anodization method. The anodic synthesis was initially developed for the preparation of aluminum oxide nanotubes and later adapted for TiO_2 nanotube arrays. This method is an electrochemical approach for the preparation of elongated TiO_2 nanostructures.

In 1999, the first anodized TiO_2 nanotube arrays were reported by Zwilling et al. [160] using electrochemical anodization of Ti-based alloys in an acidic, fluoride-based electrolyte. In the anodization cell (Fig. 1.8), a platinum (Pt) foil is used as the counter anode, and the titanium (Ti) foil acts as the cathode inside the H_2O-NHF electrolyte at 10–30 V for several hours at the room temperature [25, 95, 98, 100, 161]. The nanotubes were all oriented in the same direction, perpendicular to the surface of the electrode. Anodization of a Ti substrate in a fluoride-containing electrolyte had highly ordered and vertically oriented titania nanotubes with an open top (facing the electrolyte) and closed bottom, and a narrow distribution of tube diameter, wall thickness, spacing, and lengths.

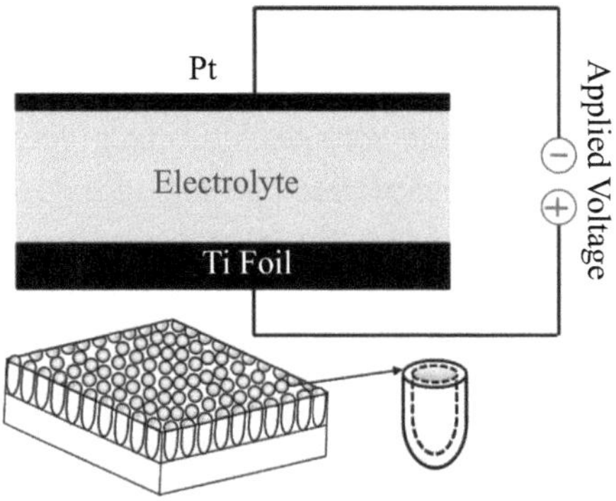

Figure 1.8 Schematic of the anodization process.

Anodization of Ti using an electrolyte of fluoride-containing polyhydric alcohols such as ethylene glycol or glycerol was done as it results in an ordered array of TiO_2 nanotubes with a smooth surface and a very high aspect ratio. It provides a scalable, easy, and inexpensive technique for future industrial application. The anodization method has a good degree of controlling the nanotube diameter, length, and wall thickness. Changes in the dimensions of TiO_2 nanotube arrays have also been studied, such as tube length, pore diameter, and tube wall, which can influence titania applications [162, 163].

The formation, growth rate, and morphology of TiO_2 nanotubes during anodization depend on various factors such as electrode areas, the distance between electrodes, temperature, synthesis time, applied voltage, and electrolyte composition. It was found that the field emission properties also affected the TiO_2 nanotube diameter and height [164]. Various electrolytes were used for titanium anodic oxidation under different anodic conditions with metallic titanium existing in the close-packed hexagonal (α phase) or body-centered cubic (β phase) forms. Anodic titanium foil consists mostly of titania, but also a certain degree of non-stoichiometric TiO_2. Different researchers have different compositions (TiO_{2-x}, where x is the oxygen vacancies) due to differences in electrolyte composition used, oxidation conditions, etc. Moreover, the anodic oxide of Ti foils may contain various amounts of elements other than titanium and oxygen, depending on Ti foils and the electrolyte. Moreover, the water content of ethylene glycol as electrolyte was found to affect the formation of TiO_2 nanotube arrays [165].

The effects of anodizing potential, anodizing temperature, concentration, anodization voltage, and time of formation of titania nanotube arrays were investigated by Xie et al. [166]. It was found that the anodizing temperature and concentration were significant parameters for the formation of nanotubes, while the applied voltage and time mainly determined the nanotube dimensions. An active surface area was increased by increasing the anodization time or voltage, which led to enhanced photovoltaic currents, and thereby an overall higher performance of the dye-sensitized solar cells [166]. Anodization time and fluoride electrolyte can be tuned for obtaining the tube length, whereas the tube diameter can be controlled by the anodization voltage. Experimental results indicate that the relationship between the anodization time and the nanotube length is present logarithmically, and the growth rate decreases with the anodization time as the length of the nanotubes increases, whereas it is present approximately linearly between the anodization voltage and nanotube diameter. Solvent type is another factor that can affect nanotube formation and growth, including the electrolyte chemistry.

After anodization formation, TiO_2 nanotube arrays are amorphous, which can be transformed into crystalline anatase, and then the crystalline anatase is transformed to crystalline rutile at elevated temperatures (see Section 1.2.3, Kinetics of TiO_2 Phase Transformation), but annealing can also affect the nanotube morphology, and then start collapsing.

Some reports indicate the presence of crystalline anatase, a mixture of anatase and rutile, or crystalline rutile in the tube wall, particularly if the anodization is carried out at higher voltages [167, 168]. This report suggests that the presence of crystalline titania could be due to HR-TEM investigation, which the electron beam can adversely affect the amorphous titania during examination in an electron microscope [169].

The common fabrication of titania nanostructure, which was obtained through this method, is nanotube (TNT), but anodized titania with various nanostructures can also be obtained by the anodization method. For example, anodization of titanium employing a electrolyte solution of NaCl 1.0 M with an applied bias voltage of 30 V for several minutes that can be used to synthesize anodic titania nanofibers [7, 170].

1.2.5.7 Electrospinning method

Anton Formhals [171] patented his first invention on the electrostatic spinning of fibers. The invention did not claim e-spinning as a technique for harnessing technology, as it took more than 60 years for e-spinning to gain momentum as an effective technique for synthesizing polymeric fibers, i.e., 1D fibers. This method is not only employed in university laboratories but is also increasingly being applied in industry and everyday life.

As shown in Fig. 1.9, the electrospun experimental setup mainly consists of a high-voltage power supply (5–30 V), a syringe pump, a syringe with a conductive needle, and a conductive collector (e.g., a flat stand collector, or a rotating drum), copper wire, and a sol-gel solution. In a typical process, a polymer and melt solution are injected with a small needle under the influence of an electric field as strong as several kV/cm [123, 172–175].

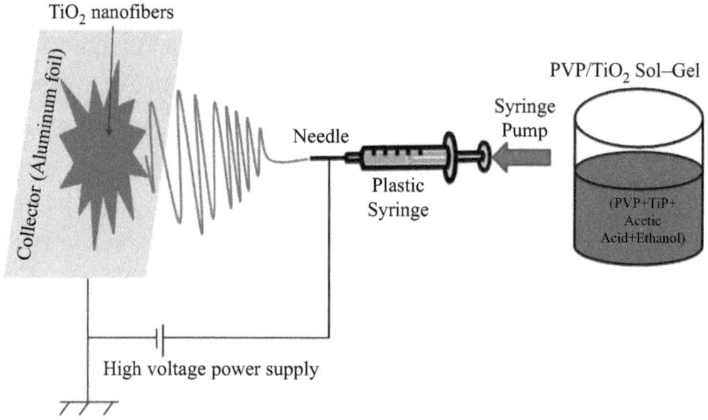

Figure 1.9 Schematic of electrospinning process for TiO_2 nanofibers.

The sol–gel solution is extruded from the microneedle tip at a constant rate by the syringe pump, forming a droplet, which is stretched into a conical shape by the applied voltage. It is named Taylor cone after the researcher who reported it for the first time. At the apex of the Taylor cone, a jet is spouted and breaks down into charged droplets that are sprayed out when the sol–gel does not contain any spinning aid. This phenomenon is known as electrospraying. The breakdown of the jet into small droplets is due to the effect of surface charges that tend to create larger surfaces in order to decrease the charge density over a larger area, while the surface tension tends to keep the droplet large, to reduce the specific surface area. At the high electric field, the effect of the surface charge is more favored by the effect of the surface tension, which leads to smaller droplets with larger specific surface area. By adding the polymeric spinning, the viscosity of sol–gel is increased to retard the formation of droplets by the chain entanglement of the polymer. When the effect of the surface tension and the viscosity at higher concentration overcomes the breakdown by the surface charge, the jet maintains its fibrous morphology. As the electric field is increased, the effect of the surface charge gets stronger, to result in thinner fibers with larger specific surface area. After a short flight from the ejection, the behavior of ejected fibers gets unstable, which can be described by three different instability modes: the Rayleigh instability, the axisymmetric instability, and the non-axisymmetric instability. In an occasion when the system is highly electrically charged by the electrospinning condition, the behavior of the jet is governed by the non-axisymmetric instability in which the jet flies in a random whipping motion at a high frequency. The whipping instability is of crucial importance in the electrospinning process because the fiber gets elongated during the whipping motion that consequently results in fiber thinning as well. The jet will elongate when the droplets become electrospray, if the molecular cohesion of the material is a suitable sol–gel [176].

Electrospinning has been recognized as an efficient method for 1D nanofiber fabrication, and the diameters of electrospun nanofiber materials can shrink from micrometers to nanometers. This amazing characteristic of very large surface area-to-volume ratio has attracted much interest due to the potential application of nanofibers in nanotechnology. Electrospinning is a relatively new and easy method for fabrication of very long nanofibers compared with other nanofibers prepared by anodization, self-assembly, and template methods. Moreover, electrospinning is a simple and cost-effective technique applicable at industrial levels for fabricating 1D nanofibers, which have numerous and diverse potential applications, including composite materials, medical materials, filtration devices, and bioengineering materials. The high surface makes the nanofibers attractive in drug delivery, catalyst support, and even in energy cells [172–174].

Electrospun nanofibers can be collected on a conductive surface to form nanowoven mats, which is one of the attractive features associated with the electrospinning method. The electrospun nanowoven mats are characterized by high surface areas and relatively small pore size, making them excellent candidates for use in filtration and membrane applications. Therefore, the electrospinning method has been used for many applications and potential applications, such as drug delivery, membrane, tissue engineering, protective clothing, wound dressing, filtration, and electronic application [174, 177, 178]. McCann et al. [179] synthesized porous electrospun TiO_2 nanofibers using miscible solvents with immiscible polymers followed by calcination of the nanofibers at an elevated temperature. The porous TiO_2 nanofibers have a much larger surface area compared to the solid TiO_2 nanofibers, which can have potential applications such as photocatalysis, energy storage, catalysis, and fuel cells.

Because of interests in the applications of electrospun nanofibers, great efforts have been made to study the effects of processing parameters. In general, the diameter, size, and morphology of electrospun nanofibers depend primarily on three processing parameters, including sol–gel solution condition, electrospinning conditions, and ambient condition [175, 180]. First, the solution properties have important influences on electrospun nanofiber diameters, which include polymer concentration, surface tension, viscosity, electrical conductivity, molecular weight distribution, and architecture of the polymer, which are related to one another [181]. The viscosity of the so-gel solution depends on the molecular weight and concentration of a polymer, solvent, and another source, where the polymer concentration is used for controlling beads and fiber diameters [181–183]. A higher TiO_2 concentration in a precursor solution will increase the TiO_2 fiber diameter [119, 123].

Second, the needle-tip-to-collector distance, needle size, flow rate, and applied voltage are electrospinning conditions [181]. The applied voltage is a significant factor for electrospinning conditions in the fiber diameters, where a high applied voltage (i.e., strong electrical repulsive forces) decreases the nanofiber diameter, resulting in elongated and highly stretched fibers [182, 184]. One of the most important roles in determining the fiber diameter and bead formation is flow rate, as it determines the amount of sol–gel solution available to be stretched into nanofibers [123, 181]. A large needle-tip-to-collector distance, an applied high-voltage setting, a low flow rate, and a comparatively low concentration of polymer solutions reduce variation in the product quality of electrospun fiber mats with a minimum number of experiments [182]. During the electrospinning process, the size and shape of the needle tip affect the formation of Taylor cone and nanofiber oscillation [185]. Ksapabutr et al. claimed that a sawtooth needle shape allowed for Taylor cones of greater length than its standard and flat counterparts [185].

Aspects of the atmospheric environment, such as temperature, pressure, and humidity, belong to the last set of parameters. The average nanofiber diameter decreases with decreasing atmospheric humidity and increasing atmospheric temperature [178].

E-spinning produces fibers and has been shown to produce many other "one-dimensional" shapes, including nanofibers, nanorods, nanowires and nanoneedles, nanobelts, and hollow tubes. Nanotubes (core-shell fibers) or hollow fibers can be produced by a special electrospinning method and complex architectures. Electrospun TiO_2 nanotubes have been synthesized by a fabrication process of core-shell nanofibers. In this technique, two syringes feed inter-separated and coaxial "inner fluid" and "outer fluid" to the spinneret. Under high voltage, the liquid of electrospinning is drawn out from the spinneret and forms a "compound Taylor cone" with a core-shell structure. The core-shell structure is built and kept in the fibers through spinning solid and can be collected from a collector.

A wide range of composites, polymers, and ceramic precursor solutions have been electrospun into fibers. The electrospinning technology enables the production of continuous polymer or metal oxide/ceramic nanofibers, which have been in the process of development since 1934 [186]. In 2002, electrospun metal oxide composite fibers were first synthesized by Dai et al. [187]. The electrospun alumina borate solution was mixed with polyvinyl alcohol (PVA) to form composite alumina nanofibers.

Electrospun TiO_2 nanofibers were the next metal oxide to be synthesized as composite nanofibers [188]. Nanofibers of TiO_2 metal oxides have been prepared from a polymer solution containing suitable TiO_2 precursors, and stabilizers. The viscosity of the TiO_2 solution can be adjusted by molecular weight and concentration of the polymer, allowing sol–gel to be electrospun into a well-defined composite amorphous titania/polymer nanofibers. Crystalline TiO_2 nanofibers are obtained by removing the carrier polymer through calcination whereby initially formed amorphous TiO_2 subsequently crystallizes to anatase and then rutile at elevated temperature.

TiO_2 nanofibers can be synthesized by a combination of the electrospinning method and the sol–gel technique [180]. In this process, a solution of polymer (binder) and TiO_2 precursor is ejected through a needle in a strong electric field (kilovolts per centimeter), whereby composite nanofibers of polymer and amorphous TiO_2 nanofibers are formed [119, 189–191]. The diameter, morphology, and grain size of the TiO_2 nanofibers have optimal values for photocatalytic activity and are directly affected by the electrospinning parameters [192]. Therefore, the electrospinning parameters such as solution concentration, applied voltage, collector distance, used solvent, and solution feed rate were extensively studied [182].

Electrospinning was proved to be a simple, low-cost, and reliable technique for TiO_2 doping to attain modified optical, electrical, and structural properties. Doping or incorporation of foreign materials into TiO_2 is often required for several titania applications such as catalysis and chemical sensors. The electrospinning method has been used for doping TiO_2 nanofibers with metal or nonmetal elements. Doped electrospun TiO_2 nanofibers brought results that indicated that doping was successfully incorporated into the crystal lattice of TiO_2 nanofibers. Meanwhile, doping could extend the visible light absorption of TiO_2 nanofibers [68]. In the electrospinning process, nanoparticles or molecules as a functional material can be easily doped or incorporated into the electrospun TiO_2 nanofibers by adding these materials to the sol–gel solution. For example, Ag-doped TiO_2 nanofibers were fabricated using the sol–gel method and electrospinning technique. A solution of titanium isopropoxide (TiP) acetic acid and ethanol was mixed into another solution that contained ethanol, polyvinyl pyrrolidone (PVP), and 0.5 N $AgNO_3$ followed by magnetic stirring for 1 day. The sol–gel solution was loaded into a syringe equipped with a 21 G stainless steel needle. A voltage of 20 kV was applied between the needle and a collector with the needle-to-collector distance of 15 cm, and a flow rate of 0.05 ml/min, which was controlled using a syringe pump [113]. In most electrospun experiments, electrospun TiO_2 nanofibers were synthesized at a constant flow rate, needle-tip-to-collector distance, and applied voltage, but with different polymer type and Ti precursors [193–195].

Various polymers have been successfully electrospun for fabrication of electrospun TiO_2 nanofibers, with typical examples, including poly(vinyl pyrrolidone) (PVP), and poly(vinyl alcohol) (PVA), poly(vinyl acetate) (PVAc), polyacrylonitrile (PAN), poly(acrylic acid) (PAA), polystyrene (PS), and poly(methyl methacrylate) (PMMA) [196]. Among of them, PVP is one of the most popular polymers because of its high solubility in water and its good compatibility with many salts containing a high molecular weight (1,300,000 g/mol), good solubility in water and alcohols, not chemical classified as hazardous or dangerous goods. PVP is used considering its safety for many applications, for example, as a binder in many pharmaceutical tablets [197], toothpaste, and shampoos, in personal care products, and as a food additive with E number 1201.

Electrospun TiO_2/PVP composite nanofibers were prepared by a solution that contained 75 wt.% titanium(diisoproproxide) bis(2,4-pentanedionate) in 2-propanol, and PVP. Diameters of 80–100 nm anatase and rutile TiO_2 nanofibers were obtained from calcination of the as-electrospun TiO_2/PVP composite nanofibers at above 300°C for 3 h in air. The results indicated a significant effect of calcination temperature on the morphology of the nanofibers, and the crystalline titania phase in the form of anatase, mixed anatase/rutile, or rutile [198]. There was a decrease in diameter size because PVP polymer, solvent, and another organic source were removed at lower calcined temperature. Upon further calcination, the nanofibers become crystalline titania nanofibers, and the polymer had been completely removed.

For Ti precursors, titanium isopropoxide is well known as a feasible parent titania material for pure titania synthesis. It is an alkoxide of titanium (IV), which is used in materials science and organic synthesis with many acquired names (see Table 1.5).

Table 1.5 Titanium isopropoxide properties

Property	Value
Chemical formula structure	$C_{12}H_{28}O_4Ti$
Molar mass (g/mol)	284.22
Appearance	Colorless to light yellow liquid
Density (at 20°C) (g/cm^3)	0.96
Melting point (°C)	14–17
Boiling point (°C)	232
Solubility in	Water, ethanol, ether, benzene, chloroform
Refractive index (RI)	1.46
Synonyms	Tetraisopropyl titanate Ti(IV) isopropoxide Ti(OiPr)4 Titanium isopropoxide Titanium(IV) isopropoxide
Reference	Pubchem Open Chemistry Database http://pubchem.ncbi.nlm.nih.gov

1.2.6 Taguchi Method

After the Second World War, the associated forces found that the quality of the Japanese telephone system was totally inappropriate and less effective for long-term communication. To improve the system, the associated command recommended that Japan established research facilities similar to the Bell Laboratories in the United States in order to develop a state-of-the-art communication system. The Japanese authority funded the Electrical Communication Laboratories (ECL) appointing Dr. Taguchi in charge of improving the R&D productivity and enhancing product quality. He observed that a great deal of money and time was invested on engineering testing and experimentation. Little importance was given to the process of creative brainstorming to minimize the expenditure of resources [199]. Dr. Taguchi started to develop new methods to improve the process of engineering experimentation. He developed techniques that are now widely known as the Taguchi method. The Taguchi approach is more than a method, which layouts experiments in the mathematical formulation of the experimental design. His method has produced a powerful and unique discipline for quality improvement that differs from traditional practices by replacing a full factorial experiment with a partial factorial, faster, less expensive, and lean experiment.

The Taguchi method is highly useful as an engineering approach for designing robust experiments and selecting optimal levels of processing parameters with minimal sensitivity to variations arising from different causes. In addition, it can also explain the effects of a large and complex number of factors on an individual and on interactive basis. An orthogonal array (OA) to simultaneously accommodate numerous experimental design factors and signal-to-noise ratio (S/N) for measuring the most robust set of operating conditions from variations within the results are the two essential tools of the Taguchi method [175, 182, 193, 199, 200].

The signal-to-noise (S/N) concept has been used in the fields of electrical, acoustics, and mechanical vibrations, and other science and engineering disciplines for several years. In a majority of cases, the quality characteristic may be a single measurable quantity such as length, weight, hours, etc., for some products with subjective measurements like "low," "bad," "good" may be used. In other examples, objective and subjective evaluations may be combined into an overall evaluation criteria (OEC).

Signal-to-noise (S/N) ratio:

$$S/N = \frac{\text{Signal}}{\text{Noise}} \tag{1.5}$$

Three characteristic formulae have been used to identify the optimum combination of factors:

Smaller-the-better equation:

$$S/N = -10\log_{10}\left[\frac{\left(\sum y_i^2\right)}{N}\right] \tag{1.6}$$

Larger-the-better equation:

$$S/N = -10\log_{10}\left[\frac{\sum\left(1/y_i^2\right)}{N}\right] \tag{1.7}$$

Nominal-the-best equation:

$$S/N = -10\log_{10}\left[\frac{\sum(y_i - m)^2}{N}\right] \tag{1.8}$$

where S/N is the signal-to-noise ratio, N is the number of trials, results target value, y_i is results of experiments, observation or quality characteristics such as titania nanofibers' diameter, inner diameter and length of titania nanotubes. The negative sign is used to indicate that the largest value gives the optimum value for the response variable and the robust design.

References

1. Augugliaro, V., Loddo, V., Palmisano, G., Pagliaro, M., and Palmisano, L. (2010). *Clean by Light Irradiation: Practical Applications of Supported TiO_2* (Royal Society of Chemistry).
2. Kim, C., Delikanli, S., Perera, S., and Zeng, H. (2010). *Dye Sensitized Solar Cells based on Free-Standing TiO_2 Nanotube* (in APS March Meeting Abstracts).

3. Bavykin, D. V. and Walsh, F. C. (2010). *Titanate and Titania Nanotubes: Synthesis, Properties and Applications* (Cambridge: RSC Publishing).
4. Dransfield, G., Guest, P., Lyth, P., McGarvey, D., and Truscott, T. (2000). Photoactivity tests of TiO_2-based inorganic sunscreens: Part 1: Non-aqueous dispersions, *J. Photochem. Photobiol. B: Biol.*, **59**, pp. 147–151.
5. Kandavelu, V., Kastien, H., and Thampi, K. R. (2004). Photocatalytic degradation of isothiazolin-3-ones in water and emulsion paints containing nanocrystalline TiO_2 and ZnO catalysts, *Appl. Catal. B: Environ.*, **48**, pp. 101–111.
6. Landmann, M., Rauls, E., and Schmidt, W. (2012). The electronic structure and optical response of rutile, anatase and brookite TiO_2, *J. Phys: Condens. Matter.*, **24**, pp. 195503–195509.
7. Low, I. M., Yam, F., and Pang, W. K. (2012). In-situ diffraction studies on the crystallization and crystal growth in anodized TiO_2 nanofibres, *Mater. Lett.*, **87**, pp. 150–152.
8. Park, J. Y., Yun, J. J., Hwang, C. H., and Lee, I. H. (2010). Influence of silver doping on the phase transformation and crystallite growth of electrospun TiO_2 nanofibers, *Mater. Lett.*, **64**, pp. 2692–2695.
9. Pfaff, G. and Reynders, P. (1999). Angle-dependent optical effects deriving from submicron structures of films and pigments, *Chem. Rev.*, **99**, pp. 1963–1982.
10. Tang, Z. R., Li, F., Zhang, Y., Fu, X., and Xu, Y. J. (2011). Composites of titanate nanotube and carbon nanotube as photocatalyst with high mineralization ratio for gas-phase degradation of volatile aromatic pollutant, *J. Phys. Chem. C*, **115**, pp. 7880–7886.
11. Weir, A., Westerhoff, P., Fabricius, L., Hristovski, K., and von Goetz, N. (2012). Titanium dioxide nanoparticles in food and personal care products, *Environ. Sci. Technol.*, **46**, pp. 2242–2250.
12. Zhang, P., Shao, C., Li, X., Zhang, M., Zhang, X., Sun, Y., and Liu, Y. (2012). In situ assembly of well-dispersed Au nanoparticles on TiO_2/ZnO nanofibers: A three-way synergistic heterostructure with enhanced photocatalytic activity, *J. Hazard. Mater.*, **237–238**, pp. 331–338.
13. Lee, C. K., Wang, C. C., Juang, L. C., Lyu, M. D., Hung, S. H., and Liu, S. S. (2008). Effects of sodium content on the microstructures and basic dye cation exchange of titanate nanotubes, *Colloids Surf. A: Physicochem. Eng. Aspects*, **317**, pp. 164–173.
14. Zhang, T., Ge, L., Wang, X., and Gu, Z. (2008). Hollow TiO_2 containing multilayer nanofibers with enhanced photocatalytic activity, *Polymer*, **49**, pp. 2898–2902.

15. Barzykin, A. and Tachiya, M. (2002). Mechanism of charge recombination in dye-sensitized nanocrystalline semiconductors: Random flight model, *J. Phys. Chem. B*, **106**, pp. 4356–4363.

16. Nalwa, H. S. (2004). *Encyclopedia of Nanoscience and Nanotechnology* (CRC Press).

17. Doh, S. J. (2008). Development of photocatalytic TiO_2 nanofibers by electrospinning and its application to degradation of dye pollutants, *J. Hazard. Mater.*, **154**, pp. 118–127.

18. Lee, S. S., Bai, H., Liu, Z., and Sun, D. D. (2012). Electrospun TiO_2/SnO_2 nanofibers with innovative structure and chemical properties for highly efficient photocatalytic H_2 generation, *Int. J. Hydrogen Energy*, **37**, pp. 10575–10584.

19. Lee, S. S., Bai, H., Liu, Z., and Sun, D. D. (2013). Novel-structured electrospun TiO_2/CuO composite nanofibers for high efficient photocatalytic cogeneration of clean water and energy from dye wastewater, *Water Res.*, **47**, pp. 4059–4073.

20. Thavasi, V., Singh, G., and Ramakrishna, S. (2008). Electrospun nanofibers in energy and environmental applications, *Energy Environ. Sci.*, **1**, pp. 205–221.

21. Weir, A., Westerhoff, P., Fabricius, L., Hristovski, K., and Goetz, N. (2012). Titanium dioxide nanoparticles in food and personal care products, *Environ. Sci. Technol.*, **46**, pp. 2242–2250.

22. Fujishima, A. and Honda, K. (1972). Photolysis-decomposition of water at the surface of an irradiated semiconductor, *Nature*, **238**, pp. 37–38.

23. Chuangchote, S., Jitputti, J., Sagawa, T., and Yoshikawa, S. (2009). Photocatalytic activity for hydrogen evolution of electrospun TiO_2 nanofibers, *ACS Appl. Mater. Interfaces*, **1**, pp. 1140–1143.

24. Liu, Z., Yan, X., Chu, W., and Li, D. (2010). Effects of impurities containing phosphorus on the surface properties and catalytic activity of TiO_2 nanotube arrays, *Appl. Surface Sci.*, **257**, pp. 1295–1299.

25. Low, I. M., Albetran, H., Prida, V., Manurung, P., and Ionescu, M. (2012). Effect of chromium-doping on the crystallization and phase stability in anodized TiO_2 nanotubes, *36th International Conference on Advanced Ceramics and Composites*, Jan 22, 2012, pp. 149–158, Daytona Beach, Florida: Wiley.

26. Carp, O., Huisman, C. L., and Reller, A. (2004). Photoinduced reactivity of titanium dioxide, *Prog. Solid State Chem.*, **32**, pp. 33–177.

27. Simons, P. and Dachille, F. (1967). The structure of TiO_2 II, a high-pressure phase of TiO_2, *Acta Crystallogr.*, **23**, pp. 334–336.

28. Latroche, M., Brohan, L., Marchand, R., and Tournoux, M. (1989). New hollandite oxides: TiO_2 (H) and K 0.06 TiO_2, *J. Solid State Chem.*, **81**, pp. 78–82.

29. Sato, H., Endo, S., Sugiyama, M., Kikegawa, T., Shimomura, O., and Kusaba, K. (1991). Baddeleyite-type high-pressure phase of TiO_2, *Science*, **251**, pp. 786–788.

30. Takahashi, Y., Kijima, N., and Akimoto, J. (2006). Synthesis, structural change upon heating, and electronic structure of ramsdellite-type TiO_2, *Chem. Mater.*, **18**, pp. 748–752.

31. Vegard, L. (1916). Results of crystal analysis–III, *London, Edinburgh, Dublin Philos. Mag. J. Sci.*, **32**, pp. 505–518.

32. Pauling, L. and Sturdivant, J. H. (1928). The crystal structure of brookite, *Z. Kristallographie-Cryst. Mater.*, **68**, pp. 239–256.

33. Cromer, D. T. and Herrington, K. (1955). The structures of anatase and rutile, *J. Am. Chem. Soc.*, **77**, pp. 4708–4709.

34. Beltran, A., Gracia, L., and Andres, J. (2006). Density functional theory study of the brookite surfaces and phase transitions between natural titania polymorphs, *J. Phys. Chem. B*, **110**, pp. 23417–23423.

35. Fujishima, A., Rao, T. N., and Tryk, D. A. (2000). Titanium dioxide photocatalysis, *J. Photochem. Photobiol. C: Photochem. Rev.*, **1**, pp. 1–21.

36. Banfield, J. (1998). Thermodynamic analysis of phase stability of nanocrystalline titania, *J. Mater. Chem.*, **8**, pp. 2073–2076.

37. Linsebigler, A. L., Lu, G., and Yates Jr, J. T. (1995). Photocatalysis on TiO_2 surfaces: Principles, mechanisms, and selected results, *Chem. Rev.*, **95**, pp. 735–758.

38. Mo, S. D. and Ching, W. (1995). Electronic and optical properties of three phases of titanium dioxide: Rutile, anatase, and brookite, *Phys. Rev. B*, **51**, pp. 13023–13032.

39. Albetran, H., Haroosh, H., Dong, Y., O'Connor, B., and Low, I. (2014). Effect of atmosphere on crystallisation kinetics and phase relations in electrospun TiO_2 nanofibres, *Ceram. Environ. Energy Appl. II: Ceram. Trans.*, **246**, pp. 125–133.

40. Bokhimi, X., Morales, A., Aguilar, M., Toledo-Antonio, J., and Pedraza, F. (2001). Local order in titania polymorphs, *Int. J. Hydrogen Energy*, **26**, pp. 1279–1287.

41. Burdett, J. K., Hughbanks, T., Miller, G. J., Richardson Jr, J. W., and Smith, J. V. (1987). Structural-electronic relationships in inorganic solids: Powder neutron diffraction studies of the rutile and anatase

polymorphs of titanium dioxide at 15 and 295 K, *J. Am. Chem. Soc.*, **109**, pp. 3639–3646.

42. Matsui, M. and Akaogi, M. (1991). Molecular dynamics simulation of the structural and physical properties of the four polymorphs of TiO_2, *Mol. Simul.*, **6**, pp. 239–244.
43. Kim, D. W., Enomoto, N., Nakagawa, Z. E., and Kawamura, K. (1996). Molecular dynamic simulation in titanium dioxide polymorphs: Rutile, brookite, and anatase, *J. Am. Ceram. Soc.*, **79**, pp. 1095–1099.
44. Wunderlich, W., Oekermann, T., Miao, L., Hue, N. T., Tanemura, S., and Tanemura, M. (2004). Electronic properties of nano-porous TiO_2 and ZnO thin films-comparison of simulations and experiments, *J. Ceram. Process Res.*, **5**, pp. 343–354.
45. Lin, H., Huang, C., Li, W., Ni, C., Shah, S. I., and Tseng, Y. H. (2006). Size dependency of nanocrystalline TiO_2 on its optical property and photocatalytic reactivity exemplified by 2-chlorophenol, *Appl. Catal. B: Environ.*, **68**, pp. 1–11.
46. Hummer, D., Heaney, P., and Post, J. (2007). Thermal expansion of anatase and rutile between 300 and 575 K using synchrotron powder X-ray diffraction, *Powder Diffr.*, **22**, pp. 352–357.
47. Reyes-Coronado, D., Rodriguez-Gattorno, G., Espinosa-Pesqueira, M., Cab, C., De Coss, R., and Oskam, G. (2008). Phase-pure TiO_2 nanoparticles: Anatase, brookite and rutile, *Nanotechnology*, **19**, pp. 145605–145615.
48. Tang, H., Prasad, K., Sanjines, R., Schmid, P., and Levy, F. (1994). Electrical and optical properties of TiO_2 anatase thin films, *J. Appl. Phys.*, **75**, pp. 2042–2047.
49. Ohno, T., Sarukawa, K., and Matsumura, M. (2001). Photocatalytic activities of pure rutile particles isolated from TiO_2 powder by dissolving the anatase component in HF solution, *J. Phys. Chem. B*, **105**, pp. 2417–2420.
50. Albetran, H., Haroosh, H., Dong, Y., Prida, V., O'Connor, B., and Low, I. M. (2014). Phase transformations and crystallization kinetics in electrospun TiO_2 nanofibers in air and argon atmospheres, *Appl. Phys. A*, **116**, pp. 161–169.
51. Meagher, E. and Lager, G. A. (1979). Polyhedral thermal expansion in the TiO, polymofphs: Refinement of the crystal structures of rutile and brookite at high temperature, *Can. Mineral.*, **17**, pp. 77–85.
52. Mills, A. and Le Hunte, S. (1997). An overview of semiconductor photocatalysis, *J. Photochem. Photobiol. A: Chem.*, **108**, pp. 1–35.

53. Eufinger, K., Poelman, D., Poelman, H., De Gryse, R., and Marin, G. (2007). Photocatalytic activity of dc magnetron sputter deposited amorphous TiO_2 thin films, *Appl. Surf. Sci.*, **254**, pp. 148–152.

54. Albetran, H., O'Connor, B., and Low, I. (2016). Effect of calcination on band gaps for electrospun titania nanofibers heated in air–argon mixtures, *Mater. Des.*, **92**, pp. 480–485.

55. Hsu, S. W., Yang, T. S., Chen, T. K., and Wong, M. S. (2007). Ion-assisted electron-beam evaporation of carbon-doped titanium oxide films as visible-light photocatalyst, *Thin Solid Films*, **515**, pp. 3521–3526.

56. Chatterjee, D. and Mahata, A. (2001). Demineralization of organic pollutants on the dye modified TiO_2 semiconductor particulate system using visible light, *Appl. Catal. B: Environ.*, **33**, pp. 119–125.

57. Anpo, M. (2000). Use of visible light. Second-generation titanium oxide photocatalysts prepared by the application of an advanced metal ion-implantation method, *Pure Appl. Chem.*, **72**, pp. 1787–1792.

58. Asahi, R., Morikawa, T., Ohwaki, T., Aoki, K., and Taga, Y. (2001). Visible-light photocatalysis in nitrogen-doped titanium oxides, *Science*, **293**, pp. 269–271.

59. Umebayashi, T., Yamaki, T., Itoh, H., and Asai, K. (2002). Band gap narrowing of titanium dioxide by sulfur doping, *Appl. Phys. Lett.*, **81**, pp. 454–456.

60. Yu, J. C., Yu, J., Ho, W., Jiang, Z., and Zhang, L. (2002). Effects of F-doping on the photocatalytic activity and microstructures of nanocrystalline TiO_2 powders, *Chem. Mater.*, **14**, pp. 3808–3816.

61. Ohno, T., Mitsui, T., and Matsumura, M. (2003). Photocatalytic activity of S-doped TiO_2 photocatalyst under visible light, *Chem. Lett.*, **32**, pp. 364–365.

62. Sakthivel, S. and Kisch, H. (2003). Daylight photocatalysis by carbon-modified titanium dioxide, *Angew. Chem. Int. Ed.*, **42**, pp. 4908–4911.

63. Yamashita, H., Harada, M., Misaka, J., Takeuchi, M., Neppolian, B., and Anpo, M. (2003). Photocatalytic degradation of organic compounds diluted in water using visible light-responsive metal ion-implanted TiO_2 catalysts: Fe ion-implanted TiO_2, *Catal. Today*, **84**, pp. 191–196.

64. Yu, J. C., Zhang, L., Zheng, Z., and Zhao, J. (2003). Synthesis and characterization of phosphated mesoporous titanium dioxide with high photocatalytic activity, *Chem. Mater.*, **15**, pp. 2280–2286.

65. Hong, X., Wang, Z., Cai, W., Lu, F., Zhang, J., Yang, Y., Ma, N., and Liu, Y. (2005). Visible-light-activated nanoparticle photocatalyst of iodine-doped titanium dioxide, *Chem. Mater.*, **17**, pp. 1548–1552.

66. Chen, D., Jiang, Z., Geng, J., Wang, Q., and Yang, D. (2007). Carbon and nitrogen co-doped TiO_2 with enhanced visible-light photocatalytic activity, *Ind. Eng. Chem. Res.*, **46**, pp. 2741–2746.

67. Lam, R. C. W., Leung, M. K. H., Leung, D. Y. C., Vrijmoed, L. L. P., Yam, W. C., and Ng, S. P. (2007). Visible-light-assisted photocatalytic degradation of gaseous formaldehyde by parallel-plate reactor coated with Cr ion-implanted TiO_2 thin film, *Sol. Energy Mater. Sol Cells*, **91**, pp. 54–61.

68. Zhang, Z., Shao, C., Zhang, L., Li, X., and Liu, Y. (2010). Electrospun nanofibers of V-doped TiO_2 with high photocatalytic activity, *J. Colloid Interface Sci.*, **351**, pp. 57–62.

69. Kamat, P. V. (2012). TiO_2 nanostructures: Recent physical chemistry advances, *J. Phys. Chem. C*, **116**, pp. 11849–11851.

70. Liu, G., Wang, L., Yang, H. G., Cheng, H. M., and Lu, G. Q. M. (2010). Titania-based photocatalysts—crystal growth, doping and heterostructuring, *J. Mater. Chem.*, **20**, pp. 831–843.

71. Zhang, R., Wu, H., Lin, D., and Pan, W. (2010). Photocatalytic and magnetic properties of the Fe- TiO_2/SnO_2 nanofiber via electrospinning, *J. Am. Ceram. Soc.*, **93**, pp. 605–608.

72. Park, J. H., Kim, S., and Bard, A. J. (2006). Novel carbon-doped TiO_2 nanotube arrays with high aspect ratios for efficient solar water splitting, *Nano Lett.*, **6**, pp. 24–28.

73. Mishra, T., Wang, L., Hahn, R., and Schmuki, P. (2014). In-situ Cr doped anodized TiO_2 nanotubes with increased photocurrent response, *Electrochim. Acta.*, **132**, pp. 410–415.

74. Zhu, J., Deng, Z., Chen, F., Zhang, J., Chen, H., Anpo, M., Huang, J., and Zhang, L. (2006). Hydrothermal doping method for preparation of Cr^{3+}- TiO_2 photocatalysts with concentration gradient distribution of Cr^{3+}, *Appl. Catal. B: Environ.*, **62**, pp. 329–335.

75. Fan, X., Chen, X., Zhu, S., Li, Z., Yu, T., Ye, J., and Zou, Z. (2008). The structural, physical and photocatalytic properties of the mesoporous Cr-doped TiO_2, *J. Mol. Catal A: Chem.*, **284**, pp. 155–160.

76. Dholam, R., Patel, N., Santini, A., and Miotello, A. (2010). Efficient indium tin oxide/Cr-doped-TiO_2 multilayer thin films for H_2 production by photocatalytic water-splitting, *Int. J. Hydrogen Energy*, **35**, pp. 9581–9590.

77. Van der Meulen, T., Mattson, A., and Österlund, L. (2007). A comparative study of the photocatalytic oxidation of propane on anatase, rutile, and mixed-phase anatase–rutile TiO_2 nanoparticles: Role of surface intermediates, *J. Catal.*, **251**, pp. 131–144.

78. Li, L. and Liu, C. Y. (2009). Facile synthesis of anatase–brookite mixed-phase N-doped TiO_2 nanoparticles with high visible-light photocatalytic activity, *Eur. J. Inorg. Chem.*, **2009**, pp. 3727–3733.

79. Xu, H. and Zhang, L. (2009). Controllable one-pot synthesis and enhanced photocatalytic activity of mixed-phase TiO_2 nanocrystals with tunable brookite/rutile ratios, *J. Phys. Chem. C*, **113**, pp. 1785–1790.

80. Zheng, Z., Liu, H., Ye, J., Zhao, J., Waclawik, E. R., and Zhu, H. (2010). Structure and contribution to photocatalytic activity of the interfaces in nanofibers with mixed anatase and TiO_2(B) phases, *J. Mol. Catal A: Chem.*, **316**, pp. 75–82.

81. Serpone, N., Maruthamuthu, P., Pichat, P., Pelizzetti, E., and Hidaka, H. (1995). Exploiting the interparticle electron transfer process in the photocatalysed oxidation of phenol, 2-chlorophenol and pentachlorophenol: Chemical evidence for electron and hole transfer between coupled semiconductors, *J. Photochem. Photobiol. A: Chem.*, **85**, pp. 247–255.

82. Marci, G., Augugliaro, V., Lopez-Munoz, M. J., Martin, C., Palmisano, L., Rives, V., Schiavello, M., Tilley, R. J., and Venezia, A. M. (2001). Preparation characterization and photocatalytic activity of polycrystalline ZnO/TiO_2 systems. 1. Surface and bulk characterization, *J. Phys. Chem. B*, **105**, pp. 1026–1032.

83. Brahimi, R., Bessekhouad, Y., Bouguelia, A., and Trari, M. (2007). Visible light induced hydrogen evolution over the heterosystem Bi_2S_3/TiO_2, *Catal. Today*, **122**, pp. 62–65.

84. Ghows, N. and Entezari, M. H. (2011). Fast and easy synthesis of core–shell nanocrystal (CdS/TiO_2) at low temperature by micro-emulsion under ultrasound, *Ultrason. Sonochem.*, **18**, pp. 629–634.

85. Li, F. T., Zhao, Y., Hao, Y. J., Wang, X. J., Liu, R. H., Zhao, D. S., and Chen, D. M. (2012). N-doped P25 TiO_2–amorphous Al_2O_3 composites: One-step solution combustion preparation and enhanced visible-light photocatalytic activity, *J. Hazard. Mater.*, **239**, pp. 118–127.

86. Shannon, R. D. and Pask, J. A. (1965). Kinetics of the anatase-rutile transformation, *J. Am. Ceram. Soc.*, **48**, pp. 391–398.

87. Eppler, R. A. (1987). Effect of antimony oxide on the anatase-rutile transformation in titanium dioxide, *J. Am. Ceram. Soc.*, **70**, pp. 64–66.

88. Akhtar, M. K., Pratsinis, S. E., and Mastrangelo, S. V. (1992). Dopants in vapor-phase synthesis of titania powders, *J. Am. Ceram. Soc.*, **75**, pp. 3408–3416.

89. Ding, X. Z., Liu, X. H., and He, Y. Z. (1996). Grain size dependence of anatase-to-rutile structural transformation in gel-derived nanocrystalline titania powders, *J. Mater Sci. Lett.*, **15**, pp. 1789–1791.

90. Gennari, F. and Pasquevich, D. (1998). Kinetics of the anatase–rutile transformation in TiO_2 in the presence of Fe_2O_3, *J. Mater. Sci.*, **33**, pp. 1571–1578.

91. Zhang, H. and Banfield, J. F. (2000). Phase transformation of nanocrystalline anatase-to-rutile via combined interface and surface nucleation, *J. Mater. Res.*, **15**, pp. 437–448.

92. Fang, D., Luo, Z., Huang, K., and Lagoudas, D. C. (2011). Effect of heat treatment on morphology, crystalline structure and photocatalysis properties of TiO_2 nanotubes on Ti substrate and freestanding membrane, *Appl. Surface Sci.*, **257**, pp. 6451–6461.

93. Liu, R., Qiang, L. S., Yang, W. D., and Liu, H. Y. (2013). The effect of calcination conditions on the morphology, the architecture and the photo-electrical properties of TiO_2 nanotube arrays, *Mater. Res. Bull.*, **48**, pp. 1458–1467.

94. Albetran, H., Alsafwan, A., Haroosh, H., Dong, Y., and Low, I. M. (2013). X-ray diffraction study on the in-situ crystallisation kinetics in electrospun PVP/TiO_2 nanofibers, *Nanostruct. Mater. Nanotechnol. VII: Ceram. Eng. Sci. Proc.*, **34**, pp. 35–49.

95. Low, I. M., Albetran, H., Prida, V. M., Vega, V., Manurung, P., and Ionescu, M. (2013). A comparative study on crystallization behavior, phase stability, and binding energy in pure and Cr-doped TiO_2 nanotubes, *J. Mater. Res.*, **28**, pp. 304–312.

96. Albetran, H., O'Connor, B., Prida, V., and Low, I. (2015). Effect of vanadium ion implantation on the crystallization kinetics and phase transformation of electrospun TiO_2 nanofibers, *Appl. Phys. A*, **120**, pp. 623–634.

97. Albetran, H. and Low, I. M. (2016). Crystallization kinetics and phase transformations in aluminum ion-implanted electrospun TiO_2 nanofibers, *Appl. Phys. A*, **122**, pp. 1044–1053.

98. Albetran, H. and Low, I. M. (2016). Effect of indium ion implantation on crystallization kinetics and phase transformation of anodized titania nanotubes using in-situ high-temperature radiation diffraction, *J. Mater. Res.*, **31**, pp. 1588–1595.

99. Albetran, H., O'Connor, B., and Low, I. M. (2017). Effect of pressure on TiO_2 crystallization kinetics using in-situ high-temperature synchrotron radiation diffraction, *J. Am. Ceram. Soc.*, **100**, pp. 3199–3207.

100. Albetran, H., Vega, V., Prida, V. M., and Low, I. M. (2018). Dynamic diffraction studies on the crystallization, phase transformation, and activation energies in anodized titania nanotubes, *Nanomaterials*, **8**, pp. 122–131.

101. Spurr, R. A. and Myers, H. (1957). Quantitative analysis of anatase–rutile mixtures with an X-ray diffractometer, *Anal. Chem.*, **29**, pp. 760–762.

102. Gamboa, J. A. and Pasquevich, D. M. (1992). Effect of chlorine atmosphere on the anatase-rutile transformation, *J. Am. Ceram. Soc.*, **75**, pp. 2934–2938.

103. Varghese, O. K., Gong, D., Paulose, M., Grimes, C. A., and Dickey, E. C. (2003). Crystallization and high-temperature structural stability of titanium oxide nanotube arrays, *J. Mater. Res.*, **18**, pp. 156–165.

104. Plugaru, R., Cremades, A., and Piqueras, J. (2004). The effect of annealing in different atmospheres on the luminescence of polycrystalline TiO_2, *J. Phys.: Condens. Matter.*, **16**, pp. 261–268.

105. Huang, J. H. and Wong, M. S. (2011). Structures and properties of titania thin films annealed under different atmosphere, *Thin Solid Films*, **520**, pp. 1379–1384.

106. Albetran, H., O'Connor, B., and Low, I. M. (2016). Activation energies for phase transformations in electrospun titania nanofibers: Comparing the influence of argon and air atmospheres, *Appl. Phys. A*, **122**, pp. 367–376.

107. Iida, Y. and Ozaki, S. (1961). Grain growth and phase transformation of titanium oxide during calcination, *J. Am. Ceram. Soc.*, **44**, pp. 120–127.

108. Cornaz, P., Van Hooff, J., Pluijm, F., and Schuit, G. (1966). Surface co-ordination of oxygen on oxygen-deficient TiO_2 and MoO_3 as revealed by esr-measurements, *Discuss. Faraday Soc.*, **41**, pp. 290–304.

109. Hanaor, D. A. and Sorrell, C. C. (2011). Review of the anatase to rutile phase transformation, *J. Mater. Sci.*, **46**, pp. 855–874.

110. Depero, L., Sangaletti, L., Allieri, B., Bontempi, E., Marino, A., and Zocchi, M. (1999). Correlation between crystallite sizes and microstrains in TiO_2 nanopowders, *J. Cryst. Growth*, **198**, pp. 516–520.

111. Depero, L., Marino, A., Allieri, B., Bontempi, E., Sangaletti, L., Casale, C., and Notaro, M. (2000). Morphology and microstructural properties of TiO_2 nanopowders doped with trivalent Al and Ga cations, *J. Mater. Res.*, **15**, pp. 2080–2086.

112. Guidi, V., Carotta, M., Ferroni, M., Martinelli, G., and Sacerdoti, M. (2003). Effect of dopants on grain coalescence and oxygen mobility

in nanostructured titania anatase and rutile, *J. Phys. Chem. B*, **107**, pp. 120–124.

113. Park, J. Y., Yun, J. J., Hwang, C. H., and Lee, I. H. (2010). Influence of silver doping on the phase transformation and crystallite growth of electrospun TiO_2 nanofibers, *Mater. Lett.*, **64**, pp. 2692–2695.

114. Rao, C., Turner, A., and Honig, J. (1959). Some observations concerning the effect of impurities on the anatase–rutile transition, *J. Phys. Chem. Solids*, **11**, pp. 173–175.

115. Yoganarasimhan, S. and Rao, C. R. (1962). Mechanism of crystal structure transformations. Part 3. Factors affecting the anatase–rutile transformation, *Trans. Faraday Soc.*, **58**, pp. 1579–1589.

116. Zhang, H. and Banfield, J. F. (1998). Thermodynamic analysis of phase stability of nanocrystalline titania, *J. Mater. Chem.*, **8**, pp. 2073–2076.

117. Zhang, H. and Banfield, J. F. (2000). Understanding polymorphic phase transformation behavior during growth of nanocrystalline aggregates: Insights from TiO_2, *J. Phys. Chem. B*, **104**, pp. 3481–3487.

118. Zhang, Q., Gao, L., and Guo, J. (2000). Effects of calcination on the photocatalytic properties of nanosized TiO_2 powders prepared by $TiCl_4$ hydrolysis, *Appl. Catal. B: Env.*, **26**, pp. 207–215.

119. Li, Q., Satur, D. J. G., Kim, H., and Kim, H. G. (2012). Preparation of sol-gel modified electrospun TiO_2 nanofibers for improved photocatalytic decomposition of ethylene, *Mater. Lett.*, **76**, pp. 169–172.

120. Zhan, S., Chen, D., Jiao, X., and Tao, C. (2006). Long TiO_2 hollow fibers with mesoporous walls: Sol–gel combined electrospun fabrication and photocatalytic properties, *J. Phys. Chem. B*, **110**, pp. 11199–11204.

121. Archana, P., Jose, R., Vijila, C., and Ramakrishna, S. (2009). Improved electron diffusion coefficient in electrospun TiO_2 nanowires, *J. Phys. Chem. C*, **113**, pp. 21538–21542.

122. Hu, P., Du, G., Zhou, W., Cui, J., Lin, J., Liu, H., Liu, D., Wang, J., and Chen, S. (2010). Enhancement of ethanol vapor sensing of TiO_2 nanobelts by surface engineering, *ACS Appl. Mater. Interfaces*, **2**, pp. 3263–3269.

123. Li, D. and Xia, Y. (2003). Fabrication of titania nanofibers by electrospinning, *Nano Lett.*, **3**, pp. 555–560.

124. Wang, J. and Lin, Z. (2009). Dye-sensitized TiO_2 nanotube solar cells with markedly enhanced performance via rational surface engineering, *Chem. Mater.*, **22**, pp. 579–584.

125. Iijima, S. (1991). Helical microtubules of graphitic carbon, *Nature*, **354**, pp. 56–58.

126. Hoyer, P. (1996). Formation of a titanium dioxide nanotube array, *Langmuir*, **12**, pp. 1411–1413.

127. Imai, H., Takei, Y., Shimizu, K., Matsuda, M., and Hirashima, H. (1999). Direct preparation of anatase TiO_2 nanotubes in porous alumina membranes, *J. Mater. Chem.*, **9**, pp. 2971–2972.

128. Michailowski, A., AlMawlawi, D., Cheng, G., and Moskovits, M. (2001). Highly regular anatase nanotubule arrays fabricated in porous anodic templates, *Chem. Phys. Lett.*, **349**, pp. 1–5.

129. Du, G., Chen, Q., Che, R., Yuan, Z., and Peng, L. M. (2001). Preparation and structure analysis of titanium oxide nanotubes, *Appl. Phys. Lett.*, **79**, pp. 3702–3704.

130. Seo, D. S., Lee, J. K., and Kim, H. (2001). Preparation of nanotube-shaped TiO_2 powder, *J. Cryst. Growth*, **229**, pp. 428–432.

131. Liu, S., Gan, L., Liu, L., Zhang, W., and Zeng, H. (2002). Synthesis of single-crystalline TiO_2 nanotubes, *Chem. Mater.*, **14**, pp. 1391–1397.

132. Tian, Z. R., Voigt, J. A., Liu, J., Mckenzie, B., and Xu, H. (2003). Large oriented arrays and continuous films of TiO_2-based nanotubes, *J. Am. Chem. Soc.*, **125**, pp. 12384–12385.

133. Yu, J., Dai, G., and Cheng, B. (2010). Effect of crystallization methods on morphology and photocatalytic activity of anodized TiO_2 nanotube array films, *J. Phys. Chem. C*, **114**, pp. 19378–19385.

134. Vega, V., Montero-Moreno, J. M., García, J., Prida, V., Rahimi, W., Waleczek, M., Bae, C., Zierold, R., and Nielsch, K. (2016). Long-range hexagonal arrangement of TiO_2 nanotubes by soft lithography-guided anodization, *Electrochimica Acta.*, **203**, pp. 51–58.

135. Vega, V., Gelde, L., González, A., Prida, V., Hernando, B., and Benavente, J. (2017). Diffusive transport through surface functionalized nanoporous alumina membranes by atomic layer deposition of metal oxides, *J. Ind. Eng. Chem.*, **52**, pp. 66–72.

136. Chen, C., Wang, N., Zhou, P., San, H., Wang, K., and Chen, X. (2016). Electrochemically reduced graphene oxide on well-aligned titanium dioxide nanotube arrays for betavoltaic enhancement, *ACS Appl. Mater. Interfaces*, **8**, pp. 24638–24644.

137. Krbal, M., Prikryl, J., Zazpe, R., Sopha, H., and Macak, J. (2017). CdS-coated TiO_2 nanotube layers: Downscaling tube diameter towards efficient heterostructured photoelectrochemical conversion, *Nanoscale*, **9**, pp. 7755–7759.

138. Licklederer, M., Cha, G., Hahn, R., and Schmuki, P. (2019). Ordered nanotubular titanium disulfide (TiS_2) structures: Synthesis and use as

counter electrodes in dye sensitized solar cells (DSSCs), *J. Electrochem. Soc.*, **166**, pp. 3009–3013.

139. Zare, M., Solaymani, S., Shafiekhani, A., Kulesza, S., Talu, S., and Bramowicz, M. (2018). Evolution of rough-surface geometry and crystalline structures of aligned TiO_2 nanotubes for photoelectrochemical water splitting, *Sci. Rep.*, **8**, pp. 10870–10881.

140. Smith, Y., Ray, R., Carlson, K., Sarma, B., and Misra, M. (2013). Self-ordered titanium dioxide nanotube arrays: Anodic synthesis and their photo/electro-catalytic applications, *Materials*, **6**, pp. 2892–2957.

141. Ozkan, S., Yoo, J., Nguyen, N. T., Mohajernia, S., Zazpe, R., Prikryl, J., Macak, J. M., and Schmuki, P. (2018). Spaced TiO_2 nanotubes enable optimized Pt atomic layer deposition for efficient photocatalytic H_2 generation, *ChemistryOpen*, **7**, pp. 797–802.

142. Mohamed, A., Barghi, S., and Rohani, S. (2018). N-and C-modified TiO_2 nanotube arrays: Enhanced photoelectrochemical properties and effect of nanotubes length on photoconversion efficiency, *Nanomaterials*, **8**, pp. 198–214.

143. Yang, D. J., Park, H., Cho, S. J., Kim, H. G., and Choi, W. Y. (2008). TiO_2-nanotube-based dye-sensitized solar cells fabricated by an efficient anodic oxidation for high surface area, *J. Phys. Chem. Solids*, **69**, pp. 1272–1275.

144. Paulose, M., Shankar, K., Yoriya, S., Prakasam, H. E., Varghese, O. K., Mor, G. K., Latempa, T. A., Fitzgerald, A., and Grimes, C. A. (2006). Anodic growth of highly ordered TiO_2 nanotube arrays to 134 μm in length, *J. Phys. Chem. B*, **110**, pp. 16179–16184.

145. Yu, C., Wei, L., Li, X., Chen, J., Fan, Q., and Yu, J. C. (2013). Synthesis and characterization of Ag/TiO_2-B nanosquares with high photocatalytic activity under visible light irradiation, *Mater. Sci. Eng. B*, **178**, pp. 344–348.

146. Rao, C. and Cheetham, A. (2001). Science and technology of nanomaterials: Current status and future prospects, *J. Mater. Chem.*, **11**, pp. 2887–2894.

147. Yang, X. L., Zhu, L., Yang, L. M., Zhou, W. Y., and Xu, Y. H. (2011). Preparation and photocatalytic activity of neodymium doping titania loaded to silicon dioxide, *Trans. Nonferrous Met. Soc. China*, **21**, pp. 335–339.

148. Kasuga, T., Hiramatsu, M., Hoson, A., Sekino, T., and Niihara, K. (1998). Formation of titanium oxide nanotube, *Langmuir*, **14**, pp. 3160–3163.

149. Sun, X. and Li, Y. (2003). Synthesis and characterization of ion-exchangeable titanate nanotubes, *Chem. A Eur. J.*, **9**, pp. 2229–2238.

150. Lin, C. H., Chao, J. H., Liu, C. H., Chang, J. C., and Wang, F. C. (2008). Effect of calcination temperature on the structure of a Pt/TiO_2(B) nanofiber and its photocatalytic activity in generating H_2, *Langmuir*, **24**, pp. 9907–9915.

151. Kuo, H. L., Kuo, C. Y., Liu, C. H., Chao, J. H., and Lin, C. H. (2007). A highly active bi-crystalline photocatalyst consisting of TiO_2(B) nanotube and anatase particle for producing H_2 gas from neat ethanol, *Catal. Lett.*, **113**, pp. 7–12.

152. Bavykin, D. V., Parmon, V. N., Lapkin, A. A., and Walsh, F. C. (2004). The effect of hydrothermal conditions on the mesoporous structure of TiO_2 nanotubes, *J. Mater. Chem.*, **14**, pp. 3370–3377.

153. Ma, R., Bando, Y., and Sasaki, T. (2004). Directly rolling nanosheets into nanotubes, *J. Phys. Chem. B*, **108**, pp. 2115–2119.

154. Hulteen, J. (1997). A general template-based method for the preparation of nanomaterials, *J. Mater. Chem.*, **7**, pp. 1075–1087.

155. Martin, C. R. (1996). Membrane-based synthesis of nanomaterials, *Chem. Mater.*, **8**, pp. 1739–1746.

156. Feng, L., Li, S., Li, H., Zhai, J., Song, Y., Jiang, L., and Zhu, D. (2002). Super-hydrophobic surface of aligned polyacrylonitrile nanofibers, *Angew. Chem.*, **114**, pp. 1269–1271.

157. Zhang, H., Li, X., and Chen, G. (2010). Fabrication of photoelectrode materials, in: Comninellis, C. and Chen, G. (eds.), *Electrochemistry for the Environment*. Springer, New York. pp. 473–513.

158. Hao, W., Pan, F., and Wang, T. (2005). Photocatalytic activity TiO_2 granular films prepared by layer-by-layer self-assembly method, *J. Mater. Sci.*, **40**, pp. 1251–1253.

159. Kim, J. H., Fujita, S., and Shiratori, S. (2006). Fabrication and characterization of TiO_2 thin film prepared by a layer-by-layer self-assembly method, *Thin Solid Films*, **499**, pp. 83–89.

160. Zwilling, V., Aucouturier, M., and Darque-Ceretti, E. (1999). Anodic oxidation of titanium and TA6V alloy in chromic media. An electrochemical approach, *Electrochim. Acta.*, **45**, pp. 921–929.

161. Gong, D., Grimes, C. A., Varghese, O. K., Hu, W., Singh, R., Chen, Z., and Dickey, E. C. (2001). Titanium oxide nanotube arrays prepared by anodic oxidation, *J. Mater. Res.*, **16**, pp. 3331–3334.

162. Macak, J. M., Albu, S. P., and Schmuki, P. (2007). Towards ideal hexagonal self-ordering of TiO_2 nanotubes, *Phys. Status Solidi (RRL) Rapid Res. Lett.*, **1**, pp. 181–183.

163. Macak, J., Tsuchiya, H., Ghicov, A., Yasuda, K., Hahn, R., Bauer, S., and Schmuki, P. (2007). TiO_2 nanotubes: Self-organized electrochemical formation, properties and applications, *Curr. Opin. Solid State Mater. Sci.*, **11**, pp. 3–18.

164. Alivov, Y., Klopfer, M., and Molloi, S. (2010). Effect of TiO_2 nanotube parameters on field emission properties, *Nanotechnology*, **21**, pp. 505706–505711.

165. Raja, K., Gandhi, T., and Misra, M. (2007). Effect of water content of ethylene glycol as electrolyte for synthesis of ordered titania nanotubes, *Electrochem. Commun.*, **9**, pp. 1069–1076.

166. Xie, Z., Adams, S., Blackwood, D., and Wang, J. (2008). The effects of anodization parameters on titania nanotube arrays and dye sensitized solar cells, *Nanotechnology*, **19**, pp. 405701–405707.

167. Marchenoir, J., Loup, J., and Masson, J. (1980). *Etude des couches poreuses formées par oxydation anodique du titane sous fortes tensions, Thin Solid Films*, **66**, pp. 357–369.

168. Delplancke, J. L., Garnier, A., Massiani, Y., and Winand, R. (1994). Influence of the anodizing procedure on the structure and the properties of titanium oxide films and its effect on copper nucleation, *Electrochim. Acta.*, **39**, pp. 1281–1289.

169. Albu, S. P., Tsuchiya, H., Fujimoto, S., and Schmuki, P. (2010). TiO_2 nanotubes–annealing effects on detailed morphology and structure, *Eur. J. Inorg. Chem.*, **2010**, pp. 4351–4356.

170. Beh, K., Yam, F., Tneh, S., and Hassan, Z. (2011). Fabrication of titanium dioxide nanofibers via anodic oxidation, *Appl. Surf. Sci.*, **257**, pp. 4706–4708.

171. Anton, F. (1934). Process and apparatus for preparing artificial threads. 1934, Google Patents.

172. Hassan, M. S., Amna, T., Yang, O. B., Kim, H., and Khil, M. (2012). TiO_2 nanofibers doped with rare earth elements and their photocatalytic activity, *Ceram. Inter.*, **38**, pp. 5925–5930.

173. Zhu., P., Nair, A. S., Shengjie, P., Shengyuan, Y., and Ramakrishna, S. (2012). Facile fabrication of TiO_2-graphene composite with enhanced photovoltaic and photocatalytic properties by electrospinning, *ACS Appl. Mater. Interfaces*, **4**, pp. 581–585.

174. Wu, H., Pan, W., Lin, D., and Li, H. (2012). Electrospinning of ceramic nanofibers: Fabrication, assembly and applications, *J. Adv. Ceram.*, **1**, pp. 2–23.

175. Albetran, H., Dong, Y., and Low, I. M. (2015). Characterization and optimization of electrospun TiO_2/PVP nanofibers using Taguchi design of experiment method, *J. Asian Ceram. Soc.*, **3**, pp. 292–300.

176. Hardick, O., Stevens, B., and Bracewell, G. (2011). Nanofibre fabrication in a temperature and humidity controlled environment for improved fibre consistency, *J. Mater. Sci.*, **46**, pp. 3890–3898.

177. Samadi, M., Shivaee, H. A., Zanetti, M., and Moshfegh, A. (2012). Visible light photocatalytic activity of novel MWCNT-doped ZnO electrospun nanofibers, *J. Mol. Catal. A: Chem.*, **359**, pp. 42–48.

178. Noor Azman, N. Z., Siddiqui, S. A., Haroosh, H. J., Albetran, H. M., Johannessen, B., and Dong, Y. (2013). Characteristics of X-ray attenuation in electrospun bismuth oxide/polylactic acid nanofibre mats, *J. Synchrotron Radiat.*, **20**, pp. 741–748.

179. McCann, J. T., Li, D., and Xia, Y. (2005). Electrospinning of nanofibers with core-sheath, hollow, or porous structures, *J. Mater. Chem.*, **15**, pp. 735–738.

180. Huang, Z. M., Zhang, Y. Z., Kotaki, M., and Ramakrishna, S. (2003). A review on polymer nanofibers by electrospinning and their applications in nanocomposites, *Compos. Sci. Technol.*, **63**, pp. 2223–2253.

181. Fridrikh, S. V., Yu, J. H., Brenner, M. P., and Rutledge, G. C. (2003). Controlling the fiber diameter during electrospinning, *Phys. Rev. Lett.*, **90**, pp. 144502–144506.

182. Patra, S., Easteal, A., and Bhattacharyya, D. (2009). Parametric study of manufacturing poly (lactic) acid nanofibrous mat by electrospinning, *J. Mater. Sci.*, **44**, pp. 647–654.

183. Inai, R., Kotaki, M., and Ramakrishna, S. (2005). Structure and properties of electrospun PLLA single nanofibres, *Nanotechnology*, **16**, pp. 208–213.

184. Kumar, A., Jose, R., Fujihara, K., Wang, J., and Ramakrishna, S. (2007). Structural and optical properties of electrospun TiO_2 nanofibers, *Chem. Mater.*, **19**, pp. 6536–6542.

185. Ksapabutr, B., Chalermkiti, T., and Panapoy, M. (2005). Effect of nozzle shapes on the formation of Taylor cone and the oscillation of fibers during electrospinning process, *Chiang Mai University J.*, **4**, pp. 115–119.

186. Dzenis, Y. A. (2004). Spinning continuous fibers for nanotechnology, *Science*, pp. 1917–1919.

187. Dai, H., Gong, J., Kim, H., and Lee, D. (2002). A novel method for preparing ultra-fine alumina-borate oxide fibres via an electrospinning technique, *Nanotechnology*, **13**, pp. 674–677.

188. Li, D., Wang, Y., and Xia, Y. (2004). Electrospinning nanofibers as uniaxially aligned arrays and layer-by-layer stacked films, *Adv. Mater.*, **16**, pp. 361–366.

189. Wessel, C., Ostermann, R., Dersch, R., and Smarsly, B. M. (2011). Formation of inorganic nanofibers from preformed TiO_2 nanoparticles via electrospinning, *J. Phys. Chem. C*, **115**, pp. 362–372.

190. Li, H., Zhang, W., and Pan, W. (2011). Enhanced photocatalytic activity of electrospun TiO_2 nanofibers with optimal anatase/rutile ratio, *J. Am. Ceram. Soc.*, **94**, pp. 3184–3187.

191. Luo, W., Hu, X., Sun, Y., and Huang, Y. (2012). Surface modification of electrospun TiO_2 nanofibers via layer-by-layer self-assembly for high-performance lithium-ion batteries, *J. Mater. Chem.*, **22**, pp. 4910–4915.

192. Li, H., Zhang, W., Li, B., and Pan, W. (2010). Diameter-dependent photocatalytic activity of electrospun TiO_2 nanofiber, *J. Am. Ceram. Soc.*, **93**, pp. 2503–2506.

193. Ghani, J., Choudhury, I., and Hassan, H. (2004). Application of Taguchi method in the optimization of end milling parameters, *J. Mater. Process. Technol.*, **145**, pp. 84–92.

194. Park, J. Y. and Kim, S. S. (2009). Effects of processing parameters on the synthesis of TiO_2 nanofibers by electrospinning, *Met. Mater. Int.*, **15**, pp. 95–99.

195. Park, J. A., Moon, J., Lee, S. J., Kim, S. H., Zyung, T., and Chu, H. Y. (2010). Structural, electrical and gas sensing properties of eletrospun TiO_2 nanofibers, *Thin Solid Films*, **518**, pp. 6642–6645.

196. Wu, C. M., Chiou, H. G., Lin, S. L., and Lin, J. M. (2012). Effects of electrostatic polarity and the types of electrical charging on electrospinning behavior, *J. Appl. Polymer. Sci.*, **126**, pp. 89–97.

197. Bühler, V. (2005). *Polyvinylpyrrolidone Excipients for Pharmaceuticals: Povidone, Crospovidone and Copovidone*. Springer Science & Business Media.

198. Nuansing, W., Ninmuang, S., Jarernboon, W., Maensiri, S., and Seraphin, S. (2006). Structural characterization and morphology of electrospun TiO_2 nanofibers, *Mater. Sci. Eng: B*, **131**, pp. 147–155.

199. Ranjit, R. (1990). *A Primer on the Taguchi Method.* (Van Nostrand Reinhold).

200. Dong, Y. and Bhattacharyya, D. (2008). Effects of clay type, clay/compatibiliser content and matrix viscosity on the mechanical properties of polypropylene/organoclay nanocomposites, *Compos Part A: Appl. Sci. Manuf.*, **39**, pp. 1177–1191.

Part II

METHODOLOGIES

Chapter 2

Material Synthesis and Methodologies

TiO_2 materials have been widely employed in diverse areas, particularly in the areas of energy, storage, and environment. Therefore, different synthesis techniques have been developed to address the needs of this material, aiming to ease the fabrication process, to reduce the cost of fabrication setup, and most importantly to increase the functionality of the TiO_2 nanostructures in the aforementioned applications.

2.1 Synthesis of TiO_2 Thin Films

TiO_2 thin films were synthesized using the solution of titanium trichloride ($TiCl_3$) and $NaHCO_3$. The process of deposition was carried out by dropwise addition of freshly prepared $NaHCO_3$ solution into $TiCl_3$ solution in a beaker glass, and then stirred until a pH of 3–4 was reached. The ITO glass substrate ($35 \times 15 \times 1$ mm^3) was hung vertically until 2.0 cm of the substrate was immersed into the solution of $TiCl_3$ and $NaHCO_3$. The solution was stirred for up to 6 h, depending on the number of deposition cycles to produce six samples, as shown in Table 2.1. The samples were then dried at 100°C for 1 h, followed by calcination at 500°C for 4 h.

X-ray diffraction data were collected using a Shimadzu 610 diffractometer. The operating conditions used were CuKα radiation

Nanostructured Titanium Dioxide in Photocatalysis
It-Meng Low, Hani Manssor Albetran, Victor Manuel de la Prida Pidal, and Fong Kwong Yam

ISBN 978-981-4877-07-7 (Hardcover), 978-1-003-14853-1 (eBook)
www.jennystanford.com

(λ = 1.54506 Å) produced at 40 kV and 30 mA. The diffraction patterns were collected over the 2θ range of 10–80°. A JEOLJSM-6360LA scanning electron microscope was used to examine the microstructures of surface and cross sections of the samples. The electrical resistivity (ρ) of TiO_2 thin films was measured using the four-point probe method.

Table 2.1 Formulations of samples as a function of deposition cycle during processing

Sample	Number of Deposition Cycles	Deposition Time per Cycle (h)
A	1	6
B	2	3
C	3	2
D	4	1.5
E	5	1.2
F	6	1

2.2 Synthesis of Electrospun TiO_2 Nanofibers

A homogenous sol–gel precursor solution was prepared by mixing titanium isopropoxide (TIP) (MW = 284.22 g/mol, 97% purity, sourced from Sigma-Aldrich), acetic acid (MW = 60.05 g/mol, 99.7% purity, Sigma-Aldrich), and ethanol (MW = 46.07 g/mol, 99.5% purity, Sigma- Aldrich) in a fixed volume ratio of 3:1:3, respectively. The sol–gel was stirred in a capped bottle for 5 min, and then, 12 wt% of polyvinyl pyrrolidone (PVP) (MW = 1,300,000 g/mol, 100% purity, Sigma-Aldrich) was dissolved in the solution at 40°C for 1 h using a stirrer. To achieve complete dissolution and mixing, the sol–gel solution precursor was stirred ultrasonically for 5 min before it was loaded into a 10 ml plastic syringe with a 25-gauge stainless steel needle. An electrospinning unit (Gamma High Voltage Research, USA) with a high-voltage power supply was used to provide 25 kV between the needle and a mesh collector covered by an aluminum foil at a distance of approximately 12 cm. A syringe pump (Chemyx Inc. USA) was used to control the solution flow rate at 2 ml/h during the process of electrospinning [1–7].

2.3 Synthesis of Anodized TiO_2 Nanotubes

Ti foils (99.6% purity) with dimensions of 10 × 10 × 0.1 mm^3 were used for electrochemical anodizing to produce self-organized and well-aligned TiO_2 nanotube arrays. The process of potentiostatic anodization was performed in a standard two-electrode electrochemical cell, with Ti as the working electrode and platinum as the counter electrode. Before anodization, Ti foils were mechanically polished, degreased by sonicating in ethanol, isopropanol, and acetone for 5 min each, followed by rinsing with deionized water, and then drying using nitrogen stream. After drying, the foils were exposed to the electrolyte, which consists of 100 mL of ethylene glycol, 2.04 mL of water, and 0.34 g of NH_4F. The electrolyte's pH was kept constant at pH = 6, and its temperature was kept at about 25°C. The anodization process was performed under an applied voltage of 60 V for 0.5 h. After that, the resulting TiO_2 structures were then rinsed in ethanol, immersed in hexamethyldililazane, and dried in air [8–11].

2.4 Synthesis of 1D TiO_2 Nanostructures

The key process to produce 1D TiO_2 nanostructure (nanobelts, nanofibers, nanorods, nanowires, nanotubes) by hydrothermal reaction and anodic alumina oxide (AAO) template-assisted deposition is reviewed. Subsequently, the important developments of anodic self-organized TiO_2 nanotube layers, especially the breakthroughs in the anodization electrolytes such as F^- and Cl^- based electrolytes are briefly reviewed. Advanced anodization techniques such as multi-step anodization to improve the ordering and create bamboo and branching nanotubes are included. Furthermore, light-driven applications such as photocatalytic water splitting, degradation of organic compounds and dye-sensitized solar cells (DSSCs) are discussed. We then extend to the modification and functionalization of TiO_2 nanotube layers, such as doping and self-doping, metal catalytic effect, integration of a semiconductor, to enhance the properties or performance of the TiO_2 nanotube layers in order to cater to the demand in diverse applications.

2.4.1 Hydrothermal: Seeded-Growth Reaction

Kasuga et al. introduced hydrothermal growth to synthesize TiO_2 nanotubes [12, 13]. The nanotubes share a similar structure with carbon nanotubes, based on a rolled atomic or molecular layer. In this approach, a TiO_2 powder was treated with NaOH to form a Ti-OH sheet-like structure, followed by an acid treatment to dehydrate the Ti-OH bonds to Ti-O-Ti, which rolls the TiO_2 sheet to form a tubular structure. The tube walls are extremely thin with small tube diameter [12, 13]. A modification of Kasuga's method was carried out by increasing the TiO_2 and NaOH treatment temperature to 120–150°C, to produce thicker multi-layered walls and larger tube diameter [14]. In addition to the nanotubes, this method may yield nanoparticles and nanoribbon, or coexistence of two of these three structures. The morphology is affected by treatment temperature, duration, precursor phase, and pH and concentration of the solution; whereas the crystallinity of the TiO_2 nanostructures is usually determined by the precursor phase [15].

Hydrothermal reaction was also applied to prepare aligned single crystalline TiO_2 nanorods on fluorine-doped tin oxide (FTO) substrate. The substrates were placed in a hydrothermal autoclave consisting of mixed HCl and titanium precursor solution, where the choices of titanium precursors include titanium butoxide [16–18], titanium isopropoxide (TTIP) [17], and titanium tetrachloride ($TiCl_4$) [19]. The autoclave was placed in an oven at temperatures 80–240°C for 1–24 h to yield rutile TiO_2 nanorods [16–19]. The nucleation and growth of nanorods are deduced as the small lattice mismatch between FTO substrate (rutile SnO_2) and rutile TiO_2 [16, 19]. This is evidenced by the unsuccessful growth of vertically aligned TiO_2 nanorods on other substrates such as glass and silicon [16, 17].

2.4.2 Templated Synthesis: Sol–Gel Deposition, Electrodeposition, Atomic Layer Deposition

Template-assisted synthesis is feasible in developing ordered structures, as the resultant structure follows exactly the dimension of the template. A highly ordered template, such as AAO porous layer, is employed as it is easy to produce and is inexpensive and commercially available. The hollow channels of the AAO porous layer are filled by the targeted material through numerous deposition techniques, including sol–gel deposition [20–22], electrodeposition [23, 24], sputtering [25], and atomic layer deposition (ALD) [25–28].

For sol–gel deposition, the TiO_2 sol was obtained by combining titanium isopropoxide with ethanol, deionized water, and acetylacetone in a certain ratio. The AAO templates were dipped in the sol solution in optimized temperature and duration, followed by a heat treatment [20–22]. It has been suggested that the nucleation process starts around the walls of the AAO channels and the subsequent growth of the nuclei forms particles, and the aggregation of particles forms nanotubes and eventually nanorods [22].

In an electrochemical deposition, an Au layer was deposited on one side of the AAO templates to serve as the working electrode. A titanium trichloride ($TiCl_3$) solution was maintained at pH 2–2.5, and the electrodeposition was carried out with different potentials against the reference electrode in a three-electrode potentiostatic system, also followed by a heat treatment [23, 24]. The growth of TiO_2 nanostructures was initiated at the Au/AAO interface. The walls of the AAO channels are positively charged due to the presence of oxygen vacancies; thus, the incoming $TiOH^{2+}$ are repelled from the walls. Hence, the growth is concentrated at the center of the AAO channels [23, 24].

The two deposition techniques mentioned above require a final-step drying and sintering to transform the amorphous to the crystalline TiO_2 nanostructure. In contrary, physical deposition such as chemical vapor deposition (CVD) and ALD produces crystalline TiO_2 nanostructure [25]. The self-limiting growth ALD process occurs in a sequential manner where the precursor vapors are pulsed for a certain duration followed by purging to remove the excess precursor. $Ti(OMe)_4$ [28] and $TiCl_4$ [25, 27] have been employed as titanium precursor, whereas H_2O as oxygen precursor, and the deposition was carried out in 130–200°C. ALD offers conformal deposition, i.e., the deposition follows closely the structure of the AAO template, with uniform thickness control according to the deposition cycles [28]. Yao et al. carried out a comparison of TiO_2 deposition within AAO template by sputtering and ALD and noted that the deposited TiO_2 was concentrated only at the surface of the AAO template, and ultimately leads to blockage of the surface when the sputtering duration was extended. Satisfactorily filled TiO_2 was achieved by the

ALD technique, despite the high-aspect-ratio nature of AAO template [25].

After TiO_2 deposition and necessary post-treatment, the AAO templates are dissolved by acid or alkaline aqueous solution to obtain the ordered TiO_2 nanorods or nanotubes [20–28]. In summary, template-assisted deposition of 1D TiO_2 nanostructures allows good control in dimension and uniformity over a large surface area.

2.4.3 Electrochemical Anodization

The electrochemical anodization method is simple and straightforward, and it is reported as the most studied technique for the synthesis of 1D TiO_2 nanotubes in the last 15 years [29]. A typical electrochemical setup is illustrated in Fig. 2.1. In the classical electrochemical oxidation of a valve metal, a barrier oxide layer is formed on the metal surface [30]. In 1984, Assefpour-Dezfuly et al. discovered that anodization in chromic acid electrolyte resulted in TiO_2 porous layers [31]. Later in 1999, Zwilling et al. added HF in the chromic acid electrolyte and produced very thin nanotube layers [32]. These early works suggest that F^- is the prime component for the formation of tubular morphology.

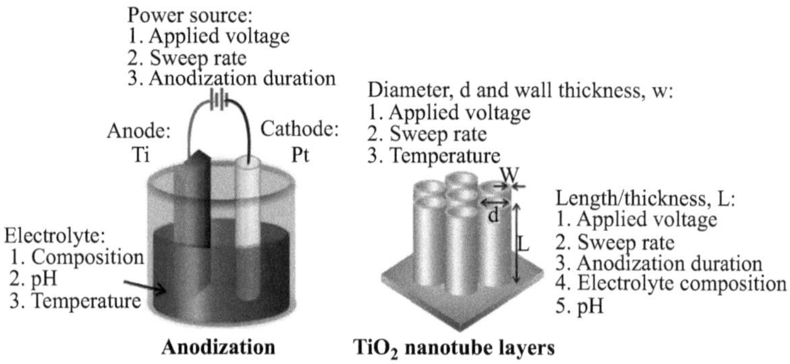

Figure 2.1 Conventional electrochemical anodization setup (left) and graphical illustration of anodic TiO_2 nanotube layers (right). The controllable parameters and their major influences on the geometry of the TiO_2 nanotube layers included (image courtesy: Ref. [33]).

Gong et al. eliminated chromic acid and employed aqueous HF electrolyte in anodization. Despite tubular structures were obtained, the TiO_2 nanotube layers were rather inhomogeneous [34]. An improvement in the morphology was proposed by Macak et al. by adding glycerol or ethylene glycol to increase the viscosity of the electrolyte, thereby reducing the growth rate to allow the growth of smoother nanotubes [35–37]. By extending the anodization durations to over 200 h, Paulose et al. reported nanotube layers up to 1000 μm [38]. The organic electrolytes typically enable the nanotube growth rate of about a few microns per hour [29, 30], which does not satisfy the needs of large-scale production. To cater to this demand, Banerjee et al. [39] added ethylenediaminetetraacetic acid (EDTA), whereas So et al. [40] adopted lactic acid in organic electrolytes to accelerate the growth rate of the nanotubes, up to tens of microns per hour.

For a long time, many works have focused on the investigation of anodization conditions in order to achieve well-defined, highly ordered, and thick anodic self-organized TiO_2 nanotube layers [29, 41, 42]. The studies include using different combinations of electrolyte [39, 43–51], controlling the pH [44, 52–54] and temperature [55, 56], as well as recycling the electrolyte [57]; varying the anodization potential [34, 44, 58–60] and duration [38, 45, 59, 61]; and employing different materials as counter electrode [62]. The effects of the conditions toward the morphologies are briefly summarized in Fig. 2.2. Surface and cross-sectional views of a conventional TiO_2 nanotube layer anodized in ethylene-glycol-based NH_4F electrolytes are depicted in Fig. 2.2(a,b). Overall, the key to the formation of tubular structure in an F^--based electrolyte is to establish a self-organization process, where a steady state is achieved by optimizing the oxide formation and chemical oxide dissolution process [29, 30].

Besides the conventional nanotube layers, other interesting morphologies such as bamboo and branching nanotubes, shown in Fig. 2.2(c,d), have been developed to increase the surface area for light-harvesting purposes [63–66]. An alternating voltage was utilized to obtain bamboo-like nanotubes with an additional surface area at the bamboo rings, as compared to the smooth-wall nanotubes [63]. On the other hand, a two-step anodization was adopted, where the first step produced the trunk and the second step intentionally decreased the anodization voltage to create the branches [64].

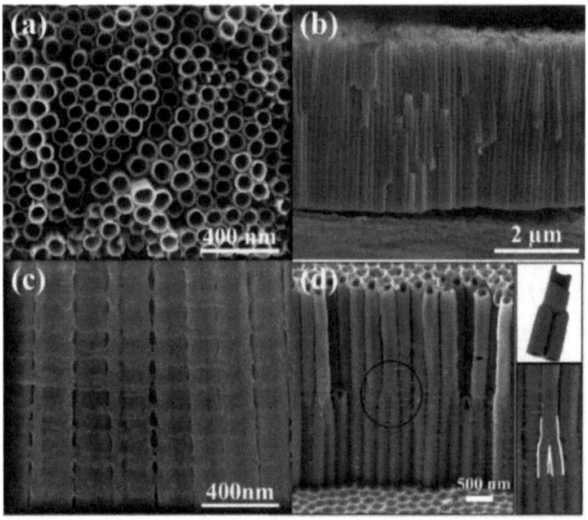

Figure 2.2 TiO_2 nanotube layers viewed by FESEM (a,b) conventional type, anodized in ethylene-glycol-based NH_4F electrolytes, surface and cross-sectional views (image courtesy: Ref. [33]); (c) bamboo type created by alternating voltages (image courtesy: Ref. [67]); (d) branching type produced by two-step anodization with different first and second anodization voltages (reprinted with permission from Ref. [64], copyright 2012, American Chemical Society).

It is worth mentioning that several studies have attempted to eliminate the F^- from the electrolyte; therefore, Cl^-, which belongs to the same group as F^-, was investigated [68–72]. Several works reported TiO_2 nanofibers from the anodization carried out in Cl^- electrolytes [68, 69]. Whereas the other works employed electrolytes containing NH_4Cl, HCl, NaCl, $HClO_4$, $NaClO_4$ and produced significantly longer nanotubes up to 30–50 μm within a short period, with very small diameter down to 10–15 nm [70–75]. During the anodization, the currents were relatively high even with the presence of high viscosity component, eventually leading to breakdown [72]. The TiO_2 nanotubes appeared as bundles scattering on the Ti surface, due to the random attack of the Ti substrate during the initial stage of anodization [70–72].

In contrast, in the F^- electrolytes, the nanotubes are perpendicularly oriented on the entire Ti substrate exposed to the electrolyte, forming a uniform oxide layer. Altogether, the anodic self-organized TiO_2 nanotube layers produced in F^- electrolytes remain as the prevalent 1D TiO_2 structure in various applications.

This leads to the effort to improve the degree of ordering of the self-organized nanotube layers, to achieve near hexagonally packed TiO_2 nanotube layers. Several works have opted to eliminate cracks and defects on the starting substrate, through mechanical, chemical, or electropolishing to obtain a smooth Ti surface prior to anodization [76–79]. Enlightened by the repetitive anodization of Al to improve the pore arrangement of AAO [80, 81], a similar approach was applied in the anodization of Ti. The first anodized layer was intentionally removed from the Ti substrate, followed by a second anodization or even repeated for a third anodization [82–85]. The removal of the first-grown layer leaves the concave textures on the Ti substrate, which serves as favorable sites for the growth of nanotubes during the subsequent anodization, hence improving the ordering of the nanotubes [84]. This idea is portrayed in the schematic diagram in Fig. 2.3. In a separate work, high-purity Ti foils (99.99%) were used as the starting substrate; the first anodization yielded rounded nanotubes, whereas the second anodization following the removal of the first layer eventually produced closely packed hexagonal nanotubes [83]. Extending the work of pre-texturing the Ti foil, nanoimprinting [86] and focused ion beam (FIB) milling [87, 88] were carried out to produce designated concaves on the Ti surface, in order to render guided-growth nanotubes.

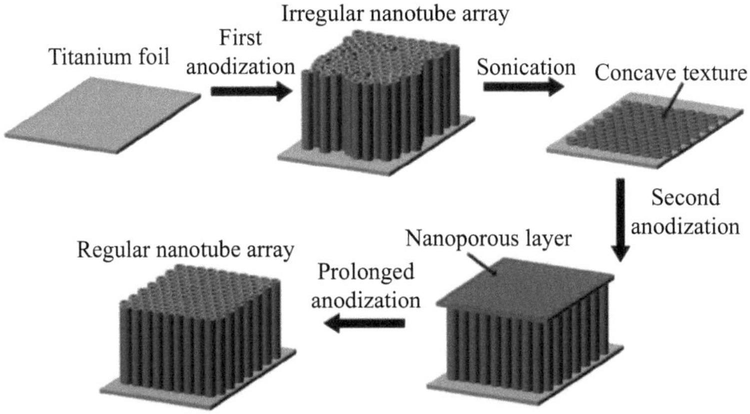

Figure 2.3 Schematic diagram illustrating the idea of two-step anodization to improve the ordering of the grown nanotubes (reprinted with permission from Ref. [84], copyright 2016, Elsevier).

2.4.4 Ion Implantation

The as-anodized TiO_2 samples were implanted with ions using the MEVVA ion implanter at the Australian Nuclear Science and Technology Organization (ANSTO). The titania nanotubes were implanted by exposing a flux of ions perpendicular to the sample surface. The average implantation energy of indium ions was 25 keV, and the peak penetration depth of indium in the titania nanotubes was close to the 70 nm stopping and range of ions in matter (SRIM) prediction based on the implantation parameters. The near-surface composition depth profiling of ion-implanted titania nanotubes was measured by Rutherford backscattering spectrometry (RBS) using He^{+1} ions at 2 MeV, and the implanted dose was $\sim 4 \times 10^{15}$ ions/cm^2.

2.5 Physical Properties of Anodic TiO_2 Nanotube Layers Annealed at Different Temperatures

In order to discuss the physical properties of the anodic self-organized TiO_2 nanotube layers, we employ a series of TiO_2 nanotube layers annealed at different temperatures as the study model. The anodization of Ti foil was carried out in an ethylene-glycol-based electrolyte consisting of 0.8 wt% NH_4F and 5 vol% solution of 0.25 M EDTA. A constant potential of 80 V was applied for 15 min with agitation throughout the anodization, to obtain TiO_2 nanotube layers [67]. These layers were annealed in air ambient for 3 h at temperatures ranging from 300 to 800°C.

2.5.1 Morphological Properties

Figure 2.4 presents the cross-sectional and surface morphology of the as-anodized and annealed TiO_2 nanotube layers imaged by field emission scanning electron microscope (FESEM). It is clearly observed that the nanotube layers treated at lower temperatures (up to 500°C) have no significant changes in morphology as compared to the as-anodized nanotube layers. The pore diameter was approximately 100 nm with very similar wall thicknesses. As the annealing temperature was increased to 600°C, a subtle increment in wall thickness, or in other words, a reduction in pore diameter was noted. For higher annealing temperatures at 700 and 800°C, a noticeable change in morphology is observed. The nanotube layers annealed at these temperatures were partially collapsed into particles form, especially for the extreme temperature at 800°C; the tubular structure was hardly distinguishable in the entire layer. Such observation suggests that high-temperature annealing destroys the structure of nanotube layers.

Based on the FESEM images in Fig. 2.4, statistical analyses of the layer thickness and pore diameter of the nanotube layers are summarized in the bottom-right corner in Fig. 2.4.

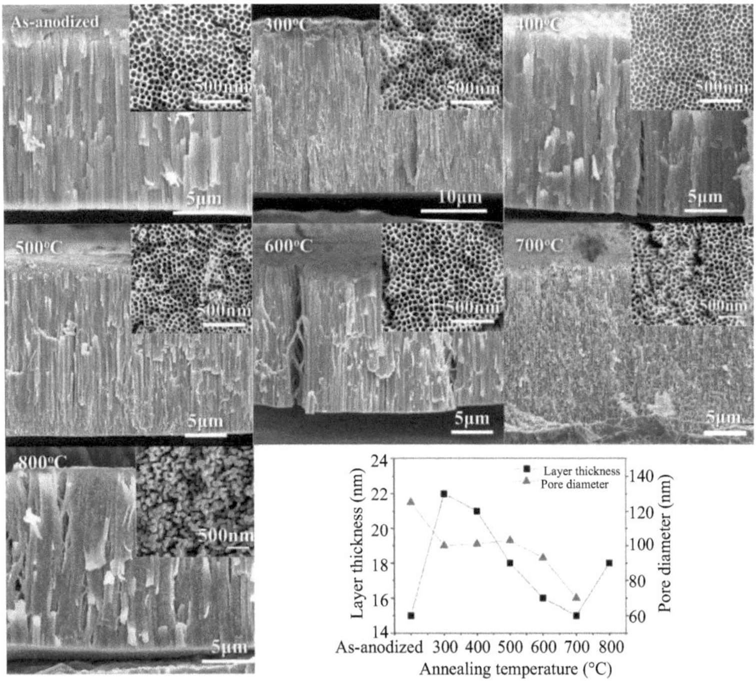

Figure 2.4 FESEM images of the cross-sectional view of TiO_2 nanotube layers anodized at 80 V for 15 min and annealed at temperatures ranging from 300 to 800°C. The insets show the top view of the nanotube layers. The bottom-right corner figure presents the layer thickness and the pore diameter of the nanotube layers as a function of annealing temperature.

2.5.2 Structural Properties

The as-anodized TiO_2 nanotube layer is amorphous, and simple thermal annealing is commonly applied to crystallize the nanotube layers [89]. Figure 2.5(a) compares the X-ray diffraction (XRD) patterns of the as-anodized and annealed TiO_2 nanotube layers, for 3 h at 300–800°C temperature in air ambient. The XRD pattern of the as-anodized layer contains only the diffraction peaks of Ti substrate [90]. Annealing temperatures at 300–600°C result in polycrystalline anatase structure, as multiple anatase diffraction peaks are observed. At 700°C, multiple rutile diffraction peaks were spotted, with more defined peaks at 800°C. This suggests that the TiO_2 nanotube (particle-like) layers annealed at 700 and 800°C comprise polycrystalline anatase and rutile structures.

The crystallite size of the TiO_2 nanotube layers was evaluated by using the Scherrer equation [91], for the dominant diffraction plane, i.e., anatase (101) and rutile (110), situated at 25.3 and 36.1°, respectively. Figure 2.5(b) shows the annealing temperature dependence of crystallite size, with an overall trend of increase in the crystallite size of both anatase and rutile with an increase in temperature, in accordance with the literature that crystallite sizes are highly dependent on the annealing temperature [90, 92]. The size of anatase crystallite lies within 25–47 nm, whereas the size of rutile crystallite (≈109–121 nm) is perceptibly larger than that of anatase.

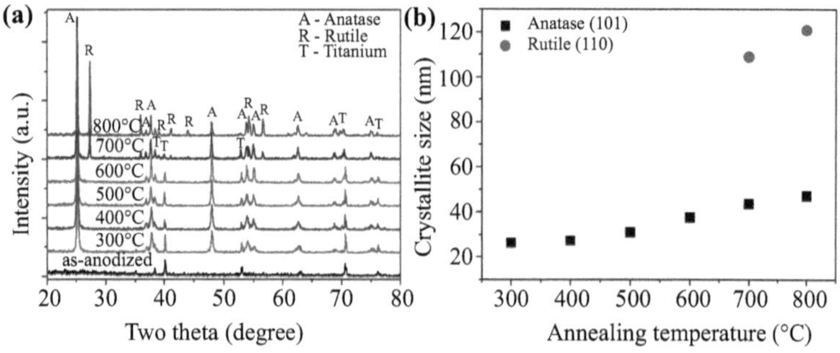

Figure 2.5 (a) XRD patterns of the as-anodized and annealed (300–800°C, 3 h, air ambient) TiO_2 nanotube layers. (b) Anatase and rutile crystallite size of the TiO_2 nanotube layers as a function of annealing temperatures, derived from the dominant diffraction plane in (a).

2.5.3 Optical Properties

The optical properties of the TiO_2 nanotube layers are investigated through reflectance measurement in the UV and visible spectral region. Figure 2.6(a) depicts the reflectance spectra of the as-anodized and annealed TiO_2 nanotube layers. The reflectance values are generally divided into two categories, with the low reflectance values for the as-anodized and low annealing temperatures up to 500°C; and high reflectance values for higher annealing temperatures 600–800°C.

The respective optical band gap is estimated based on the Kubelka–Munk function combined with the Tauc plot, applying indirect allowed transition for anatase TiO_2 [93]. This set of optical band gap values is in good agreement with the reported works between 3 and 3.4 eV [94–97].

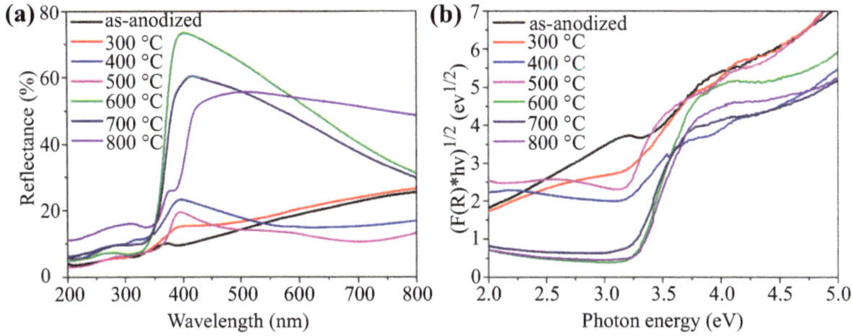

Figure 2.6 (a) Reflectance in the UV and visible spectral range and (b) Tauc plot of the Kubelka–Munk function for the determination of optical band gap, for the as-anodized and annealed (300–800°C, 3 h, air ambient) TiO_2 nanotube layers.

2.5.4 Vibrational Properties

The TiO_2 anatase crystalline structure has six active Raman modes $A_{1g}+2B_{1g}+3E_g$. The anatase modes are located at 144, 197, 639 cm^{-1} (all three are E_g), 399, 513 cm^{-1} (both are B_{1g}), and 519 cm^{-1} (A_{1g}) [98, 99]. On the other hand, the rutile crystalline structure of TiO_2 has four active modes $A_{1g}+B_{1g}+B_{2g}+E_g$. These peaks are normally observed at 232 cm^{-1} (B_{1g}), 447 cm^{-1} (E_g), 612 cm^{-1} (A_{1g}), and 826 cm^{-1} (B_{2g}) [98].

Figure 2.7(a) presents the Raman spectra of the as-anodized and annealed TiO_2 nanotube layers. The amorphous as-anodized nanotube layer does not show any active vibrational mode. Anatase modes present in all nanotube layers annealed at 300–700°C. Besides the anatase modes, for 700°C, rutile modes appear as small shoulder peak, marked in Fig. 2.7(a). These rutile peaks transform into well-defined peaks at 800°C, whereas the anatase modes diminish from sight. Overall, the Raman modes are in good agreement with the XRD patterns in Fig. 2.5. In addition, the peaks are refined at elevated temperatures, which indicate the improvement in crystallinity, similar to that in XRD observation. A complete view of the anatase E_g mode at ≈143 cm^{-1} is given in Fig. 2.7(b); by increasing the annealing temperature, the intensity of the E_g mode is increased and the full-width at half maximum (FWHM) is decreased. For 800°C, despite dominated by rutile modes, a small anatase E_g mode is recognized, shown in the inset of Fig. 2.7(b).

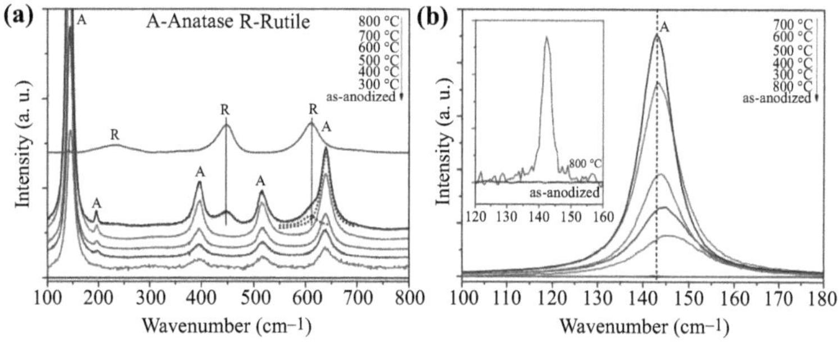

Figure 2.7 (a) Raman spectra of the as-anodized and annealed (300–800°C, 3 h, air ambient) TiO_2 nanotube layers. (b) Full anatase E_g mode at ≈143 cm^{-1} for all TiO_2 nanotube layers; inset shows magnified view for the as-anodized and 800°C annealed nanotube layers.

To enhance the Raman signal, Ag nanoparticles were introduced within the TiO_2 nanotube layers to create a surface-enhanced Raman scattering (SERS) effect [100]. Referring to the Raman spectra of the blank and Ag-nanoparticles-deposited TiO_2 nanotube layers in Fig. 2.8, the presence of Ag nanoparticles intensified each of the Raman peaks of the nanotube layer. In addition, additional Raman active modes were observed in Ag-deposited layers annealed at both 500 and 700°C, due to the surface selection rules (SSRs) in SERS [101]. To recapitulate, the Raman scattering effect occurs when photons gain or lose energy due to vibrational excitations and a shift in the frequency of the scattered light [102]. The SERS effect is a combination of electronic and chemical SERS, arises when electronic interactions between molecules and metal alter the Raman scattering process. As a result, a total scattering larger than that of the scattering of solely molecules is recorded as an enhanced Raman signal [102].

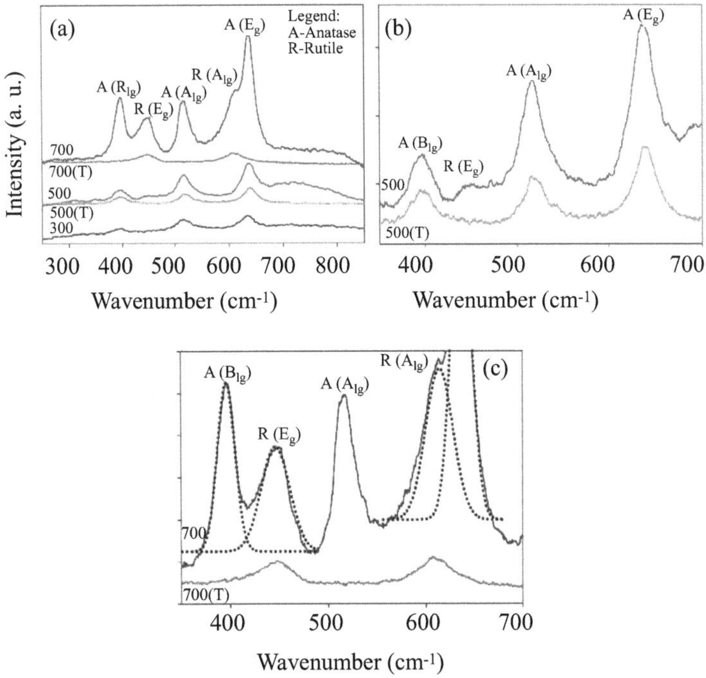

Figure 2.8 Raman spectra of the Ag-deposited and blank TiO_2 nanotube layers (a) annealed at 300, 500, and 700°C, (b,c) magnified view of layers annealed at 500 and 700°C. 500 and 500(T) denote the 500°C-annealed Ag-deposited and blank nanotube layers, respectively. The same denotation applies to the 700°C-annealed layers (image courtesy: Ref. [103]).

2.6 Modification and Functionalization of Anodic TiO_2 Nanotube Layers

As mentioned in Sections 2.4 and 2.5, self-organized anodic TiO_2 nanotube layers have a long investigation history in their fabrication and properties, with proven performances in numerous applications. Further modification or functionalization has been carried out within the anodic TiO_2 nanotube layers, to improve the specific features of the nanotube layers [104]. This includes band gap engineering, modification of surface chemistry, improvement in optical and electronic properties, creation of catalytic effects, and formation of heterojunction [33].

2.6.1 Doping

The notably large-band-gap energies (3.0–3.2 eV) of TiO_2 limit its light absorption in the visible spectral region. To exploit the solar spectrum, Asahi et al. have first reported an N-doped TiO_2 film with extended optical absorption activity up to 500 nm [105]. Since then, various dopants are introduced to modify the electronic and optical properties of TiO_2 nanotube layers. This includes lowering the conduction band, elevating the valence band, or introducing the intermediate state between conduction and valence bands [29, 106].

Nonmetal doping strategies that complement anodic TiO_2 nanotube layers are seen by adding the dopant precursor such as urea in the anodization electrolyte [107], and post-anodization electrochemical treatment of the nanotube layers in amines, urea [108], and NH_4Cl [109] solutions, to achieve N doping. Alternatively, thermal annealing of the amorphous nanotube layers was carried out in an atmosphere containing the dopant species, for instance, urea pyrolysis ambient for N and C co-doping [110], C_2H_2 [111], and CO [112] ambient both for C doping, as well as NH_3 and S ambient for N [113] and S [114] doping, respectively. These doped TiO_2 nanotube layers have shown reduced optical band gap range between 1.5 and 2.5 eV and recorded stronger visible light absorption [111–114].

In addition, high-energy ion implantation is applied to carry out N doping within TiO_2 nanotube layers and a significant reduction in band-gap energy to 2.2 eV [115]. Based on XPS measurement, a comparison against N doping via thermal annealing in NH_3 environment suggests that the N dopants implanted within the structure by this approach possess chemically bonded state [113, 115]. Although this method is claimed as the most effective way to dope TiO_2 nanotube layers, many defects are created by ion bombardment. Nevertheless, the defects can be removed by a simple heat treatment [115].

As for metal doping, a simple and direct method is through the anodization of pre-alloyed Ti-X substrates, where X is the metal dopant species. Several successful examples are Ti-Al [116], Ti-Nb [116, 117], Ti-Mo [118], Ti-Zr [119], and Ti-Al-V [120] with the different weight percentage of alloys, to obtain X-doped TiO_2 tubular structure identical to that of the undoped TiO_2 nanotube layers, as the minor content of the dopants does not alter the formation of the nanotubes. These metal-doped nanotube layers have demonstrated improved photocatalytic activities [106], electrochromic color contrast [118], and higher intercalation activity and stability [117].

2.6.2 Reduction and Self-Doping "Black TiO_2"

The presence of oxygen vacancies in TiO_2 nanotube layers affects the electronic properties [29]. Therefore, Ti^{4+} is reduced to Ti^{3+} (also known as self-doping), creates an oxygen vacancy, which acts as electron donors, increases carrier density, and hence increases conductivity [121–123]. The change in electronic properties results in a change in color; therefore, changes from yellow to blue or gray and eventually black TiO_2 may be observed [122–124].

The doping approaches discussed in Section 2.6.1 are also applied for the self-doping of TiO_2. The TiO_2 nanotube layers were annealed in reduction gas ambient, for example, H_2, Ar/H_2, or vacuum condition at elevated temperatures, to achieve the reduction effect [124, 125]. Combining Ar/H_2 with high pressure treatment has reported strong improvement in conductivity and photoconductivity as compared to conventional Ar/H_2 atmospheric pressure treatment [124].

In addition, TiO_2 nanotube layers treated by high-energy proton implantation demonstrate an advantageous length effect, with the upper ≈1 μm of the tubes being implanted by proton served as a catalytically active center and the lower part of the tubes acted as the light absorber. Such effect was not possible by a flat TiO_2 surface as the surface was partially damaged during the implantation process [126]. Nevertheless, these hydrogenated black TiO_2 nanotube layers have recorded high H_2 evolution rate, without the presence of any noble metal co-catalyst [124, 126].

A relatively simple approach, i.e., electrochemical reduction in different electrolytes, is utilized to achieve the reduction effect [121, 123, 127, 128]. An inert electrolyte such as Na_2SO_4 [123] and $(NH_4)_2SO_4$ [127] has been applied to carry out conventional cathodic reduction. Whereas for specific purposes, electrolytes such as ethylene glycol [121] and $LiClO_4$ [129] solution, which produce H^+ and Li^+ ions, respectively, are intercalated into TiO_2 nanotube layers. The hydrogenated TiO_2 or Li-ion inserted TiO_2 nanotube layers significantly improved photocurrent conversion efficiencies, performance of photo-electrocatalysts, conductivity, and capacitive properties [121, 128, 129].

2.6.3 Surface Modification

To modify the surface of the TiO_2 nanotube layers, "$TiCl_4$ treatment" is carried out to decorate the inner and outer walls of the nanotubes by TiO_2 nanoparticles [130–132]. This effort is intended to increase the specific surface area, similar to that of the bamboo morphology discussed in Section 2.3.3, for higher dye adsorption in DSSCs, as well as to improve the electronic transport properties [130–132]. Typically, the nanotube layers are immersed in a $TiCl_4$ solution, followed by thermal annealing to crystallize the TiO_2 nanoparticles. An increased photocurrent density and increased DSSC efficiency to 7–8% were documented, through anchoring TiO_2 nanoparticles on the tube walls [130, 131].

TiO_2-nanoparticles-decorated TiO_2 nanotube layers have also shown enhancement in photocurrent conversion and photocatalytic decomposition of methylene blue, due to the additional photoactive surface area within the tube walls [96]. Further improvement in both photoactivities has been attained by continuous ALD TiO_2 coating within the nanotube layers. This is attributed to the high-quality, pin-hole-free TiO_2 coating [133], which passivates the surface states and improves the charge collection efficiencies [96].

2.6.4 Incorporation of Metals and Semiconductors

The integration of secondary materials is in effort to (i) form junction with materials of suitable band position for effective band-bending, (ii) create co-catalytic effect to facilitate the charge transfer process, (iii) induce surface plasmon resonance effect, and (iv) provide a high synergistic platform [106, 134, 135].

For these purposes, numerous materials have been incorporated within the TiO_2 nanotube layers [104], which are sub-categorized into metals, oxides, chalcogenides, and graphene. Among all, chemical bath deposition with facile precursors was popularly adopted by immersing the TiO_2 nanotube layers within the precursors, with a variation of spinning, sonicating, or heating the precursors to assist the loading of secondary material within the nanotube layers. Chemical bath deposition can be easily modified into electrodeposition by applying an electric field to the precursor solution in order to control the direction and amount of the deposits. Examples of semiconductor materials deposited within TiO_2 nanotube layers by liquid deposition techniques are ZnO [136, 137], WO_3 [138, 139], Fe_2O_3 [140] , Fe_3O_4 [141], MnO_2 [142], Cu_2O [127, 143], NiO [144], CdS [145, 146], ZnS [147], CdSe [74], ZnSe [146], $CuInSe_2$ [148, 149], CdTe [150], and for the metal family are Ag [100, 151], Au [152, 153], Ni [154], and Pt [155, 156]. The challenges of both chemical bath and electrodeposition are mainly attributed to the initial distribution of uniform nuclei along the nanotube walls and the control of the deposition to prevent blockage at the surface of nanotubes.

For the decoration of noble metal particles, thermal dewetting has been carried out by radio-frequency sputtering of an ultrathin metal film followed by thermal treatment to disintegrate the metal film into metal particles [157–159]. Studying the kinetics and thermodynamics of the targeted metal film to optimize the sputtered film thickness and treatment temperature allows fine control of the size and distribution of the metal nanoparticles within a specific template [134]. This has seen success for the selective decoration of

Ag [157], Au [158], and Pt [159] nanoparticles within TiO_2 nanotube layers.

However, for the mentioned deposition techniques, the general limitation is the deposition toward the bottom of the nanotube layers at the interface with titanium substrate. Similar to the deposition of TiO_2 within AAO channels by ALD (discussed in Section 2.3.2), ALD is utilized for the deposition within TiO_2 nanotube layers, as it offers uniform and conformal coatings along the template. Uniform deposition of a range of materials has been reported, including Al_2O_3 [160–163], TiO_2 [96, 164], ZnO [165–167], WO_3 [168], In_2O_3 [169], In_2S_3 [170], CdS [171, 172], $MoSe_xO_y$ [173], as well as metal particles Pd [174] and Pt [175] within the nanotube layers.

The deposition, decoration, or incorporation of (i) metal catalysts, particularly the noble metal nanoparticles; (ii) oxides, especially transition metal oxides coating or structures; and (iii) chalcogenides has achieved the purposes described at the beginning of this section. These modified anodic TiO_2 nanotube layers have sought enhancement as a photocatalyst in DSSCs, photoelectrochemical activity, H_2 evolution, photoreduction of CO_2, direct fuel cells, photo-disintegration of pollutants, water purification, ion-intercalation batteries, supercapacitors, electrochromics, memristive devices, photodetector, chemical sensors, biosensors, drug delivery, and biomedical implants [29, 30, 41, 42, 134, 176–178].

Graphene and graphene oxide with unique physical and chemical properties have been widely studied in recent years and have been deposited within TiO_2 nanotube layers [179, 180]. Graphene oxide was separately prepared from graphite powder modified using Hummers' method. Liu et al. adopted electrodeposition followed by photoreduction process to prepare graphene sheets on the surface of TiO_2 nanotube layers [179]. On the other hand, Chen et al. [181] performed electrochemically reduced graphene oxide, while Razzaq et al. [180] employed electrophoretic deposition of graphene oxide on TiO_2 nanotube layers forming highly transparent graphene films on the surface of the nanotubes. In a separate work, Song et al. immersed TiO_2 nanotube layers in graphene oxide suspension to form graphene oxide sheets on the surface of nanotube layers [182]. The graphene-based modifications have shown enhanced visible spectral light absorption range and intensity, which promoted the photoelectrochemical and photocatalytic activities [179, 180, 182, 183].

2.7 Synthesis of Doped TiO_2 Nanostructures

The as-anodized TiO_2 nanotube samples were doped with Cr ions by exposure to a flux of ions directed perpendicular to the sample surface. The Cr ions were produced using a direct extraction high flux ion source, with a charge distribution close to the theoretical prediction (9% Cr11, 75% Cr12, 13% Cr13, 3% Cr14), corresponding to an average charge of 12.1, giving an average implantation energy of 63 keV. With these implantation parameters, the depth of Cr peak in TiO_2 is close to the stopping and range of ions in matter (SRIM) prediction of 35 nm. The near-surface composition depth profiling of ion-implanted samples was measured by RBS using He^{1+} ions at 1.8 MeV and the implanted dose was ~7×10^{14} ions/cm^2. This phase of work was conducted at the Australian Nuclear Science and Technology Organisation (ANSTO).

The synthesized TiO_2 nanofiber material was cut into 14 mm × 14 mm squares, which were then implanted with V ions using the MEVVA ion implanter at the ANSTO installation. During implantation, the samples were doped by exposing a flux of V ions directed perpendicular to the sample surface. The average implantation energy of V ions produced was 25 keV, and the penetration depth of V peak in the TiO_2 fibers was close to the stopping and range of ions in matter (SRIM) prediction of 70 nm based on these implantation parameters. The near-surface composition depth profiling of V ion-implanted TiO_2 nanofibers was measured by RBS using He^{1+} ions at 2 MeV, and the implanted dose was ~4×10^{15} ions/cm^2 [4, 5, 8–10].

References

1. Albetran, H., Haroosh, H., Dong, Y., O'Connor, B., and Low, I. (2014). Effect of atmosphere on crystallisation kinetics and phase relations in electrospun TiO_2 nanofibres, *Ceram. Environm. En. Appl. II: Ceram. Trans.*, **246**, pp. 125–133.

2. Albetran, H., Haroosh, H., Dong, Y., Prida, V., O'Connor, B., and Low, I. M. (2014). Phase transformations and crystallization kinetics in

electrospun TiO_2 nanofibers in air and argon atmospheres, *Appl. Phys. A*, **116**, pp. 161–169.

3. Albetran, H., Dong, Y., and Low, I. M. (2015). Characterization and optimization of electrospun TiO_2/PVP nanofibers using Taguchi design of experiment method, *J. Asian Ceram. Soc.*, **3**, pp. 292–300.
4. Albetran, H., O'Connor, B., Prida, V., and Low, I. (2015). Effect of vanadium ion implantation on the crystallization kinetics and phase transformation of electrospun TiO_2 nanofibers, *Appl. Phys. A*, **120**, pp. 623–634.
5. Albetran, H. and Low, I. M. (2016). Crystallization kinetics and phase transformations in aluminum ion-implanted electrospun TiO_2 nanofibers, *Appl. Phys. A*, **122**, pp. 1044–1053.
6. Albetran, H., O'Connor, B., and Low, I. M. (2016). Activation energies for phase transformations in electrospun titania nanofibers: Comparing the influence of argon and air atmospheres, *Appl. Phys. A*, **122**, pp. 367–376.
7. Albetran, H., O'Connor, B., and Low, I. (2016). Effect of calcination on band gaps for electrospun titania nanofibers heated in air–argon mixtures, *Mater. Des.*, **92**, pp. 480–485.
8. Low, I. M., Albetran, H., Prida, V., Manurung, P., and Ionescu, M. (2012). Effect of chromium-doping on the crystallization and phase stability in anodized TiO_2 nanotubes, *Dev. Strategic Mater. Comput. Des.* III, pp. 149–158.
9. Low, I. M., Albetran, H., Prida, V. M., Vega, V., Manurung, P., and Ionescu, M. (2013). A comparative study on crystallization behavior, phase stability, and binding energy in pure and Cr-doped TiO_2 nanotubes, *J. Mater. Res.*, **28**, pp. 304–312.
10. Albetran, H. and Low, I. M. (2016). Effect of indium ion implantation on crystallization kinetics and phase transformation of anodized titania nanotubes using in-situ high-temperature radiation diffraction, *J. Mater. Res.*, **31**, pp. 1588–1595.
11. Albetran, H., Vega, V., Prida, V. M., and Low, I. M. (2018). Dynamic diffraction studies on the crystallization, phase transformation, and activation energies in anodized titania nanotubes, *Nanomaterials*, **8**, pp. 122–131.
12. Kasuga, T., Hiramatsu, M., Hoson, A., Sekino, T., and Niihara, K. (1998). Formation of titanium oxide nanotube, *Langmuir*, **14**, pp. 3160–3163.
13. Kasuga, T., Hiramatsu, M., Hoson, A., Sekino, T., and Niihara, K. (1999). Titania nanotubes prepared by chemical processing, *Adv. Mater.*, **11**, pp. 1307–1311.

14. Bavykin, D. V., Parmon, V. N., Lapkin, A. A., and Walsh, F. C. (2004). The effect of hydrothermal conditions on the mesoporous structure of TiO_2 nanotubes, *J. Mater. Chem.*, **14**, pp. 3370–3377.

15. Morgan, D. L., Liu, H.-W., Frost, R. L., and Waclawik, E. R. (2009). Implications of precursor chemistry on the alkaline hydrothermal synthesis of titania/titanate nanostructures, *J. Phys. Chem. C*, **114**, pp. 101–110.

16. Liu, B. and Aydil, E. S. (2009). Growth of oriented single-crystalline rutile TiO_2 nanorods on transparent conducting substrates for dye-sensitized solar cells, *J. Am. Chem. Soc.*, **131**, pp. 3985–3990.

17. Kumar, A., Madaria, A. R., and Zhou, C. (2010). Growth of aligned single-crystalline rutile TiO_2 nanowires on arbitrary substrates and their application in dye-sensitized solar cells, *J. Phys. Chem. C*, **114**, pp. 7787–7792.

18. Ulusoy, T. G., Ghobadi, A., and Okyay, A. K. (2014). Surface engineered angstrom thick ZnO-sheathed TiO_2 nanowires as photoanodes for performance enhanced dye-sensitized solar cells, *J. Mater. Chem. A*, **2**, pp. 16867–16876.

19. Xie, Y., Wei, L., Wei, G., Li, Q., Wang, D., Chen, Y., Yan, S., Liu, G., Mei, L., and Jiao, J. (2013). A self-powered UV photodetector based on TiO_2 nanorod arrays, *Nanoscale Res. Lett.*, **8**, pp. 188.

20. Zhang, M., Bando, Y., and Wada, K. (2001). Sol-gel template preparation of TiO_2 nanotubes and nanorods, *J. Mater. Sci. Lett.*, **20**, pp. 167–170.

21. Miao, L., Tanemura, S., Toh, S., Kaneko, K., and Tanemura, M. (2004). Fabrication, characterization and Raman study of anatase-TiO_2 nanorods by a heating-sol–gel template process, *J. Cryst. Growth*, **264**, pp. 246–252.

22. Attar, A. S., Ghamsari, M. S., Hajiesmaeilbaigi, F., Mirdamadi, S., Katagiri, K., and Koumoto, K. (2009). Sol–gel template synthesis and characterization of aligned anatase-TiO_2 nanorod arrays with different diameter, *Mater. Chem. Phys.*, **113**, pp. 856–860.

23. Wang, H., Song, Y., Liu, W., Yao, S., and Zhang, W. (2013). Template synthesis and characterization of TiO_2 nanotube arrays by the electrodeposition method, *Mater. Lett.*, **93**, pp. 319–321.

24. Zhang, X., Zhang, L., Chen, W., Meng, G., Zheng, M., Zhao, L., and Phillipp, F. (2001). Electrochemical fabrication of highly ordered semiconductor and metallic nanowire arrays, *Chem. Mater.*, **13**, pp. 2511–2515.

25. Yao, Z., Wang, C., Li, Y., and Kim, N.-Y. (2015). AAO-assisted synthesis of highly ordered, large-scale TiO_2 nanowire arrays via sputtering and atomic layer deposition, *Nanoscale Res. Lett.*, **10**, pp. 166–173.

26. Foong, T. R., Shen, Y., Hu, X., and Sellinger, A. (2010). Template-directed liquid ALD growth of TiO_2 nanotube arrays: Properties and potential in photovoltaic devices, *Adv. Funct. Mater.*, **20**, pp. 1390–1396.

27. Lee, J., Kim, D. H., Hong, S. H., and Jho, J. Y. (2011). A hydrogen gas sensor employing vertically aligned TiO_2 nanotube arrays prepared by template-assisted method, *Sens. Actuators B: Chem.*, **160**, pp. 1494–1498.

28. Kemell, M., Pore, V., Tupala, J., Ritala, M., and Leskelä, M. (2007). Atomic layer deposition of nanostructured TiO_2 photocatalysts via template approach, *Chem. Mater.*, **19**, pp. 1816–1820.

29. Lee, K., Mazare, A., and Schmuki, P. (2014). One-dimensional titanium dioxide nanomaterials: Nanotubes, *Chem. Rev.*, **114**, pp. 9385–9454.

30. Macak, J. M., Tsuchiya, H., Ghicov, A., Yasuda, K., Hahn, R., Bauer, S., and Schmuki, P. (2007). TiO_2 nanotubes: Self-organized electrochemical formation, properties and applications, *Curr. Opin. Solid State Mater., Sci.*, **11**, pp. 3–18.

31. Assefpour-Dezfuly, M., Vlachos, C., and Andrews, E. (1984). Oxide morphology and adhesive bonding on titanium surfaces, *J. Mater. Sci.*, **19**, pp. 3626–3639.

32. Zwilling, V., Darque-Ceretti, E., Boutry-Forveille, A., David, D., Perrin, M. Y., and Aucouturier, M. (1999). Structure and physicochemistry of anodic oxide films on titanium and TA6V alloy, *Surf. Interface Anal.*, **27**, pp. 629–637.

33. Ng, S. W. (2018). *ZnO and MoSexOy Modified TiO_2 Nanotubes for Photoelectrochemical and Sensing Applications* (Universiti Sains Malaysia).

34. Gong, D., Grimes, C. A., Varghese, O. K., Hu, W., Singh, R., Chen, Z., and Dickey, E. C. (2001). Titanium oxide nanotube arrays prepared by anodic oxidation, *J. Mater. Res.*, **16**, pp. 3331–3334.

35. Macak, J. M., Tsuchiya, H., Taveira, L., Aldabergerova, S., and Schmuki, P. (2005). Smooth anodic TiO_2 nanotubes, *Angew. Chem. Int. Ed.*, **44**, pp. 7463–7465.

36. Macak, J. M., Tsuchiya, H., and Schmuki, P. (2005). High-aspect-ratio TiO_2 nanotubes by anodization of titanium, *Angew. Chem. Int. Ed.*, **44**, pp. 2100–2102.

37. Macak, J. M. and Schmuki, P. (2006). Anodic growth of self-organized anodic TiO_2 nanotubes in viscous electrolytes, *Electrochim. Acta.*, **52**, pp. 1258–1264.

38. Paulose, M., Prakasam, H. E., Varghese, O. K., Peng, L., Popat, K. C., Mor, G. K., Desai, T. A., and Grimes, C. A. (2007). TiO_2 nanotube arrays of 1000 μm length by anodization of titanium foil: Phenol red diffusion, *J. Phys. Chem. C*, **111**, pp. 14992–14997.

39. Banerjee, S., Misra, M., Mohapatra, S. K., Howard, C., Mohapatra, S. K., and Kamilla, S. K. (2010). Formation of chelating agent driven anodized TiO_2 nanotubular membrane and its photovoltaic application, *Nanotechnology*, **21**, pp. 145201.

40. So, S., Lee, K., and Schmuki, P. (2012). Ultrafast growth of highly ordered anodic TiO_2 nanotubes in lactic acid electrolytes, *J. Am. Chem. Soc.*, **134**, pp. 11316–11318.

41. Roy, P., Berger, S., and Schmuki, P. (2011). TiO_2 nanotubes: Synthesis and applications, *Angew. Chem. Int. Ed.*, **50**, pp. 2904–2939.

42. Liu, G., Wang, K., Hoivik, N., and Jakobsen, H. (2012). Progress on free-standing and flow-through TiO_2 nanotube membranes, *Sol. En. Mater. Sol. Cells*, **98**, pp. 24–38.

43. Cheong, Y. L., Yam, F. K., Chin, I. K., and Hassan, Z. (2013). X-ray analysis of nanoporous TiO_2 synthesized by electrochemical anodization, *Superlattices Microstruct.*, **64**, pp. 37–43.

44. Cai, Q., Paulose, M., Varghese, O. K., and Grimes, C. A. (2005). The effect of electrolyte composition on the fabrication of self-organized titanium oxide nanotube arrays by anodic oxidation, *J. Mater. Res.*, **20**, pp. 230–236.

45. Paulose, M., Shankar, K., Yoriya, S., Prakasam, H. E., Varghese, O. K., Mor, G. K., Latempa, T. A., Fitzgerald, A., and Grimes, C. A. (2006). Anodic growth of highly ordered TiO_2 nanotube arrays to 134 μm in length, *J. Phys. Chem. B*, **110**, pp. 16179–16184.

46. Mor, G., Varghese, O. K., Paulose, M., Mukherjee, N., and Grimes, C. A. (2003). Fabrication of tapered, conical-shaped titania nanotubes, *J. Mater. Res.*, **18**, pp. 2588–2593.

47. Cai, Q., Yang, L., and Yu, Y. (2006). Investigations on the self-organized growth of TiO_2 nanotube arrays by anodic oxidization, *Thin Solid Films*, **515**, pp. 1802–1806.

48. Ruan, C., Paulose, M., Varghese, O. K., and Grimes, C. A. (2006). Enhanced photoelectrochemical-response in highly ordered TiO_2 nanotube-arrays anodized in boric acid containing electrolyte, *Sol. En. Mater. Sol. Cells*, **90**, pp. 1283–1295.

49. Ruan, C., Paulose, M., Varghese, O. K., Mor, G. K., and Grimes, C. A. (2005). Fabrication of highly ordered TiO_2 nanotube arrays using an organic electrolyte, *J. Phys. Chem. B*, **109**, pp. 15754–15759.
50. Shankar, K., Mor, G. K., Fitzgerald, A., and Grimes, C. A. (2007). Cation effect on the electrochemical formation of very high aspect ratio TiO_2 nanotube arrays in formamide–water mixtures, *J. Phys. Chem. C*, **111**, pp. 21–26.
51. Yam, F. K., Beh, K. P., Ng, S. W., and Hassan, Z. (2011). The effects of morphological changes on the vibrational properties of self-organized TiO_2 nanotubes, *Thin Solid Films*, **520**, pp. 807–812.
52. Sreekantan, S., Lockman, Z., Hazan, R., Tasbihi, M., Tong, L. K., and Mohamed, A. R. (2009). Influence of electrolyte pH on TiO_2 nanotube formation by Ti anodization, *J. Alloys Compd.*, **485**, pp. 478–483.
53. Macak, J. M., Sirotna, K., and Schmuki, P. (2005). Self-organized porous titanium oxide prepared in Na_2SO_4/NaF electrolytes, *Electrochim. Acta.*, **50**, pp. 3679–3684.
54. Lockman, Z., Ismail, S., Sreekantan, S., Schmidt-Mende, L., and MacManus-Driscoll, J. L. (2009). The rapid growth of 3 μm long titania nanotubes by anodization of titanium in a neutral electrochemical bath, *Nanotechnology*, **21**, pp. 055601.
55. Wang, J. and Lin, Z. (2009). Anodic formation of ordered TiO_2 nanotube arrays: Effects of electrolyte temperature and anodization potential, *J. Phys. Chem. C*, **113**, pp. 4026–4030.
56. Chen, J., Lin, J., and Chen, X. (2010). Self-assembled TiO_2 nanotube arrays with U-shaped profile by controlling anodization temperature, *J. Nanomater.*, **2010**, pp. 38–42.
57. Prakasam, H. E., Shankar, K., Paulose, M., Varghese, O. K., and Grimes, C. A. (2007). A new benchmark for TiO_2 nanotube array growth by anodization, *J. Phys. Chem. C*, **111**, pp. 7235–7241.
58. Bestetti, M., Franz, S., Cuzzolin, M., Arosio, P., and Cavallotti, P. L. (2007). Structure of nanotubular titanium oxide templates prepared by electrochemical anodization in H_2SO_4/HF solutions, *Thin Solid Films*, **515**, pp. 5253–5258.
59. Wan, J., Yan, X., Ding, J., Wang, M., and Hu, K. (2009). Self-organized highly ordered TiO_2 nanotubes in organic aqueous system, *Mater. Charact.*, **60**, pp. 1534–1540.
60. Lockman, Z., Sreekantan, S., Ismail, S., Schmidt-Mende, L., and MacManus-Driscoll, J. L. (2010). Influence of anodisation voltage on the dimension of titania nanotubes, *J. Alloys Compd.*, **503**, pp. 359–364.

61. Tighineanu, A., Ruff, T., Albu, S., Hahn, R., and Schmuki, P. (2010). Conductivity of TiO_2 nanotubes: Influence of annealing time and temperature, *Chem. Phys. Lett.*, **494**, pp. 260–263.

62. Allam, N. K. and Grimes, C. A. (2008). Effect of cathode material on the morphology and photoelectrochemical properties of vertically oriented TiO_2 nanotube arrays, *Sol. En. Mater. Sol. Cells*, **92**, pp. 1468–1475.

63. Kim, D., Ghicov, A., Albu, S. P., and Schmuki, P. (2008). Bamboo-type TiO_2 nanotubes: Improved conversion efficiency in dye-sensitized solar cells, *J. Am. Chem. Soc.*, **130**, pp. 16454–16455.

64. Chen, B. and Lu, K. (2012). Hierarchically branched titania nanotubes with tailored diameters and branch numbers, *Langmuir*, **28**, pp. 2937–2943.

65. Rambabu, Y., Jaiswal, M., and Roy, S. C. (2015). Enhanced photoelectrochemical performance of multi-leg TiO_2 nanotubes through efficient light harvesting, *J. Phys. D: Appl. Phys.*, **48**, pp. 295302.

66. Samuel, J., Beh, K. P., Cheong, Y. L, Yusuf, W., and Yam, F. K. (2018). Novel bamboo structured TiO_2 nanotubes for energy storage/production applications, *J. Phys. Conference Series*, **995**, 012054.

67. Cheong, Y. L. (2018). *Investigation on Titanium Dioxide Nanotubes Photoanode for Dye Sensitized Solar Cell Application* (Universiti Sains Malaysia).

68. Beh, K. P., Yam, F. K., Tneh, S. S., and Hassan, Z. (2011). Fabrication of titanium dioxide nanofibers via anodic oxidation, *Appl. Surf. Sci.*, **257**, pp. 4706–4708.

69. Low, I. M., Yam, F. K., and Pang, W. K. (2012). In-situ diffraction studies on the crystallization and crystal growth in anodized TiO_2 nanofibres, *Mater. Lett.*, **87**, pp. 150–152.

70. Hahn, R., Macak, J., and Schmuki, P. (2007). Rapid anodic growth of TiO_2 and WO_3 nanotubes in fluoride free electrolytes, *Electrochem. Commun.*, **9**, pp. 947–952.

71. Ng, S. W., Yam, F. K., Beh, K. P., and Hassan, Z. (2014). Titanium dioxide nanotubes in chloride based electrolyte: An alternative to fluoride based electrolyte, *Sains Malaysiana*, **43**, pp. 947–951.

72. Cheong, Y. L., Yam, F. K., Ng, S. W., Hassan, Z., Ng, S. S, and Low, I. M. (2015). Fabrication of titanium dioxide nanotubes in fluoride-free electrolyte via rapid breakdown anodization, *J. Porous Mater.*, **22**, pp. 1437–1444.

73. Richter, C., Wu, Z., Panaitescu, E., Willey, R. J., and Menon, L. (2007). Ultra-high-aspect-ratio titania nanotubes, *Adv. Mater.*, **19**, pp. 946–948.

74. Allam, N. K., Shankar, K., and Grimes, C. A. (2008). Photoelectrochemical and water photoelectrolysis properties of ordered TiO_2 nanotubes fabricated by Ti anodization in fluoride-free HCl electrolytes, *J. Mater. Chem.*, **18**, pp. 2341–2348.

75. Chen, X., Schriver, M., Suen, T., and Mao, S. S. (2007). Fabrication of 10 nm diameter TiO_2 nanotube arrays by titanium anodization, *Thin Solid Films*, **515**, pp. 8511–8514.

76. Albertin, K., Tavares, A., and Pereyra, I. (2013). Optimized Ti polishing techniques for enhanced order in TiO_2 NT arrays, *Appl. Surf. Sci.*, **284**, pp. 772–779.

77. Lee, B. G., Hong, S.-Y., Yoo, J. E., and Choi, J. (2011). Electropolishing for the formation of anodic nanotubular TiO_2 with uniform length and density, *Appl. Surf. Sci.*, **257**, pp. 7190–7194.

78. Asgari, V., Noormohammadi, M., Ramazani, A., and Kashi, M. A. (2018). A new approach to electropolishing of pure Ti foil in acidic solution at room temperature for the formation of ordered and long TiO_2 nanotube arrays, *Corros. Sci.*, **136**, pp. 38–46.

79. Shin, Y. and Lee, S. (2008). Self-organized regular arrays of anodic TiO_2 nanotubes, *Nano Lett.*, **8**, pp. 3171–3173.

80. Lee, W., Ji, R., Gösele, U., and Nielsch, K. (2006). Fast fabrication of long-range ordered porous alumina membranes by hard anodization, *Nat. Mater.*, **5**, pp. 741.

81. Masuda, H. and Fukuda, K. (1995). Ordered metal nanohole arrays made by a two-step replication of honeycomb structures of anodic alumina, *Science*, **268**, pp. 1466–1468.

82. Zhang, G., Huang, H., Zhang, Y., Chan, H. L., and Zhou, L. (2007). Highly ordered nanoporous TiO_2 and its photocatalytic properties, *Electrochem. Commun.*, **9**, pp. 2854–2858.

83. Macak, J. M., Albu, S. P., and Schmuki, P. (2007). Towards ideal hexagonal self-ordering of TiO_2 nanotubes, *Phys. Status Solidi Rapid Res. Lett.*, **1**, pp. 181–183.

84. Cheong, Y. L., Beh, K. P., Yam, F. K., and Hassan, Z. (2016). Performance evaluation of titanium dioxide based dye-sensitized solar cells under the influence of anodization steps, nanotube length and ionic liquid-free redox electrolyte solvents, *Superlattices Microstruct.*, **94**, pp. 74–84.

85. Gui, Q., Yu, D., Li, D., Song, Y., Zhu, X., Cao, L., Zhang, S., Ma, W., and You, S. (2014). Efficient suppression of nanograss during porous anodic TiO_2 nanotubes growth, *Appl. Surf. Sci.*, **314**, pp. 505–509.

86. Kondo, T., Nagao, S., Yanagishita, T., Nguyen, N. T., Lee, K., Schmuki, P., and Masuda, H. (2015). Ideally ordered porous TiO_2 prepared by anodization of pretextured Ti by nanoimprinting process, *Electrochem. Commun.*, **50**, pp. 73–76.

87. Chen, B., Lu, K., and Tian, Z. (2011). Effects of titania nanotube distance and arrangement during focused ion beam guided anodization, *J. Mater. Chem.*, **21**, pp. 8835–8840.

88. Sopha, H., Samoril, T., Palesch, E., Hromadko, L., Zazpe, R., Skoda, D., Urbanek, M., Ng, S., Prikryl, J., and Macak, J. M. (2017). Ideally hexagonally ordered TiO_2 nanotube arrays, *ChemistryOpen*, **6**, pp. 480–483.

89. Albu, S. P., Tsuchiya, H., Fujimoto, S., and Schmuki, P. (2010). TiO_2 nanotubes: Annealing effects on detailed morphology and structure, *Eur. J. Inorg. Chem.*, **2010**, pp. 4351–4356.

90. Varghese, O. K., Gong, D., Paulose, M., Grimes, C. A., and Dickey, E. C. (2003). Crystallization and high-temperature structural stability of titanium oxide nanotube arrays, *J. Mater. Res.*, **18**, pp. 156–165.

91. Langford, J. I. and Wilson, A. (1978). Scherrer after sixty years: A survey and some new results in the determination of crystallite size, *J. Appl. Crystallogr.*, **11**, pp. 102–113.

92. Zhu, K., Neale, N. R., Halverson, A. F., Kim, J. Y., and Frank, A. J. (2010). Effects of annealing temperature on the charge-collection and light-harvesting properties of TiO_2 nanotube-based dye-sensitized solar cells, *J. Phys. Chem. C*, **114**, pp. 13433–13441.

93. López, R. and Gómez, R. (2012). Band-gap energy estimation from diffuse reflectance measurements on sol–gel and commercial TiO_2: A comparative study, *J. Sol-gel Sci. Technol.*, **61**, pp. 1–7.

94. Mor, G. K., Varghese, O. K., Paulose, M., and Grimes, C. A. (2005). Transparent highly ordered TiO_2 nanotube arrays via anodization of titanium thin films, *Adv. Funct. Mater.*, **15**, pp. 1291–1296.

95. Ly, N. T., Dao, T. H., Pham, D. L., Do, H. M., and Vu, D. L. (2013). Optical properties of TiO_2 nanotube arrays fabricated by the electrochemical anodization method, *Adv. Nat. Sci.: Nanosci. Nanotechnol.*, **5**, pp. 015004.

96. Sopha, H., Krbal, M., Ng, S., Prikryl, J., Zazpe, R., Yam, F. K., and Macak, J. M. (2017). Highly efficient photoelectrochemical and photocatalytic anodic TiO_2 nanotube layers with additional TiO_2 coating, *Appl. Mater. Today*, **9**, pp. 104–110.

97. Ng, S. W., Yam, F. K., Sohimee, S. N., Beh, K. P., Tneh, S. S., Cheong, Y. L., and Hassan, Z. (2018). Photoelectrochemical ultraviolet photodetector by anodic titanium dioxide nanotube layers, *Sensor. Actuat. A: Phys.*, **279**, pp. 263–271.

98. Zhu, K. R., Zhang, M. S., Chen, Q., and Yin, Z. (2005). Size and phonon-confinement effects on low-frequency Raman mode of anatase TiO_2 nanocrystal, *Phys. Lett. A*, **340**, pp. 220–227.

99. Zhang, W., He, Y., Zhang, M., Yin, Z., and Chen, Q. (2000). Raman scattering study on anatase TiO_2 nanocrystals, *J. Phys. D: Appl. Phys.*, **33**, pp. 912.

100. Ng, S. W., Yam, F. K., and Hassan, Z. (2012). Electrochemical impregnation of silver nanostructures in titanium dioxide nanotubes, *J. Electrochem. Soc.*, **159**, pp. 742–746.

101. Moskovits, M. and Suh, J. (1984). Surface selection rules for surface-enhanced Raman spectroscopy: Calculations and application to the surface-enhanced Raman spectrum of phthalazine on silver, *J. Phys. Chem.*, **88**, pp. 5526–5530.

102. Kneipp, K., Moskovits, M., and Kneipp, H. (2007). Surface-enhanced Raman scattering, *Phys. Today*, **60**, pp. 40.

103. Ng, S. W. (2013). *Fabrication and Characterization of Titanium Dioxide Nanotubes with Embedded Zinc Oxide and Silver Particles* (Universiti Sains Malaysia).

104. Macak, J. M. (2015). Self-organized anodic TiO_2 nanotubes: Functionalities and applications due to a secondary material, in *Electrochemically Engineered Nanoporous Materials* (Springer), pp. 65–92.

105. Asahi, R., Morikawa, T., Ohwaki, T., Aoki, K., and Taga, Y. (2001). Visible-light photocatalysis in nitrogen-doped titanium oxides, *Science*, **293**, pp. 269–271.

106. Nah, Y. C., Paramasivam, I., and Schmuki, P. (2010). Doped TiO_2 and TiO_2 nanotubes: Synthesis and applications, *Chem. Phys. Chem.*, **11**, pp. 2698–2713.

107. Mazierski, P., Nischk, M., Gołkowska, M., Lisowski, W., Gazda, M., Winiarski, M. J., Klimczuk, T., and Zaleska-Medynska, A. (2016). Photocatalytic activity of nitrogen doped TiO_2 nanotubes prepared by anodic oxidation: The effect of applied voltage, anodization time and amount of nitrogen dopant, *Appl. Catal. B: Environ.*, **196**, pp. 77–88.

108. Siuzdak, K., Szkoda, M., Sawczak, M., and Lisowska-Oleksiak, A. (2015). Novel nitrogen precursors for electrochemically driven doping of

titania nanotubes exhibiting enhanced photoactivity, *New J. Chem.*, **39**, pp. 2741–2751.

109. Li, S., Lin, S., Liao, J., Pan, N., Li, D., and Li, J. (2012). Nitrogen-doped TiO_2 nanotube arrays with enhanced photoelectrochemical property, *Int. J. Photoen.*, **2012**, pp. 7.

110. Beranek, R., Macak, J. M., Gärtner, M., Meyer, K., and Schmuki, P. (2009). Enhanced visible light photocurrent generation at surface-modified TiO_2 nanotubes, *Electrochim. Acta.*, **54**, pp. 2640–2646.

111. Hahn, R., Ghicov, A., Salonen, J., Lehto, V.-P., and Schmuki, P. (2007). Carbon doping of self-organized TiO_2 nanotube layers by thermal acetylene treatment, *Nanotechnology*, **18**, pp. 105604.

112. Park, J. H., Kim, S., and Bard, A. J. (2006). Novel carbon-doped TiO_2 nanotube arrays with high aspect ratios for efficient solar water splitting, *Nano Lett.*, **6**, pp. 24–28.

113. Macak, J. M., Ghicov, A., Hahn, R., Tsuchiya, H., and Schmuki, P. (2006). Photoelectrochemical properties of N-doped self-organized titania nanotube layers with different thicknesses, *J. Mater. Res.*, **21**, pp. 2824–2828.

114. Ni, J., Fu, S., Wu, C., Maier, J., Yu, Y., and Li, L. (2016). Self-supported nanotube arrays of sulfur-doped TiO_2 enabling ultrastable and robust sodium storage, *Adv. Mater.*, **28**, pp. 2259–2265.

115. Ghicov, A., Macak, J. M., Tsuchiya, H., Kunze, J., Haeublein, V., Frey, L., and Schmuki, P. (2006). Ion implantation and annealing for an efficient N-doping of TiO_2 nanotubes, *Nano Lett.*, **6**, pp. 1080–1082.

116. Tsuchiya, H., Berger, S., Macak, J. M., Ghicov, A., and Schmuki, P. (2007). Self-organized porous and tubular oxide layers on TiAl alloys, *Electrochem. Commun.*, **9**, pp. 2397–2402.

117. Ghicov, A., Yamamoto, M., and Schmuki, P. (2008). Lattice widening in niobium-doped TiO_2 nanotubes: Efficient ion intercalation and swift electrochromic contrast, *Angew. Chem. Int. Ed.*, **47**, pp. 7934–7937.

118. Shrestha, N. K., Nah, Y. C., Tsuchiya, H., and Schmuki, P. (2009). Self-organized nano-tubes of TiO_2–MoO_3 with enhanced electrochromic properties, *Chem. Commun.*, pp. 2008–2010.

119. Yasuda, K. and Schmuki, P. (2007). Electrochemical formation of self-organized zirconium titanate nanotube multilayers, *Electrochem. Commun.*, **9**, pp. 615–619.

120. Sarraf, M., Dabbagh, A., Razak, B. A., Mahmoodian, R., Nasiri-Tabrizi, B., Hosseini, H. R. M., Saber-Samandari, S., Kasim, N. H. A., Abdullah, H.,

and Sukiman, N. L. (2018). Highly-ordered TiO_2 nanotubes decorated with Ag_2O nanoparticles for improved biofunctionality of Ti_6Al_4V, *Surf. Coat. Technol.*, **349**, pp. 1008–1017.

121. Li, H., Chen, Z., Tsang, C. K., Li, Z., Ran, X., Lee, C., Nie, B., Zheng, L., Hung, T., and Lu, J. (2014). Electrochemical doping of anatase TiO_2 in organic electrolytes for high-performance supercapacitors and photocatalysts, *J. Mater. Chem. A*, **2**, pp. 229–236.
122. Wang, G., Ling, Y., and Li, Y. (2012). Oxygen-deficient metal oxide nanostructures for photoelectrochemical water oxidation and other applications, *Nanoscale*, **4**, pp. 6682–6691.
123. Zhou, H. and Zhang, Y. (2013). Enhancing the capacitance of TiO_2 nanotube arrays by a facile cathodic reduction process, *J. Power Sources*, **239**, pp. 128–131.
124. Liu, N., Schneider, C., Freitag, D., Hartmann, M., Venkatesan, U., Müller, J., Spiecker, E., and Schmuki, P. (2014). Black TiO_2 nanotubes: Cocatalyst-free open-circuit hydrogen generation, *Nano Lett.*, **14**, pp. 3309–3313.
125. Kim, H. S. and Kang, S. H. (2013). Effect of hydrogen treatment on anatase TiO_2 nanotube arrays for photoelectrochemical water splitting, *Bull. Korean Chem. Soc.*, **34**, pp. 2067–2072.
126. Liu, N., Häublein, V., Zhou, X., Venkatesan, U., Hartmann, M., Mačković, M., Nakajima, T., Spiecker, E., Osvet, A., and Frey, L. (2015). "Black" TiO_2 nanotubes formed by high-energy proton implantation show noble-metal-co-catalyst free photocatalytic H_2-evolution, *Nano Lett.*, **15**, pp. 6815–6820.
127. Macak, J. M., Gong, B. G., Hueppe, M., and Schmuki, P. (2007). Filling of TiO_2 nanotubes by self-doping and electrodeposition, *Adv. Mater.*, **19**, pp. 3027–3031.
128. Kim, C., Lee, S., Kim, S., and Yoon, J. (2016). Effect of annealing temperature on the capacitive and oxidant-generating properties of an electrochemically reduced TiO_2 nanotube array, *Electrochim. Acta.*, **222**, pp. 1578–1584.
129. Kang, U. and Park, H. (2013). Lithium ion-inserted TiO_2 nanotube array photoelectrocatalysts, *Appl. Catal. B: Environ.*, **140**, pp. 233–240.
130. So, S., Hwang, I., and Schmuki, P. (2015). Hierarchical DSSC structures based on "single walled" TiO_2 nanotube arrays reach a back-side illumination solar light conversion efficiency of 8%, *En. Environ. Sci.*, **8**, pp. 849–854.
131. Wang, J. and Lin, Z. (2009). Dye-sensitized TiO_2 nanotube solar cells with markedly enhanced performance via rational surface engineering, *Chem. Mater.*, **22**, pp. 579–584.

132. Roy, P., Kim, D., Paramasivam, I., and Schmuki, P. (2009). Improved efficiency of TiO_2 nanotubes in dye sensitized solar cells by decoration with TiO_2 nanoparticles, *Electrochem. Commun.*, **11**, pp. 1001–1004.

133. Roelofs, K. E., Pool, V. L., Bobb-Semple, D. A., Palmstrom, A. F., Santra, P. K., Van Campen, D. G., Toney, M. F., and Bent, S. F. (2016). Impact of conformality and crystallinity for ultrathin 4 nm compact TiO_2 layers in perovskite solar cells, *Adv. Mater. Interfaces.*, **3**, pp. 1600580.

134. Altomare, M., Nguyen, N. T., and Schmuki, P. (2016). Templated dewetting: Designing entirely self-organized platforms for photocatalysis, *Chem. Sci.*, **7**, pp. 6865–6886.

135. Zhang, J., Xiao, G., Xiao, F.-X., and Liu, B. (2017). Revisiting one-dimensional TiO_2 based hybrid heterostructures for heterogeneous photocatalysis: A critical review, *Mater. Chem. Frontiers*, **1**, pp. 231–250.

136. Momeni, M. M. and Ghayeb, Y. (2015). Visible light-driven photoelectrochemical water splitting on $ZnO–TiO_2$ heterogeneous nanotube photoanodes, *J. Appl. Electrochem.*, **45**, pp. 557–566.

137. Ng, S. W., Yam, F. K., and Hassan, Z. (2015). Rapid formation of zinc oxide nanosheets on titanium dioxide nanotubes through electrochemical method, *Optoelectron. Adv. Mater. Rapid Commun.*, **9**, pp. 1429–1434.

138. Song, Y. Y., Gao, Z. D., Wang, J. H., Xia, X. H., and Lynch, R. (2011). Multistage coloring electrochromic device based on TiO_2 nanotube arrays modified with WO_3 nanoparticles, *Adv. Funct. Mater.*, **21**, pp. 1941–1946.

139. Martins, A. S., Cordeiro-Junior, P. J. M., Nuñez, L., and de Vasconcelos Lanza, M. R. (2017). A simple method for the electrodeposition of WO_3 in TiO_2 nanotubes: Influence of the amount of tungsten on photoelectrocatalytic activity, *Electrocatalysis*, **8**, pp. 115–121.

140. Mohapatra, S. K., Banerjee, S., and Misra, M. (2008). Synthesis of Fe_2O_3/TiO_2 nanorod–nanotube arrays by filling TiO_2 nanotubes with Fe, *Nanotechnology*, **19**, pp. 315601.

141. Kupcik, R., Rehulka, P., Bilkova, Z., Sopha, H., and Macak, J. M. (2017). New interface for purification of proteins: One-dimensional TiO_2 nanotubes decorated by Fe_3O_4 nanoparticles, *ACS Appl. Mater. Interfaces*, **9**, pp. 28233–28242.

142. Zhou, H., Zou, X., and Zhang, Y. (2016). Fabrication of TiO_2@ MnO_2 nanotube arrays by pulsed electrodeposition and their application for high-performance supercapacitors, *Electrochim. Acta.*, **192**, pp. 259–267.

143. Zhang, S., Zhang, S., Peng, F., Zhang, H., Liu, H., and Zhao, H. (2011). Electrodeposition of polyhedral Cu_2O on TiO_2 nanotube arrays for enhancing visible light photocatalytic performance, *Electrochem. Commun.*, **13**, pp. 861–864.

144. Shrestha, N. K., Yang, M., Nah, Y.-C., Paramasivam, I., and Schmuki, P. (2010). Self-organized TiO_2 nanotubes: Visible light activation by Ni oxide nanoparticle decoration, *Electrochem. Commun.*, **12**, pp. 254–257.

145. Sun, W. T., Yu, Y., Pan, H. Y., Gao, X. F., Chen, Q., and Peng, L. M. (2008). CdS quantum dots sensitized TiO_2 nanotube-array photoelectrodes, *J. Am. Chem. Soc.*, **130**, pp. 1124–1125.

146. Nguyen, V., Li, W., Pham, V., Wang, L., Sheng, P., Cai, Q., and Grimes, C. (2016). A CdS/ZnSe/TiO_2 nanotube array and its visible light photocatalytic activities, *J. Colloid Interface Sci.*, **462**, pp. 389–396.

147. Lin, Z. Q., Lai, Y. K., Hu, R. G., Li, J., Du, R. G., and Lin, C. J. (2010). A highly efficient ZnS/CdS@ TiO_2 photoelectrode for photogenerated cathodic protection of metals, *Electrochim. Acta.*, **55**, pp. 8717–8723.

148. Wang, Q., Zhu, K., Neale, N. R., and Frank, A. J. (2009). Constructing ordered sensitized heterojunctions: Bottom-up electrochemical synthesis of p-type semiconductors in oriented n-TiO_2 nanotube arrays, *Nano Lett.*, **9**, pp. 806–813.

149. Das, S., Sopha, H., Krbal, M., Zazpe, R., Podzemna, V., Prikryl, J., and Macak, J. M. (2017). Electrochemical infilling of $CuInSe_2$ within TiO_2 nanotube layers and subsequent photoelectrochemical studies, *ChemElectroChem*, **4**, pp. 495–499.

150. Seabold, J. A., Shankar, K., Wilke, R. H., Paulose, M., Varghese, O. K., Grimes, C. A., and Choi, K. S. (2008). Photoelectrochemical properties of heterojunction CdTe/TiO_2 electrodes constructed using highly ordered TiO_2 nanotube arrays, *Chem. Mater.*, **20**, pp. 5266–5273.

151. Paramasivam, I., Macak, J., Ghicov, A., and Schmuki, P. (2007). Enhanced photochromism of Ag loaded self-organized TiO_2 nanotube layers, *Chem. Phys. Lett.*, **445**, pp. 233–237.

152. Paramasivam, I., Macak, J., and Schmuki, P. (2008). Photocatalytic activity of TiO_2 nanotube layers loaded with Ag and Au nanoparticles, *Electrochem. Commun.*, **10**, pp. 71–75.

153. Noothongkaew, S., Han, J. K., Lee, Y. B., Thumthan, O., and An, K. S. (2017). Au NPs decorated TiO_2 nanotubes array candidate for UV photodetectors, *Prog. Nat. Sci: Mater. Int.*, **27**, pp. 641–646.

154. Qamar, M., Kim, S., and Ganguli, A. (2009). TiO_2-based nanotubes modified with nickel: Synthesis, properties, and improved photocatalytic activity, *Nanotechnology*, **20**, pp. 455703.

155. Kang, Q., Yang, L., and Cai, Q. (2008). An electro-catalytic biosensor fabricated with Pt–Au nanoparticle-decorated titania nanotube array, *Bioelectrochemistry*, **74**, pp. 62–65.

156. Anitha, V., Goswami, A., Sopha, H., Nandan, D., Gawande, M. B., Cepe, K., Ng, S., Zboril, R., and Macak, J. M. (2018). Pt nanoparticles decorated TiO_2 nanotubes for the reduction of olefins, *Appl. Mater. Today*, **10**, pp. 86–92.

157. Nguyen, N. T., Altomare, M., Yoo, J., and Schmuki, P. (2015). Efficient photocatalytic H_2 evolution: Controlled dewetting–dealloying to fabricate site-selective high-activity nanoporous Au particles on highly ordered TiO_2 nanotube arrays, *Adv. Mater.*, **27**, pp. 3208–3215.

158. Yoo, J., Lee, K., and Schmuki, P. (2013). Dewetted Au films form a highly active photocatalytic system on TiO_2 nanotube-stumps, *Electrochem. Commun.*, **34**, pp. 351–355.

159. Yoo, J., Altomare, M., Mokhtar, M., Alshehri, A., Al-Thabaiti, S. A., Mazare, A., and Schmuki, P. (2016). Photocatalytic H_2 generation using dewetted Pt-decorated TiO_2 nanotubes: Optimized dewetting and oxide crystallization by a multiple annealing process, *J. Phys. Chem. C*, **120**, pp. 15884–15892.

160. Kim, J.-Y., Lee, K.-H., Shin, J., Park, S. H., Kang, J. S., Han, K. S., Sung, M. M., Pinna, N., and Sung, Y.-E. (2014). Highly ordered and vertically oriented TiO_2/Al_2O_3 nanotube electrodes for application in dye-sensitized solar cells, *Nanotechnology*, **25**, pp. 504003.

161. Zazpe, R., Knaut, M., Sopha, H., Hromadko, L., Albert, M., Prikryl, J., Gartnerova, V., Bartha, J. W., and Macak, J. M. (2016). Atomic layer deposition for coating of high aspect ratio TiO_2 nanotube layers, *Langmuir*, **32**, pp. 10551–10558.

162. Sopha, H., Salian, G. D., Zazpe, R., Prikryl, J., Hromadko, L., Djenizian, T., and Macak, J. M. (2017). ALD Al_2O_3-coated TiO_2 nanotube layers as anodes for lithium-ion batteries, *ACS Omega*, **2**, pp. 2749–2756.

163. Gui, Q., Xu, Z., Zhang, H., Cheng, C., Zhu, X., Yin, M., Song, Y., Lu, L., Chen, X., and Li, D. (2014). Enhanced photoelectrochemical water splitting performance of anodic TiO_2 nanotube arrays by surface passivation, *ACS Appl. Mater. Interfaces*, **6**, pp. 17053–17058.

164. Tupala, J., Kemell, M., Härkönen, E., Ritala, M., and Leskelä, M. (2012). Preparation of regularly structured nanotubular TiO_2 thin films on ITO

and their modification with thin ALD-grown layers, *Nanotechnology*, **23**, pp. 125707.

165. Ng, S., Kuberský, P., Krbal, M., Prikryl, J., Gärtnerová, V., Moravcová, D., Sopha, H., Zazpe, R., Yam, F. K., and Jäger, A. (2018). ZnO coated anodic 1D TiO_2 nanotube layers: Efficient photo-electrochemical and gas sensing heterojunction, *Adv. Eng. Mater.*, **20**, pp. 1700589.

166. Cai, H., Yang, Q., Hu, Z., Duan, Z., You, Q., Sun, J., Xu, N., and Wu, J. (2014). Enhanced photoelectrochemical activity of vertically aligned ZnO-coated TiO_2 nanotubes, *Appl. Phys. Lett.*, **104**, pp. 053114.

167. Jeong, J. S., Choe, B.-H., Lee, J. H., Lee, J. J., and Choi, W.-Y. (2014). ZnO-coated TiO_2 nanotube arrays for a photoelectrode in dye-sensitized solar cells, *J. Electron. Mater.*, **43**, pp. 375–380.

168. Turkevych, I., Kosar, S., Pihosh, Y., Mawatari, K., Kitamori, T., Ye, J., and Shimamura, K. (2014). Synergistic effect between TiO_2 and ubiquitous metal oxides on photocatalytic activity of composite nanostructures, *J. Ceram. Soc. Jpn.*, **122**, pp. 393–397.

169. Macak, J. M., Prikryl, J., Sopha, H., and Strizik, L. (2015). Antireflection In_2O_3 coatings of self-organized TiO_2 nanotube layers prepared by atomic layer deposition, *Phys. Status Solidi Rapid Res Lett.*, **9**, pp. 516–520.

170. Sarkar, S. K., Kim, J. Y., Goldstein, D. N., Neale, N. R., Zhu, K., Elliott, C. M., Frank, A. J., and George, S. M. (2010). In_2S_3 atomic layer deposition and its application as a sensitizer on TiO_2 nanotube arrays for solar energy conversion, *J. Phys. Chem. C*, **114**, pp. 8032–8039.

171. Zheng, X., Yu, D., Xiong, F. Q., Li, M., Yang, Z., Zhu, J., Zhang, W. H., and Li, C. (2014). Controlled growth of semiconductor nanofilms within TiO_2 nanotubes for nanofilm sensitized solar cells, *Chem. Commun.*, **50**, pp. 4364–4367.

172. Krbal, M., Prikryl, J., Zazpe, R., Sopha, H., and Macak, J. (2017). CdS-coated TiO_2 nanotube layers: Downscaling tube diameter towards efficient heterostructured photoelectrochemical conversion, *Nanoscale*, **9**, pp. 7755–7759.

173. Ng, S., Krbal, M., Zazpe, R., Prikryl, J., Charvot, J., Dvořák, F., Strizik, L., Slang, S., Sopha, H., and Kosto, Y. (2018). MoSexOy-coated 1D TiO_2 nanotube layers: Efficient interface for light-driven applications, *Adv. Mater. Interfaces*, **5**, pp. 1701146.

174. Assaud, L., Brazeau, N., Barr, M. K., Hanbücken, M., Ntais, S., Baranova, E. A., and Santinacci, L. (2015). Atomic layer deposition of Pd nanoparticles on TiO_2 nanotubes for ethanol electrooxidation:

Synthesis and electrochemical properties, *ACS Appl. Mater. Interfaces*, **7**, pp. 24533–24542.

175. Yoo, J., Zazpe, R., Cha, G., Prikryl, J., Hwang, I., Macak, J. M., and Schmuki, P. (2018). Uniform ALD deposition of Pt nanoparticles within 1D anodic TiO_2 nanotubes for photocatalytic H_2 generation, *Electrochem. Commun.*, **86**, pp. 6–11.

176. Indira, K., Mudali, U. K., Nishimura, T., and Rajendran, N. (2015). A review on TiO_2 nanotubes: Influence of anodization parameters, formation mechanism, properties, corrosion behavior, and biomedical applications, *J. Bio- Tribo-Corrosion*, **1**, pp. 28.

177. Xiao, F. X., Miao, J., Tao, H. B., Hung, S. F., Wang, H. Y., Yang, H. B., Chen, J., Chen, R., and Liu, B. (2015). One-dimensional hybrid nanostructures for heterogeneous photocatalysis and photoelectrocatalysis, *Small*, **11**, pp. 2115–2131.

178. Fu, Y. and Mo, A. (2018). A review on the electrochemically self-organized titania nanotube arrays: Synthesis, modifications, and biomedical applications, *Nanoscale Res. Lett.*, **13**, pp. 187.

179. Liu, C., Teng, Y., Liu, R., Luo, S., Tang, Y., Chen, L., and Cai, Q. (2011). Fabrication of graphene films on TiO_2 nanotube arrays for photocatalytic application, *Carbon*, **49**, pp. 5312–5320.

180. Razzaq, A., Grimes, C. A., and In, S.-I. (2016). Facile fabrication of a noble metal-free photocatalyst: TiO_2 nanotube arrays covered with reduced graphene oxide, *Carbon*, **98**, pp. 537–544.

181. Chen, C., Wang, N., Zhou, P., San, H., Wang, K., and Chen, X. (2016). Electrochemically reduced graphene oxide on well-aligned titanium dioxide nanotube arrays for betavoltaic enhancement, *ACS Appl. Mater. Interfaces*, **8**, pp. 24638–24644.

182. Song, P., Zhang, X., Sun, M., Cui, X., and Lin, Y. (2012). Graphene oxide modified TiO_2 nanotube arrays: Enhanced visible light photoelectrochemical properties, *Nanoscale*, **4**, pp. 1800–1804.

183. Zubair, M., Kim, H., Razzaq, A., Grimes, C. A., and In, S. I. (2018). Solar spectrum photocatalytic conversion of CO_2 to CH_4 utilizing TiO_2 nanotube arrays embedded with graphene quantum dots, *J. CO_2 Utilization*, **26**, pp. 70–79.

Chapter 3

Characterization Techniques

3.1 Scanning Electron Microscopy

The morphology of TiO_2 nanofibers was examined using scanning electron microscopy (SEM) (EVO 40XVP). The samples were coated with platinum to avoid charging, and images were taken using secondary electrons. Energy dispersive X-ray spectroscopy (EDS) at an acceleration voltage of 10 kV was used to qualitatively analyze the elemental compositions [1–3].

The morphologies of the electrospun TiO_2 nanofibers were examined using an EVO 40XVP scanning electron microscope with an accelerating voltage of 15 kV. Prior to the microstructure observations, the samples were coated with platinum to avoid charging. The SEM images of the as-spun nanofibers were taken using secondary and backscattered electrons at working distances of 4 and 4.5 mm, respectively. The samples were characterized again at the same accelerating voltage after thermal annealing in air and argon atmospheres. Associated EDS with an acceleration voltage of 10 kV, magnification of x606, silicon detector and zero electronic beam tilt was used to analyze the elemental compositions of the SEM-imaged materials [1, 2].

Nanostructured Titanium Dioxide in Photocatalysis
It-Meng Low, Hani Manssor Albetran, Victor Manuel de la Prida Pidal, and Fong Kwong Yam

ISBN 978-981-4877-07-7 (Hardcover), 978-1-003-14853-1 (eBook)
www.jennystanford.com

3.2 High-Resolution Transmission Electron Microscopy

A small piece of implanted nanofiber material was ground and suspended with ethanol. Two drops of suspension were mounted onto carbon-coated copper grids for transmission electron microscopy (TEM) imaging. High-resolution TEM (HR-TEM) was carried out with a JEOL3000F TEM operating at 300 kV and fitted with a Gatan Orius 10-MP charge-coupled device (CCD) camera [4, 5].

3.3 X-Ray Photoelectron Spectroscopy

X-ray photoelectron spectroscopy (XPS) analyses were performed using a Phoibos 100MCD5 system (SPECS Nanotechnology, Delft, Netherlands). The X-ray source was MgKα (1253.6 eV) radiation operating at 200 W. The operating pressure of the spectrometer was 1×10^{-9} mbar, and the analyzed area was ~500 μm in diameter. XPS measurements were carried out in two stages. First, a survey spectrum was measured from 1200 to 0 eV using an energy pass of 90 eV and an energy step of 1 eV to obtain a global measurement. Then, high-resolution spectra were acquired with an energy pass of 30 eV and an energy step of 0.1 eV for more detailed analysis [4, 6].

3.4 Electron Backscatter Diffraction

For samples heated in air under normal pressure, electron backscatter diffraction (EBSD) analyses were performed by SEM (Tescan, Mira3) with a Nordlys Nano EBSD detector. Samples were prepared using an epoxy resin to encase the powder particles. The specimen surface was polished using a series of successively finer grades of diamond paste. The polishing stage was necessary to expose an unaltered, reasonably flat sample cross section. Specimens were coated with an 8 nm thick carbon layer and then mounted with conductive carbon sticky tabs onto stubs to avoid charging. Samples were tilted to 70°C from horizontal using an SEM-protected holder. The operating conditions for EBSD collection were as follows: an acceleration voltage of 20 kV, a spot size of 23 nm, a working distance of 19.7 mm, and a beam current of 842 pA. Amorphous and crystalline areas for the polished cross-sectional TiO_2 samples were studied, and phase identification for crystalline titania was determined using EBSD software (AZtec Software, Oxford). An electron beam was directed at a point of interest on tilted samples. An optimization pattern tool was used to identify amorphous and crystalline areas; an image was produced on a phosphor screen by detecting a Kikuchi pattern, which is generated by electron-beam interaction with only the crystal structure in the sample surface. Kikuchi patterns were captured by a high-speed and high-sensitivity CCD camera behind a phosphor screen and transferred to a computer for further processing [7].

3.5 In Situ High-Temperature X-Ray and Synchrotron Radiation Diffraction

The in situ crystallization behavior of electrospun TiO_2 nanofibers was evaluated using the high-temperature synchrotron radiation diffraction (SRD) in air and argon. The SRD measurements were conducted at the powder diffraction beamline of the Australian Synchrotron. The specimens were mounted on an Anton Parr HTK16 hot platinum stage and heated with an Anton Paar HTK20 furnace. The SRD data were collected at an incident angle of 3° and a wavelength of 0.1126 nm. The SRD patterns were acquired initially at ambient temperature and then in steps of 100°C from 200°C to 900°C at a rate of 10°C/min with a data acquisition time of 10 min per pattern. Each pattern was measured over the angular range of 2θ = 5–84°. In situ isothermal high-temperature SRD in air (powder diffraction beamline, Australian Synchrotron, Melbourne, Australia; 3° incident angle, 0.1126 nm wavelength) was used to study the high-temperature crystallization kinetics of the anodized titania nanotube arrays. The specimen was mounted and heated using an Anton Parr HTK 16 hot platinum stage and an MYTHEN II microstrip detector (Australian Synchrotron, Melbourne, VIC, Australia). The heating protocol (10°C/min; data acquisition of 2 min/pattern; ambient temperature, 100, 200, and 300°C non-isothermally; 50°C steps from 400 to 650°C isothermally at 400, 450, 500, 550, 600, and 650°C and 50°C steps from 650 to 900°C non-isothermally) for the SRD studies of the titania nanotube arrays is given in Fig. 3.1. The residence times from 400 to 650°C are shown in Table 3.1. Each pattern of the anodized titania nanotube arrays was measured over of 2θ = 5–84°.

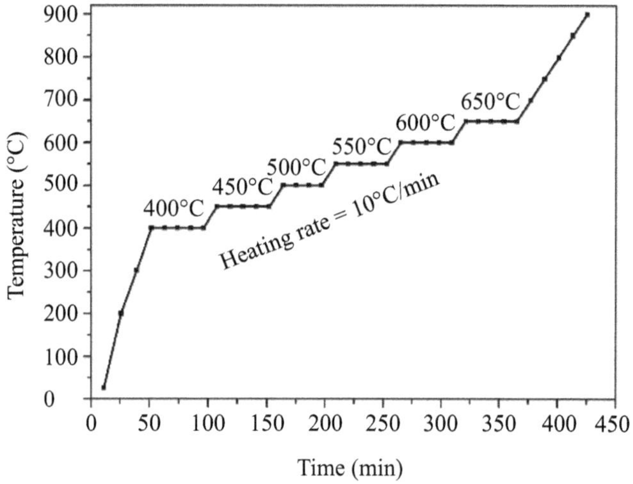

Figure 3.1 SRD heating protocol at 10°C/min heating, 2 min/pattern acquired initially at ambient temperature, 100, 200, and 300°C non-isothermally; 50°C steps from 400 to 650°C isothermally at 400, 450, 500, 550, 600, and 650°C, and 50°C steps from 650 to 900°C non-isothermally.

Table 3.1 Heating schedule during in situ isothermal high-temperature SRD

Temperature (°C)	Residence Time (min)
400	44
450	44
500	34
550	44
600	44
650	44

The SRD data were analyzed using Rietveld pattern fitting with the Rietica program (version 2.1) [8], to compute the relative crystalline phase abundances, and the lattice parameters of anatase and rutile at each temperature [9]. The parameters optimized in the Rietveld refinements were the pattern background, 2θ-zero, and for each phase the scale factor, lattice parameters, and the peak shape parameters. The Rietveld refinements were conducted with the crystal structures of anatase (ICSD 202242) and rutile (ICSD64987). The Jade (version 6) program was used to perform peak broadening analyses in order to obtain the crystalline size (L) and the strain (ε) for each of the two phases as a function of temperature, using Williamson–Hall (W–H) plots [10].

3.6 Analysis of Absolute Phase Compositions

Rietveld analysis provided only the relative levels of anatase and rutile. In order to determine the absolute levels for the amorphous and crystalline components, a novel technique was developed using the temperature dependence of the SRD pattern background levels. The background-temperature plots were analyzed by assuming that the background $B(T)$ at temperature T is given by a component $B_A(T)$ due to amorphous scatter and a secondary component B_S attributed to atomic X-ray scatter (Compton and Rayleigh). On this basis, the weight percentage of amorphous TiO_2 ($wt_{Am}\%$) was calculated using:

$$wt_{Am}\% = \frac{B(T) - B_f}{B_o - B_f} \tag{3.1}$$

where $B(T)$ is the background intensity at temperature T, B_o is the background intensity prior to the onset of the amorphous-to-crystalline TiO_2 transformation, and B_f is the estimated background intensity when amorphous TiO_2 has been completely transformed into crystalline TiO_2 [11].

3.7 Estimation of Activation Energies

The SRD results were used to examine the crystallization kinetics of TiO_2 with the equation of Matusita et al. for non-isothermal heating conditions [12]:

$$ln[-ln(1-x)] = -nln\alpha - 1.052mE/RT + \text{constant} \tag{3.2}$$

where x is the volume fraction of transformed material; n is a numerical factor depending on the nucleation process, where $n = m + 1$, with $m = 1, 2, 3$ being the number of dimensional growth; α is the constant heating rate; R is the ideal gas constant (=8.3145 J/Kmol); T is the temperature in Kelvin; and E is the activation energy in kJ/mol.

To obtain the activation energies of amorphous-to-crystalline TiO_2 (anatase and rutile) transformation or anatase-to-rutile transformation at a heating rate of 10°C/min, the weight percentages (wt) of crystalline TiO_2 or rutile were used rather than the volume fractions because anatase and rutile have almost the same density, and $m = 1$ for one-dimensional nanofibers [12].

If Eq. 3.2 is written as:

$$\ln\left[-\ln\left(1-wt\right)\right] = -1.052E/RT + \text{constant} \quad (3.3)$$

A plot of ln [−ln (1 − wt)] for crystalline TiO_2 (total of anatase and rutile) or rutile against $1/T$ is expected to be linear, and the activation energy for each transformation can be estimated from the gradient. The activation energies were also calculated using a modified version of the Arrhenius equation:

$$\ln wt = -E/RT + \ln wt_o \quad (3.4)$$

where wt is the weight percentage of the transformed phase(s) (crystalline TiO_2 or rutile), and wt_o is the initial weight percentage [11].

3.8 Estimation of Crystallite Size and Strain

The average crystallite size (L) and the rms-microstrain (ε) for each phase at each temperature were estimated using Williamson–Hall (W–H) plots [13]:

$$\beta\cos\theta = \lambda/L + 4\varepsilon\sin\theta \quad (3.5)$$

where λ is the X-ray wavelength, θ is the Bragg angle, and β is the integral breadth of each peak. Care was taken to confine the peaks used in the analysis to well-defined peaks with minimal peak overlap. The five peaks used for the rutile plots were 110 (2θ = 19.91°), 011 (26.07°), 111 (29.74°), 120 (31.72°), and 220 (40.44°). Only two peaks could be used for the anatase plots: 011 (18.41°) and 020

(34.52°). The integral widths were determined using the program X'Pert HighScore Plus (version 3.0).

The integral widths were large (in the $\Delta 2\theta$ range 0.18–0.43° for rutile and 0.26–1.08° for anatase) compared with the broadening due to the instrument. Instrument broadening corrections were applied according to the optics of the SRD instrument. According to Eq. 3.6, a near-parallel beam is incident on the sample and then diffracted by the sample [14, 15]:

$$\mathrm{FWHM}\,(^\circ) = \left[\varphi_V^2\left(\frac{2\tan\theta}{\tan\theta_m}-1\right)^2+\delta^2\right]^{1/2} \tag{3.6}$$

where φ_V is a vertical divergence of the incident beam (= 0.02°), θ is Bragg angle for the SRD line, θ_m is Bragg angle for the Si (111) double-crystal monochromator (= 10.3435°), and δ is the divergence angle for the diffracted beam, which is calculated using:

$$\delta = \frac{\left(d_S^2 + d_{IP}^2\right)^{1/2}}{R} \tag{3.7}$$

where d_S is the width of the incident beam at the specimen, in the direction of the incident beam (=0.2 mm), d_{1P} is the spatial resolution of the imaging plate (=0.053 mm) and angular resolution = 0.004° for the MYTHEN detector, and R is the radius of the instrument, or the specimen-to-detector distance (=760 mm). On this basis, the instrument width was 0.020° for the data used for W–H plots [4].

References

1. Albetran, H., Haroosh, H., Dong, Y., Prida, V., O'Connor, B., and Low, I. M. (2014). Phase transformations and crystallization kinetics in electrospun TiO_2 nanofibers in air and argon atmospheres, *Appl. Phys. A*, **116**, pp. 161–169.

2. Albetran, H., Dong, Y., and Low, I. M. (2015). Characterization and optimization of electrospun TiO_2/PVP nanofibers using Taguchi design of experiment method, *J. Asian Ceram. Soc.*, **3**, pp. 292–300.

3. Albetran, H., Vega, V., Prida, V. M., and Low, I.-M. (2018). Dynamic diffraction studies on the crystallization, phase transformation, and activation energies in anodized titania nanotubes, *Nanomaterials*, **8**, pp. 122.

4. Albetran, H., O'Connor, B., Prida, V., and Low, I. (2015). Effect of vanadium ion implantation on the crystallization kinetics and phase transformation of electrospun TiO_2 nanofibers, *Appl. Phys. A*, **120**, pp. 623–634.
5. Albetran, H. and Low, I. M. (2016). Crystallization kinetics and phase transformations in aluminum ion-implanted electrospun TiO_2 nanofibers, *Appl. Phys. A*, **122**, pp. 1044.
6. Low, I. M., Albetran, H., Prida, V. M., Vega, V., Manurung, P., and Ionescu, M. (2013). A comparative study on crystallization behavior, phase stability, and binding energy in pure and Cr-doped TiO_2 nanotubes, *J. Mater. Res.*, **28**, pp. 304–312.
7. Albetran, H. and Low, I. (2019). Crystallization kinetics study of In-doped and (In-Cr) co-doped TiO_2 nanopowders using in-situ high-temperature synchrotron radiation diffraction, *Arab. J. Chem.* https://doi.org/10.1016/j.arabjc.2019.04.003.
8. Pang, W. K., Low, I. M., and Sun, Z. M. (2010). In situ high-temperature diffraction study of the thermal dissociation of Ti_3AlC_2 in vacuum, *J. Am. Ceram. Soc.*, **93**, pp. 2871–2876.
9. Hummer, D., Heaney, P., and Post, J. (2007). Thermal expansion of anatase and rutile between 300 and 575 K using synchrotron powder X-ray diffraction, *Powder Diffr.*, **22**, pp. 352–357.
10. Albetran, H., O'Connor, B., and Low, I. M. (2017). Effect of pressure on TiO_2 crystallization kinetics using in situ high-temperature synchrotron radiation diffraction, *J. Am. Ceram. Soc.*, **100**, pp. 3199–3207.
11. Albetran, H., O'Connor, B., and Low, I. M. (2016). Activation energies for phase transformations in electrospun titania nanofibers: Comparing the influence of argon and air atmospheres, *Appl. Phys. A*, **122**, pp. 367.
12. Matusita, K., Komatsu, T., and Yokota, R. (1984). Kinetics of non-isothermal crystallization process and activation energy for crystal growth in amorphous materials, *J. Mater. Sci.*, **19**, pp. 291–296.
13. Williamson, G. K. and Hall, W. H. (1953). X-ray line broadening from filed aluminium and wolfram, *Acta Metall.*, **1**, pp. 22–31.
14. Cox, D., Toby, B., and Eddy, M. (1988). Acquisition of powder diffraction data with synchrotron radiation, *Aust. J. Phys.*, **41**, pp. 117–132.
15. O'Connor, B. H., Van Riessen, A., Carter, J., Burton, G. R., Cookson, D. J., and Garrett, R. F. (1997). Characterization of ceramic materials with bigdiff: A synchrotron radiation debye-scherrer powder diffractometer, *J. Am. Ceram. Soc.*, **80**, pp. 1373–1381.

Part III

Materials Characterization

Chapter 4

In Situ Isothermal High-Temperature Diffraction Studies on the Crystallization, Phase Transformation, and Activation Energies in Anodized Titania Nanotubes

The influence of calcination time on the phase transformation and crystallization kinetics of anodized titania nanotube arrays was studied using in situ isothermal and non-isothermal synchrotron radiation diffraction (SRD) from room temperature to 900°C. Crystalline anatase first appeared at 400°C, and crystalline rutile appeared at 550°C. Isothermal heating of the anodized titania nanotubes by an increase in the calcination time at 400, 450, 500, 550, 600, and 650°C resulted in a slight reduction in anatase abundance, but an increase in the abundance of rutile because of an anatase-to-rutile transformation. Isothermal in situ synchrotron radiation diffraction data analysis using the Avrami equation was used to model the titania crystallization mechanism, and the Arrhenius equation was used to estimate the activation energies of the titania phase transformation. The corresponding activation energies were 22 (10) kJ/mol for the titanium-to-anatase transformation, and 207 (17) kJ/mol for the anatase-to-rutile transformation.

Nanostructured Titanium Dioxide in Photocatalysis
It-Meng Low, Hani Manssor Albetran, Victor Manuel de la Prida Pidal, and Fong Kwong Yam

ISBN 978-981-4877-07-7 (Hardcover), 978-1-003-14853-1 (eBook)
www.jennystanford.com

4.1 Introduction

Titania has attracted considerable attention as a promising and attractive raw material for use in various applications such as hydrogen production, gas sensors, photoelectrochemical cells, and dye-sensitized solar cells because of its stability, nontoxicity, photodurability, affordability, and excellent photocatalytic activity [1–5].

Anatase, rutile, and brookite are three common naturally occurring titania crystal structures [6–9]. The thermal treatment of amorphous titania at elevated temperature yields anatase, rutile, or a mixture of these materials as the most common titania phases, whereas brookite formation is rare [10]. Titania phase transformation is influenced by several conditions, such as calcining temperature, heating rate, calcination time, synthesis method, dopants, the level of impurities, grain size, and atmospheric condition and type [11–16].

The effect of calcination time on the titania phase transformation has not been investigated intensively and is not yet clear and fully understood. Some studies of this effect exist in the literature, for example, titanium dioxide nanotube arrays have been heated at 500°C and 10°C/min with a calcination time of 1, 2, 4, 8, and 12 h [17]. The XRD patterns show that only the (101) anatase peak of the crystallographic plane appears after 1 h of calcination, and (101) and (200) anatase crystallographic planes appear after 2 h. An anatase-to-rutile transformation was observed after 4 h of calcination, and the anatase phase was transformed completely into the rutile phase when the calcination time was 12 h. This effect was studied by heating anatase in air at 950°C for 4, 24, and 48 h [18]. Analysis of X-ray diffraction (XRD) measurements shows that an increase in calcination time increased the rutile percentage and decreased the anatase percentages. The rutile percentages were 5 wt% for 4 h, 65 wt% for 24 h, and 92 wt% for 48 h.

There is much interest in synthesizing one-dimensional titania nanostructures such as nanofibers, nanobelts, nanorods, nanowires, and nanotubes, as they deliver unique electronic characteristics, such as quantum-confinement effects, a high surface-area-to-volume ratio, a high mechanical strength, and a high electron mobility [3, 4, 19, 20].

Fabrication methods that have been used to synthesize one-dimensional nanostructured titania include the sol–gel, electrospinning, hydrothermal, and anodization methods. Electrochemical anodization is currently one of the most common inexpensive and simple methods used to fabricate one-dimensional nanotube titania nanostructures [2, 19, 21]. At room temperature, electrochemical anodization of the titanium-based alloy as a cathode in an electrolyte with a platinum grid as an anode produced anodized titania nanotube arrays [2, 22, 23].

In this study, the effect of calcination time on phase transformations, crystallization kinetics, and activation energies of anodized titania nanotube arrays was investigated using isothermal in situ SRD from room temperature to 900°C. The sample was characterized using field emission scanning electron microscopy (FESEM) and associated energy dispersive spectroscopy (EDS).

4.2 Results and Discussion

4.2.1 Microstructural Imaging

After anodizing in an ammonium fluoride/ethylene glycol electrolyte (25°C, 20 h), a white homogeneous layer resulted on the titanium foil (Fig. 4.1). Images in Figs. 4.1b and 4.1c present typical secondary-electron FESEM micrographs for the anodized titania nanotube arrays, following the SRD experiments with a low and a high magnification, respectively. Some grains were distributed randomly among the titania nanotube. The grains are crystalline rutile, which result from the transformation of metastable anatase to stable crystalline rutile at a high temperature (see Section 4.2.2, Crystallization Kinetics).

The EDS spectra of arrays of anodized titania nanotube are shown in Fig. 4.2, with strong signatures for titanium, oxygen, and platinum, and with the platinum content resulting because of the platinum coating. Quantitative EDS analysis (conducted assuming that the platinum coating peaks are negligible, and with AZtec Software, Oxford, UK) indicates that 66.18% O Kα and 33.82% Ti Kα exist, and this confirms that titania is present [24]. The flat samples were studied by quantitative EDS, and an approximate analysis was obtained for the nanotubes. The nanotube EDS was obtained from a small square to reduce problems related to the sample not being flat.

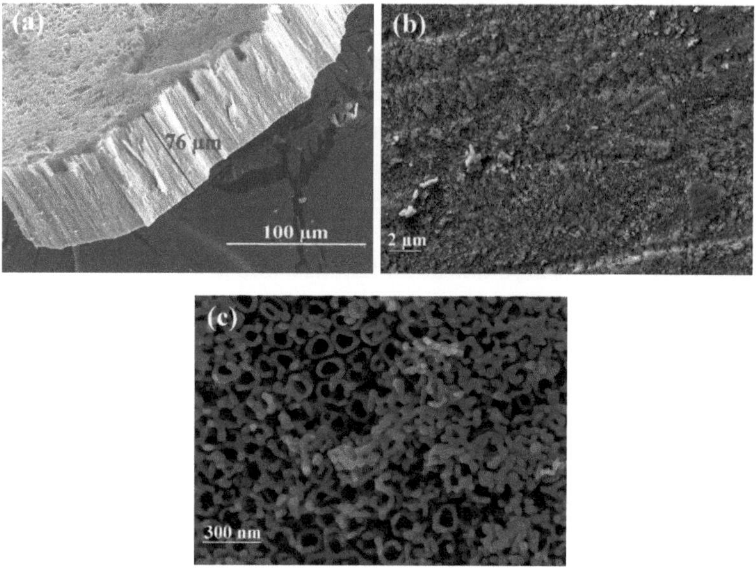

Figure 4.1 FESEM micrographs for anodized titania nanotube arrays (a) after the anodization process (60 V, 25°C, 20 h), and following SRD experiments and viewed with (b) low and (c) high magnification.

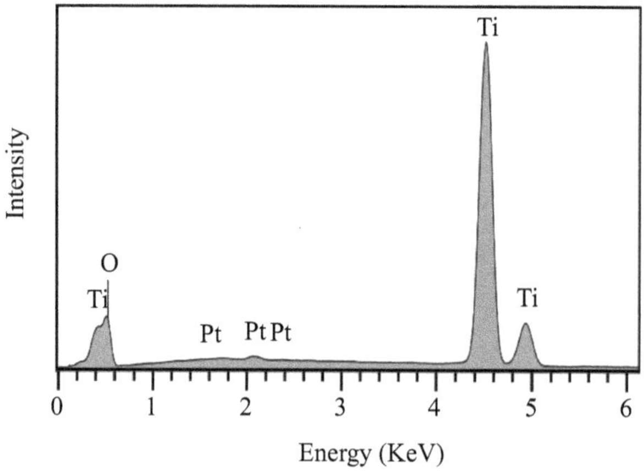

Figure 4.2 EDS spectra for heat treatment (to 900°C) and cooling to room temperature for nanotubes of anodized titania.

4.2.2 Crystallization Kinetics

Figure 4.3 shows the stacked SRD plots during thermal calcination in air with an experimental SRD time, at ambient temperature; 100, 200, and 300°C; and with 50°C steps from 400 to 900°C. Diffraction analysis indicates that the initial arrays of titania nanotube were amorphous, with only Ti_{α} peaks, but at 400°C, crystalline anatase formed. The small crystalline rutile peak indicates that the transformation of anatase to rutile commenced at 550°C. These results are similar to the interpretation of the SRD plot for pure anodized titania nanofiber by Low et al. [25]. The metastable Ti_{α} transformed into another metastable Ti_{β} at 700°C, and both these materials transformed into titania phases at elevated temperature.

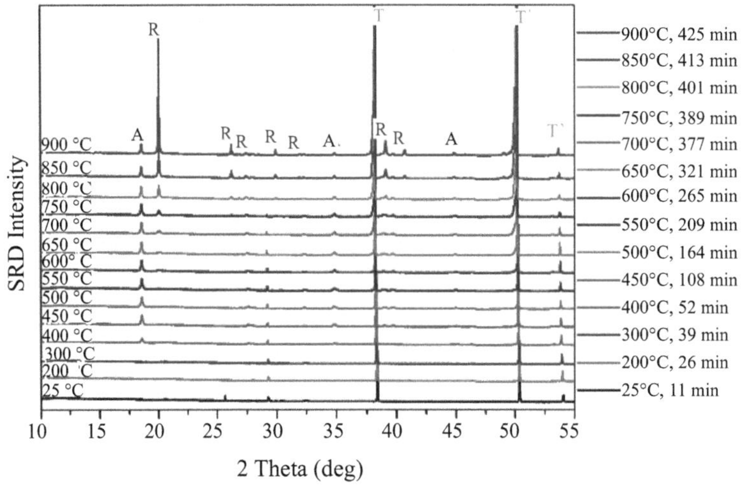

Figure 4.3 Stacked in situ high-temperature SRD plot for as-anodized titania nanotube arrays obtained at ambient temperature; in 100°C steps from 100 to 300°C; and 50°C steps from 400 to 900°C at 10°C/min, 2 min/pattern data-acquisition time [Ti: titanium, A: anatase, and R: rutile].

Relative analysis was used to calculate the relative phase levels of titanium-alpha, titanium-beta, crystalline anatase, and crystalline rutile in the anodized nanotube arrays. The residual values for the Topas refinements R_{wp} and R_{exp} were ~9 and 2, respectively. The goodness-of-fit, χ^2, ranged from 4 to 5, which indicates that the quality of the refinements was acceptable [26–28]. Figure 4.4 shows typical SRD Rietveld residual plots at 900°C with goodness-of-fit values of 4.26 for this pattern.

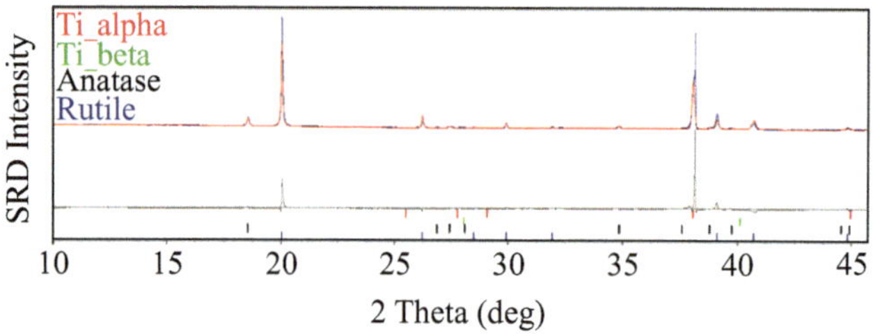

Figure 4.4 SRD Rietveld difference plots for anodized titania nanotube arrays data measured at 900°C. Black crosses show measured patterns, and solid red lines show calculated patterns. Difference between measured and calculated patterns is shown by the gray residual plot. Red, green, black, and blue bars show peak positions for Tiα, Tiβ, anatase, and rutile, respectively. The plots were computed with TOPAS (Version 4.2).

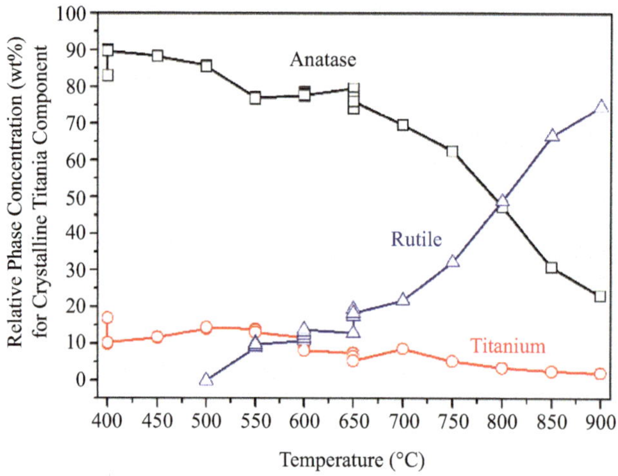

Figure 4.5 Phase abundances of titanium, crystalline anatase, and crystalline rutile for anodized titania nanotube arrays calcined at 400–900°C.

Figure 4.5 shows the phase abundances of $Ti_{(\alpha+\beta)}$, crystalline anatase, and crystalline rutile for the titania nanotube arrays from 400 to 900°C. At 400°C, and from 52 to 96 min after initiation of the experiment, the Ti_{α} transformed rapidly into anatase. Anatase grew progressively after the formation, whereas the Ti_{α} concentration was reduced steadily beyond 900°C because of its transformation to anatase. At 550°C, the anatase-to-rutile transformation did not commence immediately, but required 44 min for its transformation, which occurred rapidly from 600 to 650°C. The calcination time affected the anatase-to-rutile transformation, which increased moderately with increasing calcination time at 550, 660, and 650°C.

4.2.3 Activation Energies

The activation energies of $Ti_{(\alpha+\beta)}$-to-anatase and anatase-to-rutile were obtained by using the Avrami and Arrhenius equations. Figure 4.6 shows the Avrami plot that was used to determine the average temperature-dependent rate constant at 400, 450, 500, 550, 600, and 650°C. Linear regressions of the Arrhenius equation were used to calculate the activation energies from the isothermal SRD data, with the assumed sequential transformations being titanium-to-anatase and then anatase-to-rutile. The corresponding activation energies for this process were 22 (10) kJ/mol for the titanium-to-anatase transformation, and 207 (17) kJ/mol for the anatase-to-rutile transformation. A comparison of the isothermal SRD

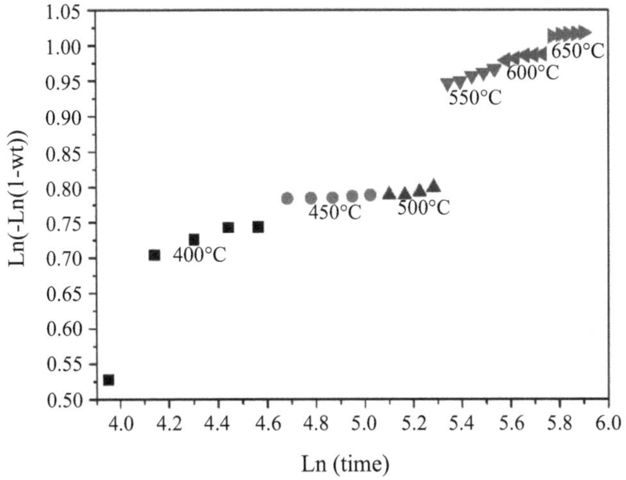

Figure 4.6 The Avrami plot used to determine the average temperature-dependent rate constant at 400, 450, 500, 550, 600, and 650°C.

anatase-to-rutile activation energies for the anodized titania nanotube arrays in this study indicates that these values are close to those of the activation energy reported in the literature, such as 213 kJ/mol for powder, and membranes [29, 30].

4.3 Conclusion

The effect of calcination time, crystallization kinetics, and activation energies of anodized titania nanotube arrays was investigated using in situ high-temperature SRD from 25 to 900°C isothermally at 400, 450, 500, 550, 600, and 650°C. Only anodized titanium diffraction peaks were visible for calcination from room temperature to 300°C, which indicates the formation of amorphous titania nanotubes. The titanium-to-anatase transformation started at 400°C, whereas the anatase-to-rutile transformation occurred at 550°C. The analysis shows that an increase in calcination time increased the rutile percentage but decreased the anatase percentage. The activation energies from the analysis of the isothermal SRD data using the Avrami and Arrhenius equations provided estimates of 22 (10) kJ/mol for the titanium-to-anatase transformation and 207 (17) kJ/mol for the anatase-to-rutile transformation.

Acknowledgments

The authors acknowledge financial support from the Australian Synchrotron (Powder Diffraction Beamline, PD-3611). The authors thank E/Prof. B.H. O'Connor and Dr. Matthew Rowles at the Department of Physics and Astronomy, Curtin University for TOPAS assistance in the data analysis.

References

1. Senna, M., Myers, N., Aimable, A., Laporte, V., Pulgarin, C., Baghriche, O., and Bowen, P. (2013). Modification of titania nanoparticles for photocatalytic antibacterial activity via a colloidal route with glycine and subsequent annealing. *J. Mater. Res.*, **28**: 354–361.
2. Macak, J. M., Tsuchiya, H., Ghicov, A., Yasuda, K., Hahn, R., Bauer, S., and Schmuki, P. (2007). TiO_2 nanotubes: Self-organized electrochemical

formation, properties and applications. *Curr. Opin. Solid State Mater. Sci.*, **11**: 3–18.

3. Xiong, H., Slater, M. D., Balasubramanian, M., Johnson, C. S., and Rajh, T. (2011). Amorphous TiO_2 nanotube anode for rechargeable sodium ion batteries. *J. Phys. Chem. Lett.*, **2**: 2560–2565.
4. Yanga, D., Parka, H., Choa, S., Kima, H., and Choi, W. (2008). TiO_2-nanotube-based dye-sensitized solar cells fabricated by an efficient anodic oxidation for high surface area. *J. Phys. Chem. Solids*, **69**: 1272–1275.
5. Albetran, H., O'Connor, B., and Low, I. (2016). Effect of calcination on band gaps for electrospun titania nanofibers heated in air–argon mixtures. *Mater. Des.*, **92**: 480–485.
6. Hanaor, D. A. and Sorrell, C. C. (2011). Review of the anatase to rutile phase transformation. *J. Mater. Sci.*, **46**: 855–874.
7. Chuangchote, S., Jitputti, J., Sagawa, T., and Yoshikawa, S. (2009). Photocatalytic activity for hydrogen evolution of electrospun TiO_2 nanofibers. *ACS Appl. Mater. Interfaces*, **1**: 1140–1143.
8. Kim, D., Enomoto, N., Nakagawa, Z., and Kawamura, K. (1996). Molecular dynamic simulation in titanium dioxide polymorphs: Rutile, brookite, and anatase. *J. Am. Ceram. Soc.*, **79**: 1095–1099.
9. Liu, G., Wang, L., Yang, H. G., Cheng, H. M., and Lu, G. Q. (2010). Titania-based photocatalysts-crystal growth, doping and heterostructuring. *J. Mater. Chem.*, **20**: 831–843.
10. Beltran, A., Gracia, L., and Andres, J. (2006). Density functional theory study of the brookite surfaces and phase transitions between natural titania polymorphs. *J. Phys. Chem. B*, **110**: 23417–23423.
11. Ghicov, A., Tsuchiya, H., Macak, J. M., and Schmuki, P. (2006). Annealing effects on the photoresponse of TiO_2 nanotubes. *Phy. Status Solidi A*, **203**: 28–30.
12. Liu, Z., Yan, X., Chu, W., and Li, D. (2010). Effects of impurities containing phosphorus on the surface properties and catalytic activity of TiO_2 nanotube arrays. *Appl. Surf. Sci.*, **257**: 1295–1299.
13. Antony, R. P., Mathews, T., Ajikumar, P. K., Krishna, D. N., Dash, S., and Tyagi, A. K. (2012). Electrochemically synthesized visible light absorbing vertically aligned N-doped TiO_2 nanotube array films. *Mater. Res. Bull.*, **47**: 4491–4494.
14. Iida, Y. and Ozaki, S. (1961). Grain growth and phase transformation of titanium oxide during calcination. *J. Am. Ceram. Soc.*, **44**: 120–127.

15. Shannon, R. D. and Pask, J. A. (1965). Kinetics of the anatase-rutile transformation. *J. Am. Ceram. Soc.*, **48**: 391–398.

16. Low, I. M., Albetran, H., Prida, V. M., Vega, V., Manurung, P., and Ionescu, M. (2013). A comparative study on crystallization behavior, phase stability, and binding energy in pure and Cr-doped TiO_2 nanotubes. *J. Mater. Res.*, **28**: 304–312.

17. Liu, R., Qiang, L. S., Yang, W. D., and Liu, H. Y. (2013). The effect of calcination conditions on the morphology, the architecture and the photo-electrical properties of TiO_2 nanotube arrays. *Mater. Res. Bull.*, **48**: 1458–1467.

18. Gamboa, J. A. and Pasquevich, D. M. (1992). Effect of chlorine atmosphere on the anatase–rutile transformation. *J. Am. Ceram. Soc.*, **75**: 2934–2938.

19. Li, H., Cao, L., Liu, W., Su, G., and Dong, B. (2012). Synthesis and investigation of TiO_2 nanotube arrays prepared by anodization and their photocatalytic activity. *Ceram. Int.*, **38**: 5791–5797.

20. Varghese, O. K., Gong, D., Paulose, M., Grimes, C. A., and Dickey, E. C. (2003). Crystallization and high-temperature structural stability of titanium oxide nanotube arrays. *J. Mater. Res.*, **18**: 156–165.

21. Tan, A. W., Pingguan-Murphy, B., Ahmad, R., and Akbar, S. A. (2012). Review of titania nanotubes: Fabrication and cellular response. *Ceram. Int.*, **38**: 4421–4435.

22. Arunchandran, C., Ramya, S., George, R. P., and Mudali, U. K. (2013). Corrosion inhibitor storage and release property of TiO_2 nanotube powder synthesized by rapid breakdown anodization method. *Mater. Res. Bull.*, **48**: 635–639.

23. Liao, J., Lin, S., Pan, N., Li, D., Li, S., and Li, J. (2012). Free-standing open-ended TiO_2 nanotube membranes and their promising through-hole applications. *Chem. Eng. J.*, **211**: 343–352.

24. Albetran, H. and Low, I. M. (2016). Effect of indium ion implantation on crystallization kinetics and phase transformation of anodized titania nanotubes using in-situ high-temperature radiation diffraction. *J. Mater. Res.*, **31**: 1588–1595.

25. Low, I. M., Yam, F., and Pang, W. K. (2012). In situ diffraction studies on the crystallization and crystal growth in anodized TiO_2 nanofibres. *Mater. Lett.*, **87**: 150–152.

26. Albetran, H., Haroosh, H., Dong, Y., Prida, V. M., O'Connor, B. H., and Low, I. M. (2014). Phase transformations and crystallization kinetics in electrospun TiO_2 nanofibers in air and argon atmospheres. *Appl. Phys. A*, **116**: 161–169.

27. Albetran, H., O'Connor, B., and Low, I. (2016). Activation energies for phase transformations in electrospun titania nanofibers: Comparing the influence of argon and air atmospheres. *Appl. Phys. A*, **122**: 367.

28. Albetran, H., O'Connor, B. H., and Low, I. M. (2015). Effect of vanadium ion implantation on the crystallization kinetics and phase transformation of electrospun TiO_2 nanofibers. *Appl. Phys. A*, **120**: 623–634.

29. Kumar, K. N. and Engell, J. (1995). Pore-structure stabilization by controlling particle coordination. *J. Mater. Sci. Lett.*, **14**: 1784–1788.

30. Kumar, K. N. P., Keizer, K., and Burggraaf, A. J. (1993). Textural stability of titania–alumina composite membranes. *J. Mater. Chem.*, **3**: 917–922.

Chapter 5

Effect of Calcination on Band Gap for Electrospun Titania Nanofibers Heated in Air–Argon Mixtures

The relationship between the band gap in electrospun titania nanofibers at ambient temperature and the nature of the air–argon atmosphere in which the material has been heated non-isothermally to 900°C was investigated by ultraviolet–visible absorption spectrometry at room temperature. The results for heating in different selected air–argon mixtures show that the UV-region band gap found in unheated as-spun amorphous nanofibers, 3.33 eV, may be shifted well into the visible region by calcining in the different air–argon mixtures. The band gap value found for heating in air, 3.09 eV, reduces systematically when the material is heated in an air–argon mixture, with the gap in pure argon being 2.18 eV. The progressive lowering of the band gap is attributed to the development of crystallinity in the fibers as the material is calcined and the associated development of oxygen vacancies when heated in argon and, therefore, to the formation of defect states below the conduction band.

Nanostructured Titanium Dioxide in Photocatalysis
It-Meng Low, Hani Manssor Albetran, Victor Manuel de la Prida Pidal, and Fong Kwong Yam

ISBN 978-981-4877-07-7 (Hardcover), 978-1-003-14853-1 (eBook)
www.jennystanford.com

5.1 Introduction

Titanium dioxide (TiO_2) or titania has received much attention as a photocatalytic and photovoltaic semiconductor because it is relatively inexpensive and ecofriendly and has long-term photostability [1–2]. It is a wide-band-gap semiconductor ceramic with an optical band gap for amorphous titania being in the range of 3.30–3.5 eV [3]. This value decreases for two common crystalline titania forms: crystalline anatase (~3.20 eV) and crystalline rutile (~3.02 eV) [4, 5].

Although rutile has a smaller band gap than crystalline anatase, the latter is most favored because it has a higher photoactivity than crystalline rutile, whereas a mixture of crystalline anatase and rutile has been reported to exhibit superior photocatalytic ability [6–8]. Thus, many studies have been conducted on the photocatalytic activity of crystalline titania because crystallinity is an important parameter that influences band gap narrowing in titania. Titania powder, which is used as a commercial photocatalytic material (Degussa, P25), may include small amounts of amorphous phase, a minor quantity of crystalline rutile, and more than 70% crystalline anatase [9]. Some studies have focused on photocatalytic activities and the band gap for amorphous titania or a mixture of amorphous and crystalline titania [3].

An optimum combination of titania phases with lower band gap can be synthesized by thermal treatment of amorphous titania, which transforms to crystalline anatase at relatively low temperature, and then the crystalline anatase transforms to crystalline rutile at higher temperature. The amorphous-to-crystalline transformation of electrospun titania nanofibers has been studied by the authors from room temperature to 900°C in 100% air and 100% argon atmospheres [10]. After thermal treatment, a sample heated in 100% argon contains more crystalline content (anatase and rutile) than that heated in 100% air due to the influence of oxygen vacancies created under argon. To the best of the author's knowledge, the effect of degree of crystallinity on the titania band gap for a mixture of amorphous and crystalline titania has not been reported previously.

When chemically pure, titania must be exposed to ultraviolet (UV) radiation to become photoactive. Various treatments, such as an organic dye attachment, hydrogen plasma reduction, and extrinsic chemical doping [11, 12], have been investigated to reduce the band gap of titania in order to achieve photocatalytic activity under visible light (1.77–3.10 eV).

Impurity doping with various anions and cations has been used to create oxygen vacancies and thus modify the electronic structure to achieve visible light absorption [13–19]. A major disadvantage with chemical doping, however, is that dopant chemicals may be corrosive and pollute the environment (i.e., secondary pollution). A reduction in the band gap without chemical doping is, therefore, an attractive proposition.

The introduction of oxygen vacancies can result in Ti^{3+} states, which may be used as intrinsic donors to narrow the band gap of titania such that they can act as n-type donors. The introduction of oxygen vacancy defects into non-doped titania can influence the photocatalytic behavior significantly. Here the defects act as recombination centers for photo-induced electron–hole pairs, thereby reducing the band gap and enabling photoactivity under visible light [20]. Oxygen vacancies have been created by annealing titania in hydrogen, argon, or in a vacuum [20–22]. The oxygen vacancy concentration increases with increase in calcination temperature in these atmospheres, whereas it remains constant in air.

Using titania in an electrospun nanofiber form is attractive for photocatalysis applications since titania nanofibers have a high ratio of surface area to volume, which increases the decomposition rate of air and water pollutants and this allows for photocatalytic reactions to occur more rapidly on the photocatalyst surface [23–30].

In this work, a novel strategy was used to create oxygen vacancies in a mixture of amorphous titania, crystalline anatase, and crystalline rutile in electrospun titania nanofibers to narrow the band gap. Amorphous titania nanofibers were synthesized by electrospinning, and their morphology and band gap were characterized before and after calcination in air–argon mixtures, using field emission scanning electron microscopy (FESEM), energy dispersive spectroscopy (EDS), and UV–visible spectrometry.

5.2 Results and Discussion

5.2.1 Microstructure Imaging

Figure 5.1 shows secondary electron FESEM images of electrospun titania nanofibers recorded at ambient temperature: (a) as-electrospun prior to heating, and after non-isothermal thermal processing from 25 to 900°C at 10°C/min in (b) 100% air and (c) 100% argon atmospheres. The heated nanofibers have uneven surfaces compared with the as-spun amorphous fibers because of the development of crystalline grains of anatase and rutile [25].

Figure 5.1 Secondary electron FESEM micrographs of electrospun titania nanofibers recorded at ambient temperature: (a) as-electrospun prior to heating, and after non-isothermal heating from 25°C to 900°C at 10°C/min in (b) 100% air and (c) 100% argon atmospheres.

Table 5.1 shows that the crystalline grains were larger in 100% argon than in 100% air, which is consistent with oxygen vacancies created in argon at elevated temperature affecting titania grain growth positively because of enhanced solid-state diffusion. Atomic diffusion or the cooperative rearrangement of titania atoms and grain boundary movement during titania phase transformations occurs more easily in 100% argon than in 100% air because of the higher mobility of titania atoms in the presence of oxygen vacancies. This can lead to higher grain growth and larger titania grains [31, 32].

Table 5.1 Nanofiber and grain size measurements

Calcination Conditions	Nanofiber Size (nm)	Grain Size (nm)
As-spun (Not calcined)	665 ± 330	Amorphous
100% Air	351 ± 145	21 ± 5
100% Argon	328 ± 113	118 ± 49

Note: The ± symbol refers to the size range for size assessment. The number of measurements for each entry was 40.

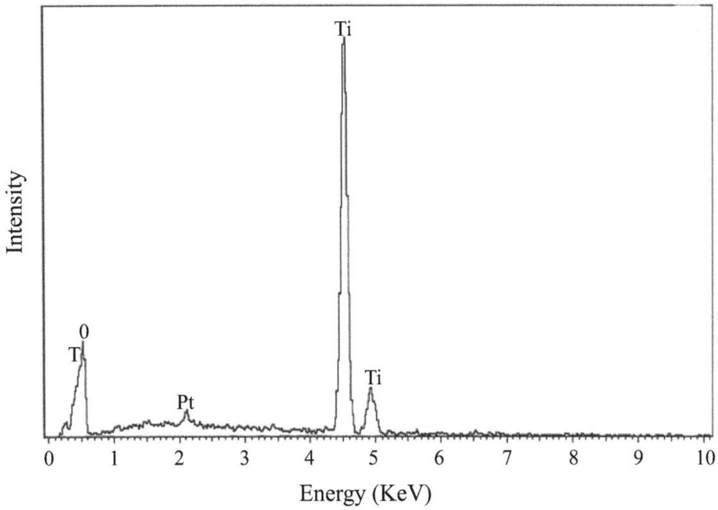

Figure 5.2 EDS spectrum of electrospun titania nanofibers recorded at ambient temperature after heating from 25°C to 900°C at 10°C/min in 100% argon.

The EDS spectrum of the nanofiber sample was recorded after thermal treatment in 100% argon (Fig. 5.2). The spectrum is similar to that for as-spun nanofiber (not shown here) besides an absence of the C peak from loss of organic substance and PVP polymer. The EDS spectrum shows strong Ti and O signatures and a weak Pt signature from the platinum coating on the sample, which are essentially the same when the sample was heated in 100% air. The complete loss of organic material and PVP polymer from the electrospun titania nanofibers occurred at ~450°C; see our preliminary results from in situ high-temperature synchrotron radiation diffraction (SRD) (see Fig. 5.4) and thermal gravimetric analysis (TGA) [10]. The titania nanofibers were initially amorphous by the pronounced amorphous humps in the SRD patterns for 100% air and 100% argon, but these humps disappeared by 400°C. The TGA results showed that the loss of organic material (ethanol and acetic acid) occurred from room temperature to ~150°C, and the PVP polymer decomposition occurred from ~150 to 450°C.

5.2.2 Influence of Calcining Atmosphere

Figure 5.3 shows the progressive color change in titania from white after thermal treatment in 100% air to progressively darker shades of gray after thermal treatment in air–argon compositions from 100% air to 100% argon. This color change is attributed to oxygen loss from the titania to form oxygen vacancy defects, and the color intensification is considered to result from an increase in oxygen vacancy concentration with increasing argon concentration. A similar color change effect has been reported by Gamboa et al. [33] in a calcining experiment in which the effect of low levels of chlorine addition to air and argon was studied. This color change was attributed to the formation of oxygen vacancies when heating in argon.

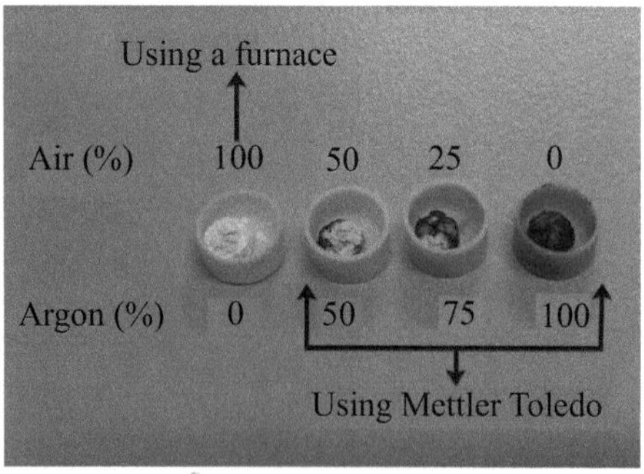

Figure 5.3 Color changes in electrospun titania nanofibrous material observed at room temperature following non-isothermal heating from 25°C to 900°C at 10°C/min in argon–air mixtures.

5.2.3 Phase Compositions

Figure 5.4 shows the stacked SRD plots measured in 100% air and 100% argon over the temperature range of 25–900°C and noting that the peaks at ~27.7° and 35.6° occur because of the Pt heating holder. Table 5.2 shows the variation in absolute phase

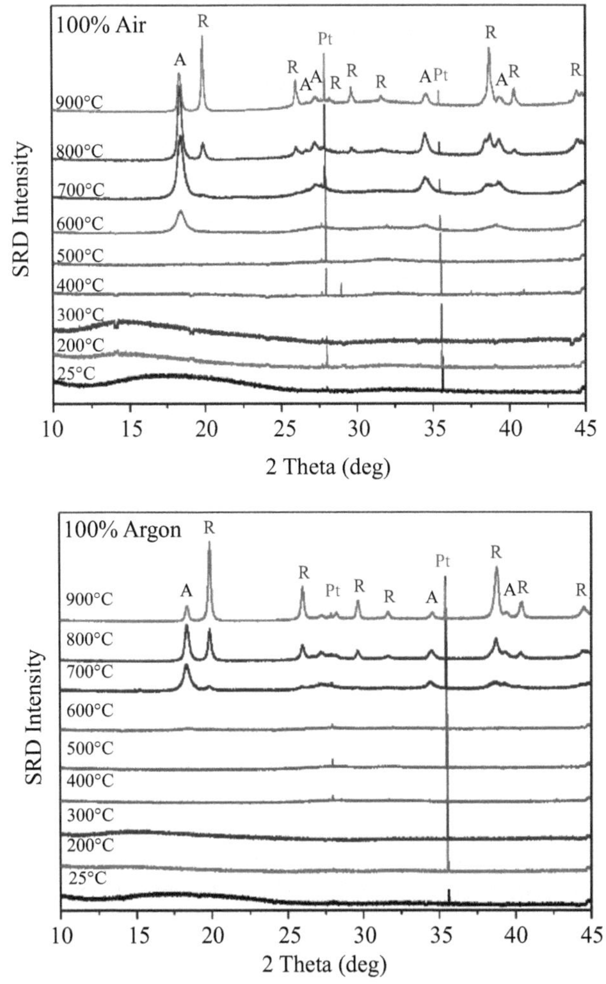

Figure 5.4 Stacked SRD plots for electrospun titania nanofiber material when heated in 100% air and 100% argon, from 25 to 900°C [A: anatase, R: rutile, Pt: platinum]. Plots taken from Albetran et al. [25].

concentrations following SRD for the 100% argon and 100% air atmospheres. The presence of oxygen vacancies makes the anatase-to-rutile transformation in 100% argon more efficient than in 100% air because some relaxation of the Ti binding environment occurs. Details of absolute levels of amorphous titania and crystalline anatase and rutile at each temperature extracted from the SRD data are given in the preliminary novel study by the authors [10].

Table 5.2 Phase compositions and band gaps for titania nanofibers at room temperature after calcining to 900°C in 100% air, 100% argon, and air–argon mixtures

Calcination Conditions	Amorphous (%)	Anatase (%)	Rutile (%)	Crystallinity (%)	Band Gap (eV)
As-spun (not calcined)	100	0	0	0	3.33
100% Air	40.5	20.9	38.6	59.5	3.09
50% Air–50% Argon	NA*	NA*	NA*	NA*	2.94
25% Air–75% Argon	NA*	NA*	NA*	NA*	2.91
100% Argon	26.6	10.3	63.1	73.4	2.18

*Not available

5.2.4 UV–Visible Spectral Analysis

Figure 5.5 compares the UV–visible diffuse reflectance spectra of non-calcined as-electrospun nanofibers with those after thermal treatment in the argon–air mixtures followed by cooling to room temperature. The absorption thresholds, band gap, and absorbance were determined from an absorbance versus wavelength graph. The absorbance plots for unheated fibers, and those after heating in 100%, 50% air, and 25% air are similar in form. The absorbance falls from ~0.7–0.8 to ~0.1–0.2 with all showing the expected precipitous drop in absorbance when the photon energy exceeds the band energy gap. A systematic red shift occurs as the level of argon increases, which is consistent with the development of oxygen vacancy levels within the band gap. The red shift seen here resembles that reported by various authors such as Buha [34] who reported titania thin-film absorbance spectra with different levels of carbon doping, in which an increase in C-doping level increased the magnitude of absorption across the entire visible range. A similar substantial red shift in absorption of Ag-doped titania nanosquares and Fe-doped titania nanoparticles was observed under visible light irradiation [15, 17].

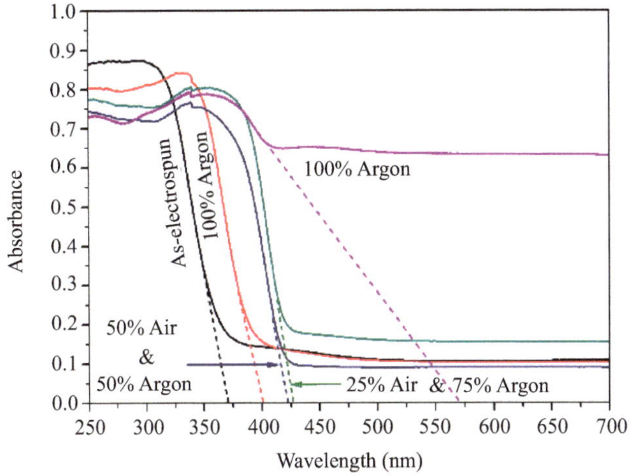

Figure 5.5 UV–visible spectra for as-electrospun titania nanofibrous material after non-isothermal heating from 25°C to 900°C at 10°C/min in 100% air, 50% air–50% argon, 25% air–75% argon, and 100% argon, and then cooling to room temperature.

The absorbance–wavelength plot for the sample heated in 100% argon is substantially different from those for the other samples discussed above, with there being a reduced absorbance range, from *ca.* 0.80 to 0.65, across the wavelength band investigated. This effect is consistent with accounts in the literature that the development of substantial oxygen vacancy defects is responsible for such changes in optical properties. The absorbance change resembles that for C-doped titania thin films for which the absorbance fell from 0.7 to 0.5 over the range 300 nm to 800 nm [34]. Following the present study, the measurement of additional absorbance spectra would be required for argon contents of between 75% and 100% to reveal the underlying band structure.

5.2.5 Influence of Calcining Atmosphere on Band-Gap Structure

Figure 5.6 shows a plot of band gap versus argon content corresponding to the absorbance spectra in Fig. 5.5, and Table 5.2 lists the band gap values for each sample type and also specifies the sample phase compositions [10, 25, 30]. Table 5.3 compares band gap results from this study with literature data for titania materials with different forms, mixtures, and atmospheric treatment.

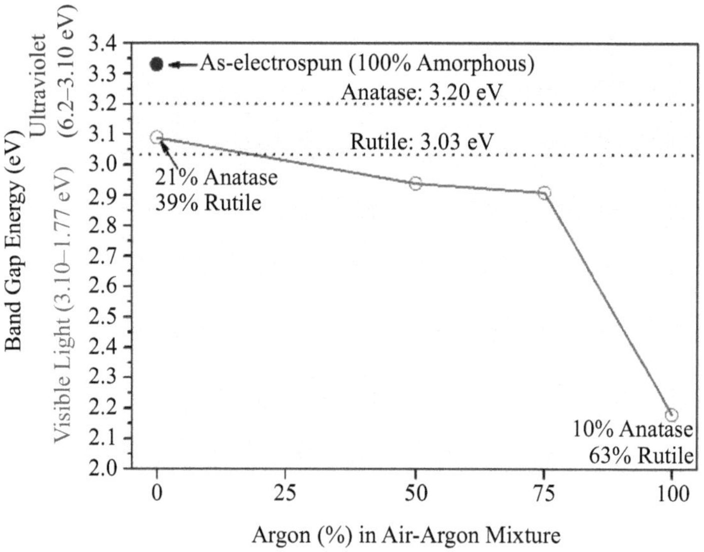

Figure 5.6 Band gap values of as-electrospun titania nanofibrous material, extracted from UV–visible spectra in Fig. 5.4. Band gaps in eV at room temperature following non-isothermal heating from 25°C to 900°C at 10°C/min in air–argon mixtures, and then cooling to room temperature.

The effect on band gap for the as-produced amorphous fiber sample through calcining in 100% air was to reduce it from the amorphous value of 3.33 eV to 3.09 eV, i.e., from within the UV to the visible region. As earlier work on the same sample has indicated that the development of vacancies on heating in air is minimal, the reduction to 3.09 eV is attributed entirely to the development of crystallinity in the fibers, noting the phase levels of about 21% anatase and 39% rutile on calcining amorphous material in 100% air [10]. It is noted that the measured gap value 3.09 eV agrees with

Table 5.3 Comparison of band gap results from present study with data from literature for titania materials

Study	Titania Nanomaterial	Treatment	Band Gap (eV)	Am: A: R	Comment
This study	Fiber	As-spun	3.33	100: 0: 0	
		100% Air	3.09	40: 21: 39	
		100% Argon	2.18	27: 10: 63	Presence of oxygen vacancies
Potlog et al. [1]	Films	Vacuum	3.34	0: 62: 38	
		100% H_2	3.22	0:100: 0	
Tang et al. [4]	Films	Argon and O_2	3.20	0: 100: 0	
			3.00	0: 0: 100	
Chen et al. [38]	Powder	100% Air	3.15	0: 100: 0	
Rosseler et al. [39]	Sol–gel promoted template	100% Air	3.02	0: 50: 50	TiO_2–polyoxyethylene cetylether
			2.99	0: 40: 60	TiO_2–CTAB
			2.97	0: 30: 70	TiO_2–polyvinyl alcohol
			2.94	0: 5: 95	TiO_2–polyethylene glycol
Coronado et al. [40]	Particles	100% Air	3.21	0: 100: 0	
			3.00	0: 0: 100	

Am: amorphous, A: anatase, and R: rutile

the concentration-weighted gap for the two phases: 3.20 eV for pure anatase [5] and 3.03 eV for pure rutile [4, 5]. This appears to be consistent with literature reports of band gaps for mixtures of anatase and rutile [5].

Two factors appear to drive the further reduction in band gap as argon is introduced: (i) changes in the titania phase mix and (ii) the progressive development of oxygen vacancies. The measured gap of 2.18 eV for heating in 100% argon is related to the titania phase composition (10% anatase and 63% rutile) for which the concentration-weighted gap would be 3.05 eV, which is ~0.87 eV greater than the measured gap of 2.18 eV. The 0.87 eV difference is attributed to the oxygen vacancy development. The presence of oxygen vacancies provides more surface-active sites and allows for a larger charge carrier density. This is consistent with the electronic structure of titania material being changed by an absent oxygen atom, resulting in the localization of one or two electrons in an oxygen state. The formation of oxygen vacancies results in the formation of unpaired electrons or Ti^{3+} centers, which form oxygen vacancy states below the conduction band and act as n-type donors [35, 36].

The development of oxygen vacancies as the concentration of argon increases was discussed by the authors [10] in terms of non-stoichiometric effects of titania heated in 100% argon. Oxygen is lost when the sample is heated in argon due to the formation of non-stoichiometric anatase (TiO_{2-x}). This is consistent with the work of Andersson et al. who investigated the composition range of argon-annealed $TiO_{1.9-2}$ using X-ray powder diffraction [37].

Figure 5.7 shows a schematic depicting the assumed band gap states and the apparent influence of oxygen vacancies after calcining to 900°C and cooling to room temperature in air–argon mixtures. The introduction of oxygen vacancies in titania results with the presence of localized new oxygen vacancy states existing between the valence band (VB) and conduction band (CB) in the titania band structure. Electrons may be excited from the VB to the oxygen vacancy states even under visible light. As the argon level increases from 25% to 100%, the effective energy gap red shifts systematically and the sample becomes active in visible light, with the band gap narrowing significantly to 2.94, 2.91, and then 2.18 eV for 50% air–50% argon, 25% air–75% argon, and 100% argon, respectively.

Thus, the combined influence of crystallinity development and oxygen vacancy formation has extended the light excitation of electrospun titania nanofibers from the UV to well into the visible light range. This has been achieved without chemical doping. Finally, a comparison of band gap results, as shown in Table 5.3, indicates that the current band gap results are consistent with other data for nano-titania materials reported in the literature.

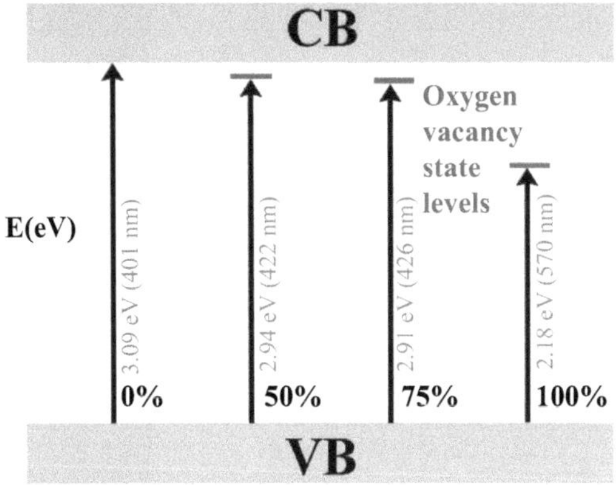

Figure 5.7 Band gap states and the assumed influence of oxygen vacancies with change in air–argon mixture. Argon (%) in air–argon mixtures [CB = conduction band, VB = valence band].

5.3 Conclusion

This chapter focuses on the use of heat treatment of amorphous titania nanofibers, rather than chemical doping, to develop titania photocatalysts that are photoactive in visible light. The results have significance for the development of waste treatment technologies using photocatalytic titania nanofibers, which may function in the visible light. The results for heating in different selected air–argon mixtures show that the UV-region band gap found in as-electrospun amorphous nanofibers, 3.33 eV, shifted well into the visible region by calcining in air, argon, and argon–air mixtures. The band gap value observed for heating in air, 3.09 eV, reduces systematically when the material is heated in an air–argon mixture, with the gap for heating in 100% argon being 2.18 eV. The progressive lowering of the band gap is attributed to the development of crystallinity in the fibers as the material is calcined and the associated development of oxygen vacancies when heated in argon.

The narrowed band gap is attributed to the heterostructure titania phases (anatase and rutile) and the formation of oxygen vacancies, which creates a state below the conduction band. Increasing the percentage of argon in the air–argon mixtures reduces the state to below the titania conduction band, which thus increases the oxygen vacancy concentration and the subsequent band gap narrowing.

Acknowledgments

The authors thank the following Curtin University colleagues: Dr. Y. Dong of Mechanical Engineering for access to the electrospinning equipment; E/Prof. B.H. O'Connor and Ms. E. Miller of the John de Laeter Centre for assistance with Rietveld modelling and electron microscopy, respectively; and Prof. M. Tade, Ms. K. Haynes, and Mr. A. Chan of Chemical Engineering for access to the TGA/DSC Mettler Toledo and UV–Visible spectrometer instruments.

References

1. T. Potlog, P. Dumitriu, M. Dobromir, A. Manole, and D. Luca, Nb-doped TiO_2 thin films for photovoltaic applications, *Mater. Des.* **85** (2015): 558–563.
2. D. Wu and M. Long, Realizing visible-light-induced self-cleaning property of cotton through coating N-TiO_2 film and loading AgI particles, *ACS. Appl. Mater. Interfaces* **3** (2011): 4770–4774.
3. K. Eufinger, D. Poelman, H. Poelman, R. D. Gryse, and G. B. Marin, Photocatalytic activity of dc magnetron sputter deposited amorphous TiO_2 thin films, *Appl. Surf. Sci.* **254** (2007): 148–152.
4. H. Tang, K. Prasad, R. Sanjines, P. E. Schmid, and F. Levy, Electrical and optical properties of TiO_2 anatase thin films, *J. Appl. Phys.* **75** (1994): 2042–2047.
5. D. O. Scanlon, C. W. Dunnill, J. Buckeridge, S. A. Shevlin, A. J. Logsdail, S. M. Woodley, C. R. Catlow, M. J. Powell, R. G. Palgrave, I. P. Parkin, G. W. Watson, T. W. Keal, P. Sherwood, A. Walsh, and A. A. Sokol, Band alignment of rutile and anatase TiO_2, *Nat. Mater.* **12** (2013): 798–801.

6. A. Natoli, A. Cabeza, A. G. Torre, M. A. Aranda, and I. Santacruz, Colloidal processing of macroporous TiO_2 materials for photocatalytic water treatment, *J. Am. Ceram. Soc.* **95** (2012): 502–508.
7. S. Kment, P. Kluson, V. Stranak, P. Virostko, J. Krysa, M. Cada, J. Pracharova, M. Kohout, M. Morozova, P. Adamek, and Z. Hubicka, Photo-induced electrochemical functionality of the TiO_2 nanoscale films, *Electrochim. Acta* **54** (2009): 3352–3359.
8. H. Li, W. Zhang, and W. Pan, Enhanced photocatalytic activity of electrospun TiO_2 nanofibers with optimal anatase/rutile ratio, *J. Am. Ceram. Soc.* **94** (2011): 3184–3187.
9. B. Ohtani, O. O. Mahaney, D. Li, and R. Abe, What is degussa (evonik) P25? Crystalline composition analysis, reconstruction from isolated pure particles and photocatalytic activity test, *J. Photochem. Photobiol. A: Chem.* **216** (2010): 179–182.
10. H. Albetran, B. H. O'Connor, and I. M. Low, Activation energies for phase transformations in electrospun titania nanofibers: Comparing the influence of argon and air atmospheres, *Appl. Phys. A* **122** (2016): 367.
11. S. W. Hsu, T. S. Yang, T. K. Chen, and M. S. Wong, Ion-assisted electron-beam evaporation of carbon-doped titanium oxide films as visible-light photocatalyst, *Thin Solid Films* **515** (2007): 3521–3526.
12. J. Yi, X. Yuan, H. Wang, H. Yu, and F. Peng, Preparation of $Bi_2Ti_2O_7/TiO_2$ nanocomposites and their photocatalytic performance under visible light irradiation, *Mater. Des.* **86** (2015): 152–155.
13. V. J. Babu, A. S. Nair, Z. Peining, and S. Ramakrishna, Synthesis and characterization of rice grains like nitrogen-doped TiO_2 nanostructures by electrospinning–photocatalysis, *Mater. Lett.* **65** (2011): 3064–3068.
14. A. Ghicov, J. M. Macak, H. Tsuchiya, J. Kunze, V. Haeublein, L. Frey, and P. Schmuki, Ion implantation and annealing for an efficient N-doping of TiO_2 nanotubes, *Nano. Lett.* **6** (2006): 1080–1082.
15. C. Yu, L. Wei, X. Li, J. Chen, Q. Fan, and J. C. Yu, Synthesis and characterization of Ag/TiO_2-B nanosquares with high photocatalytic activity under visible light irradiation, *Mater. Sci. Eng. B* **178** (2013): 344–348.
16. M. Takeuchi, H. Yamashita, M. Matsuoka, M. Anpo, T. Hirao, N. Itoh, and N. Iwamoto, Photocatalytic decomposition of No under visible light irradiation on the Cr-ion-implanted TiO_2 thin film photocatalyst, *Catal. Lett.* **67** (2000): 135–137.
17. S. K. Patel, S. Kurian, and N. S. Gajbhiye, Room-temperature ferromagnetism of Fe-doped TiO_2 nanoparticles driven by oxygen vacancy, *Mater. Res. Bull.* **48** (2013): 655–660.

18. D. Li, N. Ohashi, S. Hishita, T. Kolodiazhnyi, and H. Haneda, Origin of visible-light-driven photocatalysis: A comparative study on N/F-doped and N–F-codoped TiO_2 powders by means of experimental characterizations and theoretical calculations, *J. Solid. State. Chem.* **178** (2005): 3293–3302.

19. R. P. Antony, T. Mathews, P. K. Ajikumar, D. N. Krishna, S. Dash, and A. K. Tyagi, Electrochemically synthesized visible light absorbing vertically aligned N-doped TiO_2 nanotube array films, *Mater. Res. Bull.* **4** (2012): 4491–4497.

20. J. H. Huang and M. S. Wong, Structures and properties of titania thin films annealed under different atmosphere, *Thin Solid Films* **520** (2011): 1379–1384.

21. A. Ghicov, H. Tsuchiya, J. M. Macak, and P. Schmuki, Annealing effects on the photoresponse of TiO_2 nanotubes, *Physica. Status. Solidi. A* **203** (2006): 28–30.

22. Q. Wu and R. V. Krol, Selective photoreduction of nitric oxide to nitrogen by nanostructured TiO_2 photocatalysts: Role of oxygen vacancies and iron dopant, *J. Am. Chem. Soc.* **134** (2012): 9369–9375.

23. J. Y. Park, J. J. Yun, C. H. Hwang, and I. H. Lee, Influence of silver doping on the phase transformation and crystallite growth of electrospun TiO_2 nanofibers, *Mater. Lett.* **64** (2010): 2692–2695.

24. I. M. Low, F. K. Yam, and W. K. Pang, In-situ diffraction studies on the crystallization and crystal growth in anodized TiO_2 nanofibres, *Mater. Lett.* 87 (2012): 150–152.

25. H. Albetran, H. Haroosh, Y. Dong, V. M. Prida, B. H. O'Connor, and I. M. Low, Phase transformations and crystallization kinetics in electrospun TiO_2 nanofibers in air and argon atmospheres, *Appl. Phys. A* **116** (2014): 161–169.

26. Q. Li, D. J. G. Satur, H. Kim, and H. G. Kim, Preparation of sol–gel modified electrospun TiO_2 nanofibers for improved photocatalytic decomposition of ethylene, *Mater. Lett.* **76** (2012): 169–172.

27. H. Li, W. Zhang, B. Li, and W. Pan, Diameter-dependent photocatalytic activity of electrospun TiO_2 nanofiber, *J. Am. Ceram. Soc.* **93** (2010): 2503–2506.

28. S. Chuangchote, J. Jitputti, T. Sagawa, and S. Yoshikawa, Photocatalytic activity for hydrogen evolution of electrospun TiO_2 nanofibers, *ACS. Appl. Mater. Interfaces* **1** (2009): 1140–1143.

29. R. Chandrasekar, L. Zhang, J. Y. Howe, N. E. Hedin, Y. Zhang, and H. Fong, Fabrication and characterization of electrospun titania nanofibers, *J. Mater. Sci.* **44** (2009): 1198–1205.

30. H. Albetran, B. H. O'Connor, and I. M. Low, Effect of vanadium ion implantation on the crystallization kinetics and phase transformation of electrospun TiO_2 nanofibers, *Appl. Phys. A* **120** (2015): 623–634.

31. Y. Iida and S. Ozaki, Grain growth and phase transformation of titanium oxide during calcination, *J. Am. Ceram. Soc.* **44** (1961): 120–127.

32. X. Ding and X. Liu, Correlation between anatase-to-rutile transformation and grain growth in nanocrystalline titania powders, *J. Mater. Res.* **13** (1998): 2556–2559.

33. J. A. Gamboa and D. M. Pasquevich, Effect of chlorine atmosphere on the anatase-rutile transformation, *J. Am. Ceram. Soc.* **75** (1992): 2934–2938.

34. J. Buha, Solar absorption and microstructure of C-doped and H-co-doped TiO_2 thin films, *J. Phys. D: Appl. Phys.* **45** (2012): 385305.

35. I. Nakamura, N. Negishi, S. Kutsuna, T. Ihara, S. Sugihara, and K. Takeuchi, Role of oxygen vacancy in the plasma-treated TiO_2 photocatalyst with visible light activity for No removal, *J. Mole. Cataly. A: Chem.* **161** (2000): 205–212.

36. H. Seo, L. R. Baker, A. Hervier, J. Kim, J. Whitten, and G. A. Somorjai, Generation of highly n-type titanium oxide using plasma fluorine insertion, *Nano. Lett.* **11** (2010): 751–756.

37. S. Andersson, B. Collen, U. Kuylenstierna, and A. Magneli, Phase analysis studies on the titanium-oxygen system, *Acta. Chem. Scand.* **11** (1957): 1641–1652.

38. D. Chen, Z. Jiang, J. Geng, Q. Wang, and D. Yang, Carbon and nitrogen co-doped TiO_2 with enhanced visible-light photocatalytic activity, *Ind. Eng. Chem. Res.* **46** (2007): 2741–2746.

39. O. Rosseler, M. V. Shankar, M. K. Du, L. Schmidlin, N. Keller, and V. Keller, Solar light photocatalytic hydrogen production from water over Pt and Au/TiO_2 (anatase/rutile) photocatalysts: Influence of noble metal and porogen promotion, *J. Catal.* **269** (2010): 179–190.

40. 40. D. Reyes-Coronado, G. Rodriguez-Gattorno, M. E. Espinosa-Pesqueira, C. Cab, R. de Coss, and G. Oskam, Phase-pure TiO_2 nanoparticles: Anatase, brookite and rutile, *Nanotechnology* **19** (2008): 145605.

Chapter 6

Characterization and Optimization of Electrospun TiO_2/PVP Nanofibers Using Taguchi Design of Experiment Method

TiO_2 nanofibers were prepared within polyvinyl pyrrolidone (PVP) polymer using a combination of sol–gel and electrospinning techniques. Based on a Taguchi design of experiment (DoE) method, the effects of sol–gel and electrospinning on the diameter of TiO_2/PVP nanofibers, including titanium isopropoxide (TiP) concentration, flow rate, needle-tip-to-collector distance, and applied voltage, were evaluated. The analysis of DoE for nanofiber diameters demonstrated that TiP concentration was the most significant factor. An optimum combination to obtain smallest diameters was also determined with a minimum variation for electrospun TiO_2/PVP nanofibers. The optimum combination was determined to be 60% TiP concentration, at the flow rate of 1 ml/h, with the needle-tip-to-collector distance at 11 cm (position *a*), and the applied voltage of 18 kV. This combination was further validated by conducting a confirmation experiment that used two differently sized needles to study the effect of needle size. The average nanofiber diameter size was approximately the same for both needle sizes in good accordance with the optimum performance estimated by the Taguchi DoE method.

Nanostructured Titanium Dioxide in Photocatalysis
It-Meng Low, Hani Manssor Albetran, Victor Manuel de la Prida Pidal, and Fong Kwong Yam

ISBN 978-981-4877-07-7 (Hardcover), 978-1-003-14853-1 (eBook)
www.jennystanford.com

6.1 Introduction

The electrospinning technique has attracted considerable attention as a relatively new, cheap, and simple synthesis method for one-dimensional nanostructures [1–4]. The unique nanofibers prepared by electrospinning generally exhibit high surface-area-to-volume ratios, high porosity, nanosized effects, and excellent mechanical strength [5]. They have been suggested for many applications, such as membrane separation, drug delivery, protective clothing, wound dressings, filtration, tissue engineering, and electronic applications [6–9].

Titanium dioxide (TiO_2), also known as titania, has emerged as a promising photocatalyst in the current market. It has the advantages of being photocatalytically stable, reasonably inexpensive, and relatively easy to produce and use. It is a human and environment-friendly photocatalyst used to treat polluted air and water, and to split water to generate hydrogen [10]. A high surface-area-to-volume ratio (SA/vol) of electrospun nanofibers can significantly improve the photocatalytic performance of TiO_2 because of small fiber size, and the high SA/vol provides further means for quick charge transfer to the dynamics of hole–electron (e^-/h^+) recombination on a large specific surface area of TiO_2 nanofibers [6, 11].

TiO_2 nanofibers can be synthesized by combining electrospinning with a TiO_2 sol–gel technique [2, 3, 11–14]. The components of the electrospun experiment to synthesize TiO_2 nanofibers comprise a syringe pump, a syringe with a conductive needle, a high-voltage supply, a conductive collector, copper wires, and a sol–gel of polymer (binder), and a TiO_2 precursor [4]. In general, electrospun nanofiber diameters depend primarily on three processing parameter sets [15] mentioned below.

The first set of adjusted parameters involves the sol–gel solution. Its properties include electrical conductivity, viscosity, surface tension, polymer concentration, and molecular weight. They are related to one another, and these relationships have important influences on electrospun nanofiber diameters [16]. The viscosity of the TiO_2 solution depends on the molecular weight and concentration of the material solution, such as polymer, solvent, and TiO_2 source. The polymer concentration is one of the most significant factors in controlling beads and fiber diameters [16–18]. The TiO_2 fiber diameter increases with a higher TiO_2 concentration in a precursor solution [12, 19].

Electrospinning conditions are the second set of parameters, which consist of applied voltage, flow rate, needle size, and needle-tip-to-collector distance [16]. Optimally applied voltage is a significant factor in the fiber diameters [17, 19–21]. High applied voltage (i.e., strong electrical repulsive forces) reduces the nanofiber diameter, resulting in highly stretched and elongated fibers [20] [3]. Flow rate plays one of the most important roles in determining the fiber diameter and bead formation, because it determines the amount of sol–gel solution available to be stretched into nanofibers [16, 19, 20]. During the electrospinning process, the shape and size of the needle tip affect the formation of Taylor cone and nanofiber oscillation [22]. Ksapabutr et al. [22] claimed that a sawtooth needle shape allowed for Taylor cones of greater length than standard and flat counterparts. A needle-tip-to-collector distance with sufficient field gradient produces fibers with less bead defects, but the more considerable distance increases the nanofiber diameter owing to subsequent decreases in the electric field gradient [17]. Applied high-voltage setting, a large needle-tip-to-collector distance, a comparatively low concentration of polymer solutions, and a low flow rate reduce the variation in product quality of electrospun fiber mats with a minimum number of experiments [17]. The relationship between electrospun variables and associated fibrous structures is still not well understood to achieve ultrafine bead-free nanofibers with good dimensional stability.

Aspects of the atmospheric environment, such as humidity, pressure, and temperature, belong to the last set of parameters. The average nanofiber diameter decreases with increasing atmospheric temperature and decreasing atmospheric humidity [23].

Electrospun TiO_2/PVP nanofibers were synthesized using different titanium oxide precursors with a constant flow rate, needle-tip-to-collector distance, and applied voltage in most electrospun experiments [2, 8, 28–31]. The fibers have a smooth and uniform surface with a random orientation, and the average fiber diameters range from 132 to 2280 nm [24–27]. Kumar et al. used a different applied voltage and flow rate with a constant needle-to-collector distance. At ~10 cm, the average diameter of the as-spun TiO_2/PVP nanofiber was 450 nm at 10 kV and 1 ml/h, 262 nm at 20 kV and 1 ml/h, and 145 nm at 20 kV and 0.5 ml/h [20].

The Taguchi method for robust experimental design is a useful engineering approach to select the optimal levels of processing parameters with the minimal sensitivity to different causes of variations. Furthermore, such a method can also elucidate the effects of a large and complex number of factors on an individual and interactive basis. In general, two essential tools are required, namely an orthogonal array (OA) to simultaneously accommodate several experimental design factors, and signal-to-noise ratio (S/N) to measure the most robust set of operating conditions from variations within the results [17, 28–31].

In this study, TiO_2 nanofibers were fabricated with PVP polymer as precursor using both sol–gel and electrospinning techniques. An optimum combination of parameters obtained from TiP concentration, flow rate, needle-tip-to-collector distance, and applied voltage in response to minimizing diameter size and its variation for TiO_2/PVP nanofibers was determined by means of the Taguchi DoE method. Such optimum condition was further implemented to explore the effect of needle size on fiber diameters accordingly.

6.2 Theory and Fundamentals

6.2.1 Taguchi DoE

Four factors of sol–gel and electrospinning parameters have been selected for this experiment, which are given by TiP concentration, flow rate, collector distance, and applied voltage at three different levels (Table 6.1). The full-factorial experiment of 81 (3^4) trials can be completed in just 27 runs due to the slope collector, but that entails a large number of tests, which are significant in both experimental cost and time. As a result, Taguchi DoE layouts are more applicable when compared to a traditional full-factorial counterpart. This is because it reduced the number of tests to a practical level, thus significantly saving the experiment time and associated costs as opposed to the conduct of four factors individually. The L_9 DoE orthogonal array was selected with the assumption of no factorial interactions, resulting in nine trials, as illustrated in Table 6.2.

Table 6.1 Four factors and three levels selected in the DoE study for electrospun TiO_2/PVP nanofibers

	Factor Description	Level		
		1	2	3
A	TiP concentration (wt%)	40	50	60
B	Flow rate (ml/h)	0.5	1	2
C	Needle-tip-to-collector distance (cm)	~11	~8	~9
D	Applied voltage (kV)	14	18	25

Table 6.2 Taguchi orthogonal array with nine trials (L_9)

Trial	Factor			
	A: TiP concentration (wt%)	B: Flow rate (ml/h)	C: Needle-tip-to-collector distance	D: Applied voltage (kV)
T1	40	0.5	a	14
T2	40	1	b	18
T3	40	2	c	25
T4	50	0.5	b	25
T5	50	1	c	14
T6	50	2	a	18
T7	60	0.5	c	18
T8	60	1	a	25
T9	60	2	b	14

6.2.2 Analysis of Variance

As mentioned earlier, the Taguchi DoE method replaces the full-factorial experiments with only a simple orthogonal array of nine trials. To determine the significant factors, and optimum combination of factors, an analysis of variance (ANOVA) was utilized in order to offer a measure of confidence by determining and analyzing the data variance [31]. In ANOVA, the total variation (S_T), the sum of squares of each factor (Si), and the percentage contribution (%) were computed, respectively [31].

6.2.3 Total Variation (S_T)

The total variation (S_T) is the sum of squares of all trial results, which is expressed in the following form:

$$S_T = \left[\sum_{i=1}^{N} \overline{Y}_i^2\right] - \left[\frac{\left(\sum_{i=1}^{N} \overline{Y}_i\right)^2}{N}\right] \tag{6.1}$$

where $\overline{Y}_i$ is the mean fiber diameter and N is the number of trials in Taguchi DoE study.

6.2.4 Total Variance of Each Factor (Si)

$$S_A = \frac{A_{40}^2}{3} + \frac{A_{50}^2}{3} + \frac{A_{60}^2}{3} - C.F \tag{6.2}$$

$$S_B = \frac{B_{0.5}^2}{3} + \frac{B_1^2}{3} + \frac{B_2^2}{3} - C.F \tag{6.3}$$

$$S_C = \frac{C_a^2}{3} + \frac{C_b^2}{3} + \frac{C_c^2}{3} - C.F \tag{6.4}$$

$$S_D = \frac{D_{14}^2}{3} + \frac{D_{18}^2}{3} + \frac{D_{25}^2}{3} - C.F \tag{6.5}$$

where S_A, S_B, S_C, and S_D are the sum squares for four factors of TiP concentration, flow rate, collector distance, and applied voltage at three different levels, respectively. $C.F$ denotes the correction factor, which is a term similar to $\left[\frac{\left(\sum_{i=1}^{N} \overline{Y}_i\right)^2}{N}\right]$ in Eq. 6.1 and it remains constant for all factors. The correction factor ($C.F$) is used for calculation of all sums of squares [31].

6.2.5 Percentage Contribution (%)

The percentage contribution of four factors (P_A, P_B, P_C, or P_D) is the ratio of the total variance of each factor (S_A, S_B, S_C, or S_D) to total variation (S_T) as given by:

$$P_i = \frac{S_i}{S_T} \times 100 \quad (6.6)$$

where i is the number of factors ($i = 4$ for this study).

6.2.6 Signal-to-Noise Ratio (S/N) of Electrospun TiO_2 Nanofiber Diameter

A "smaller the better" characteristic formula [17, 28, 29, 31] has been used to identify the optimum combination of factors to reduce both the fiber diameter and its variation in electrospun TiO_2/PVP nanofibers as indicated below:

$$S/N = -10\log\left(\frac{1}{n}\sum_{i=1}^{n} y_i^2\right) \quad (6.7)$$

where S/N is the signal-to-noise ratio, n is the number of measurements, and y is the diameter of electrospun TiO_2/PVP nanofibers. Mathematically, the greater the value of S/N, the smaller the variance for electrospun TiO_2/PVP nanofibers.

6.3 Results and Discussion

6.3.1 Nanofiber Morphology and Diameter

Scanning electron microscopy (SEM) micrographs of electrospun nanofiber morphology for the L_9 DoE are illustrated in Fig. 6.1. The randomly oriented nanofibers had smooth surfaces along with most of large beads as typical defects, except only minor bead defects were noticed in Figs. 6.1e–g. Figure 6.2 shows a typical EDS spectrum of the DoE study with elemental signals for Ti, O, Pt, and C, where the Ti and O elements are ascribed to TiO_2; C and Pt elements are due to the PVP polymer and the platinum coating, respectively [14].

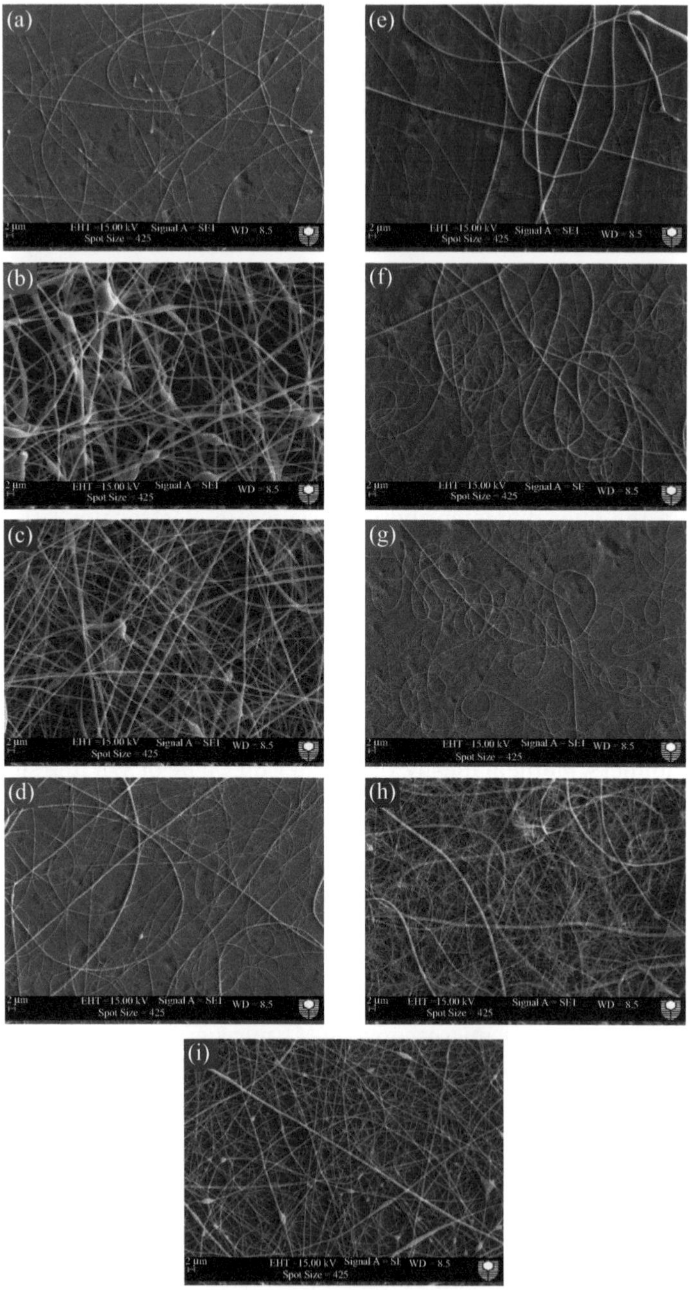

Figure 6.1 SEM micrographs of electrospun TiO_2/PVP nanofibers used in DoE study: (a) T1, (b) T2, (c) T3, (d) T4, (e) T5, (f) T6, (g) T7, (h) T8, (i) T9. All scale bars represent 2 μm.

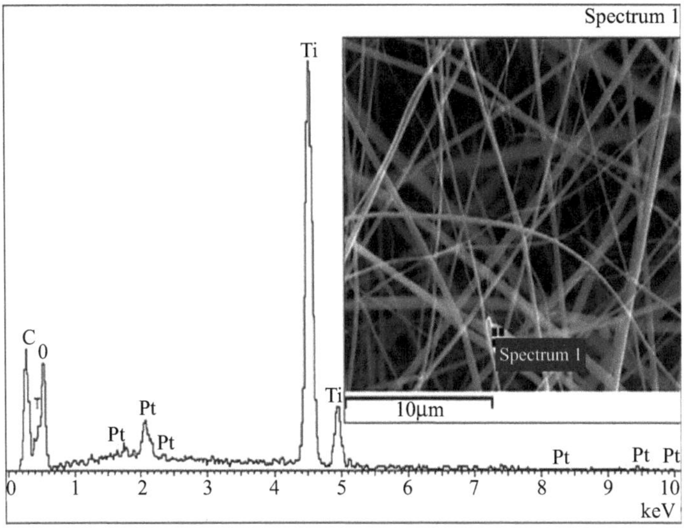

Figure 6.2 EDS spectra of electrospun TiO_2/PVP nanofibers.

Figure 6.3 shows the frequency contribution diagrams for the DoE study in the diameter range of 25–1050 nm. The corresponding average fiber diameters and standard deviations are illustrated in Table 6.3. T2 in the DoE study yielded the highest average diameter (305 ± 165 nm) with the highest range of variation in diameter (125–1050 nm). The highest viscosity level of the sol–gel solution with 40% TiP concentration and the lowest needle-tip-to-collector distance (position *b*) were the main reason for the increased nanofiber diameter. However, the nanofiber diameter decreased slightly for T1, T3, and T9, where the average diameters and the standard deviations were 218 ± 126 nm, 271 ± 203 nm, and 227 ± 57 nm, respectively. Both T1 and T3 had the same sol–gel concentrations as opposed to T2 with 40 wt% TiP, but with different electrospinning parameters. The needle-tip-to-collector distances for T1 (position *a*) and T3 (position *c*) were higher than that for T2 (position *b*), which was one reason for decreasing fiber diameter. Other reasons to achieve this were to reduce the flow rate or increase the applied voltage. T1 had the lowest level of flow rate (0.5 ml/h), but it also had the lowest level of applied voltage (14 kV). T3 had the highest level of applied voltage (25 kV) but also had the highest level of flow rate (2 ml/h). For T9, the main reason for reduction in diameter size was the TiP concentration of 60 wt%, despite its lowest distance (position *b*), lowest applied voltage (14 kV), and highest flow rate (2 ml/h). The TiP concentration of T4, T5, and T6 was 50 wt%, but with different combinations of electrospinning factors. Most of the diameter variation was from 75 to 275 nm, and the average diameters were under 200 nm. For both T7 and T8, the highest level of TiP concentration at 60 wt% was used. The average fiber diameters and the standard deviations were 85 ± 18 nm for T7, and 70 ± 18 for T8 due to the different electrospinning conditions. Both T7 and T8 had a minimum variance of fiber diameter, where about 80% of the frequency contribution was detected in the fiber diameter range of 50–100 nm.

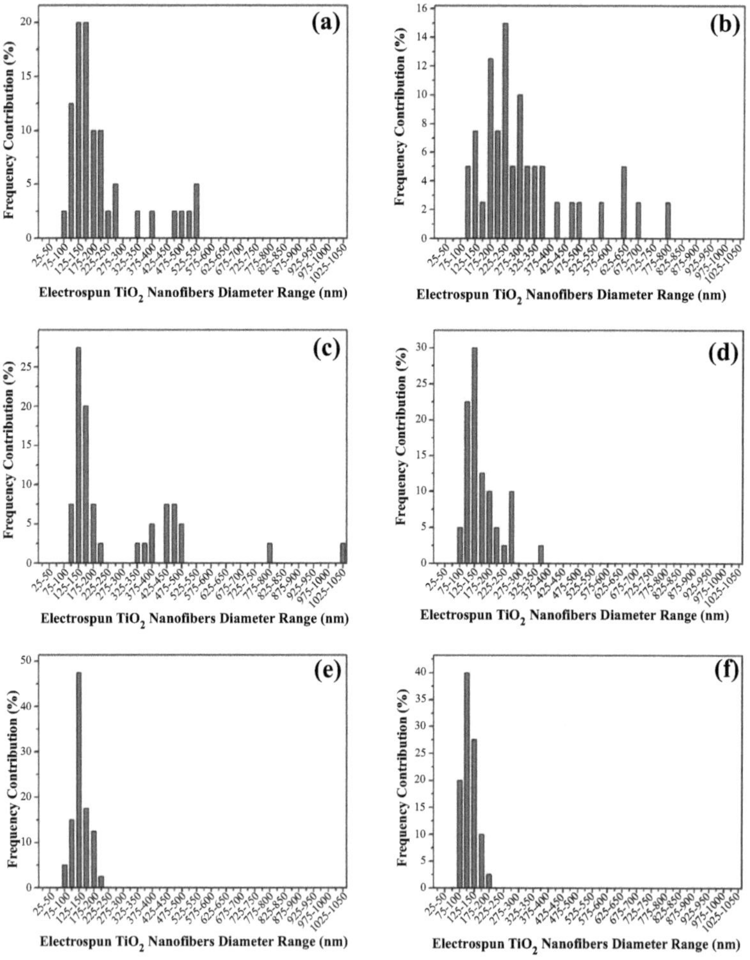

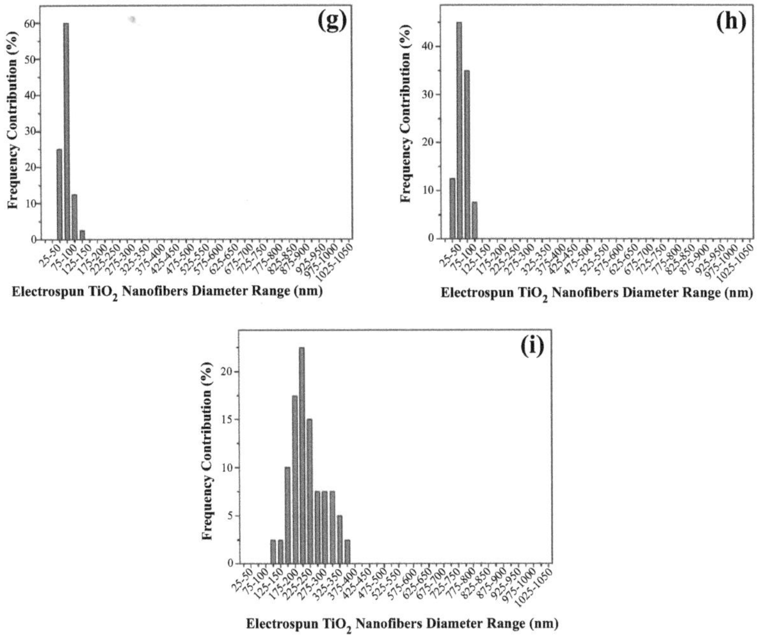

Figure 6.3 Frequency contributions to electrospun TiO_2/PVP nanofiber diameter range in DoE study: (a) T1, (b) T2, (c) T3, (d) T4, (e) T5, (f) T6, (g) T7, (h) T8, (i) T9.

Table 6.3 Changes in nanofiber diameters of electrospun TiO_2/PVP

Trial	Combination of Factors	Average Nanofiber Diameter and Standard Deviation (nm)	*S/N*
T1	$A_{40}\,B_{0.5}\,C_a\,D_{14}$	218±126	-47.99
T2	$A_{40}\,B_1\,C_b\,D_{18}$	305±165	-50.78
T3	$A_{40}\,B_2\,C_c\,D_{25}$	271±203	-50.56
T4	$A_{50}\,B_{0.5}\,C_b\,D_{25}$	162±59	-44.73
T5	$A_{50}\,B_1\,C_c\,D_{14}$	142±28	-43.21
T6	$A_{50}\,B_2\,C_a\,D_{18}$	118±25	-41.61
T7	$A_{60}\,B_{0.5}\,C_c\,D_{18}$	85±18	-38.81
T8	$A_{60}\,B_1\,C_a\,D_{25}$	70±18	-37.22
T9	$A_{60}\,B_2\,C_b\,D_{14}$	227+57	-47.37

Figure 6.4 shows the effect of TiP concentration, flow rate, needle-tip-to-collector distance, and applied voltage on the average nanofiber diameter of electrospun TiO_2/PVP. It has been found that the fiber diameters decreased with increasing TiP concentration (decrease in viscosity due to the low concentration of PVP polymer in the TiP/PVP solution), needle-tip-to-collector distance (position *a*), and applied voltage setting while decreasing the flow rate, which is in good agreement with previous literature [11, 17, 20]. The nanofiber diameters decreased relatively sharply for the TiP concentration from 265 nm with the 40 wt% TiP to 127 nm with 60 wt% TiP. This finding suggests that the TiP concentration is the most significant factor for achieving the small nanofiber diameter of electrospun TiO_2/PVP with the minimum variance (further detail is included in the Taguchi analysis).

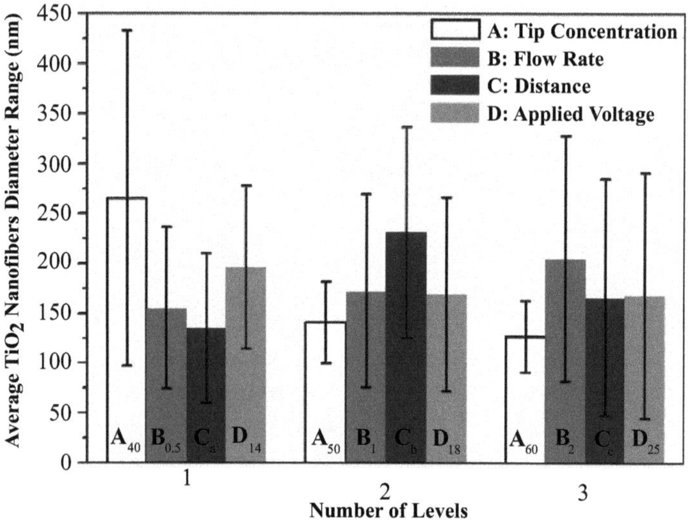

Figure 6.4 The effect of TiP concentration, flow rate, needle-tip-to-collector distance, and applied voltage on the average nanofiber diameter of electrospun TiO_2/PVP.

6.3.2 Analysis of Variance (ANOVA)

The relative percentage contributions of electrospinning factors were determined using Eqs. 6.1–6.6. The ANOVA diagram depicted in Fig. 6.5 demonstrates the effect of processing parameters on nanofiber diameters of electrospun TiO_2/PVP by the percentage contributions for selected L_9 DoE factors. The TiP concentration (factor A) was a significant variable to control the nanofiber diameter with a percentage contribution of 63.45%. This provides further evidence that increasing the TiP concentration decreases the diameter size. The second and third most prevalent factors in minimizing nanofiber diameters were the needle-tip-to-collector distance (factor C) and the flow rate (factor B) with the percentage of contributions of 26.59% and 7.23% accordingly. The applied voltage (factor D) appeared to be insignificant at 2.72% only. It is implied that the applied voltage has a trivial impact on the nanofiber diameter, which can be maintained at the workable experimental range.

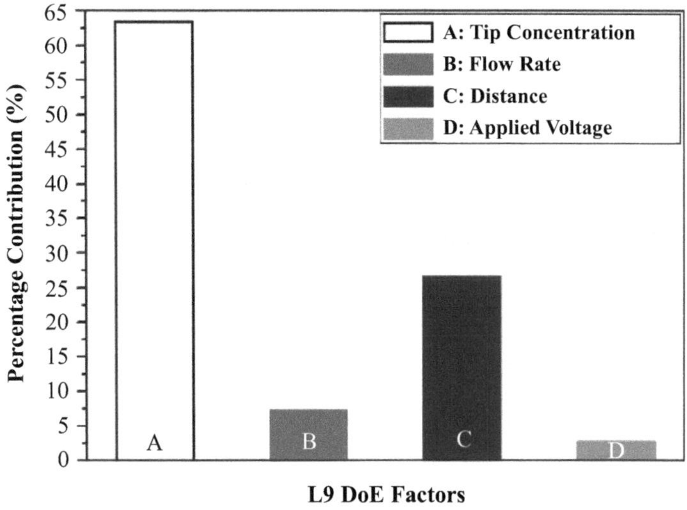

Figure 6.5 The ANOVA diagram for the determination of significant factors to influence electrospun TiO_2/PVP nanofiber diameters.

6.3.3 Optimum Combination of Factors

The S/N ratios for electrospun TiO_2/PVP nanofibers were calculated using the "smaller the better" Eq. 6.7 for the nine Taguchi DoE study (Table 6.3). Since the purpose of this research work was to determine the smallest nanofiber with the minimum variance, the highest value of *S*/*N* ratio yields the optimum condition from the Taguchi DoE work [17, 28, 31]. As illustrated in Table 6.3, T8 ($A_{60}B_1C_aD_{25}$) was determined to be the best candidate for inclusion in the optimal combination, which has the maximum *S*/*N* value of −37.22 with the average nanofiber diameter and standard deviation at 70 ± 18 nm. However, T7 ($A_{60}B_{0.5}C_cD_{18}$) with the *S*/*N* value of −38.81 and the average fiber diameter and standard deviation at 85 ± 18 nm is another strong candidate. The average fiber diameter for T8 is slightly smaller than that for T7, but the fiber variance for T7 is less than that for T8, and the standard deviation is ±18 nm in both cases. In Fig. 6.3, 60% of the frequency contributions of the fibers in the nanofiber diameter range from 75 to 100 nm for T7, whereas for T8, it is 45% in the diameter range of 50–75 nm, and 35% in the 75–100 nm range.

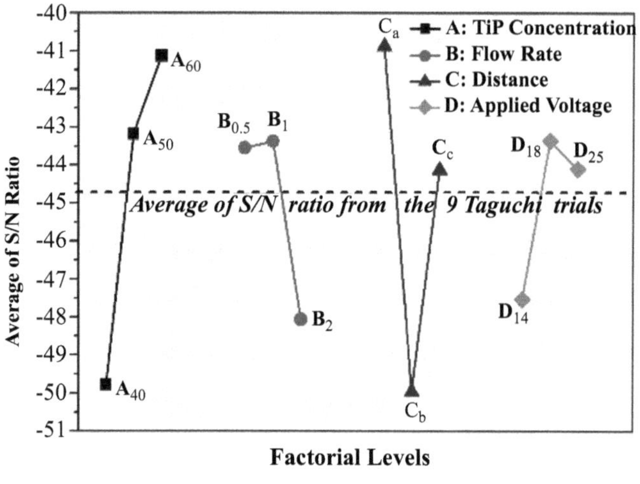

Figure 6.6 Average *S*/*N* ratio diagram for the determination of the optimum combination of factors for electrospun TiO_2/PVP nanofibers with the smallest diameter size and the minimum variation.

Therefore, the average *S*/*N* ratio was calculated to determine the best level factor for the optimum contribution to minimizing the diameter and variation of the electrospun TiO_2/PVP nanofiber as shown in Fig. 6.6. It suggests that 60 wt% TiP concentration, a flow rate of 1 ml/h, needle-tip-to-collector distance (position *a*), and applied voltage of 18 kV ($A_{60}B_1C_aD_{18}$) constitute the optimum combination of factors to obtain small electrospun TiO_2/PVP nanofiber with the minimum variation. It is further proven that the TiP concentration and the needle-tip-to-collector distance are the two most significant factors, whereas the flow rate and the applied voltage appear to be relatively insignificant.

6.3.4 Confirmation Experiment to Optimum Conditions

The confirmation experiment for the optimum combination of factors was subsequently carried out as a necessary and important post-step in the Taguchi DoE method [31]. The optimum combination ($A_{60}B_1C_aD_{18}$) was tested with the same needle size of 0.514 mm and a new needle size of 0.819 mm to study the effect on the nanofiber diameter of electrospun TiO_2/PVP. Figure 6.7 shows the typical SEM images of the confirmation experiments with both needle sizes. The randomly oriented nanofibers had smooth surfaces without bead defects in both images. The mean diameter and range are 79 ± 11 nm and 84 ± 13 nm for the confirmation experiments with needle sizes of 0.514 mm and 0.819 mm, respectively. It is clear that needle size is a non-significant factor in the nanofiber diameter of electrospun TiO_2/PVP. The diameters of these nanofibers for both needle sizes compare closely with the best Taguchi candidate values obtained in T7 (85 ± 18) and T8 (70 ± 18). T8 gives the smallest nanofiber diameter, but the optimum combination ($A_{60}B_1C_aD_{18}$) gives comparatively small electrospun TiO_2/PVP nanofiber diameter with the smallest standard deviation.

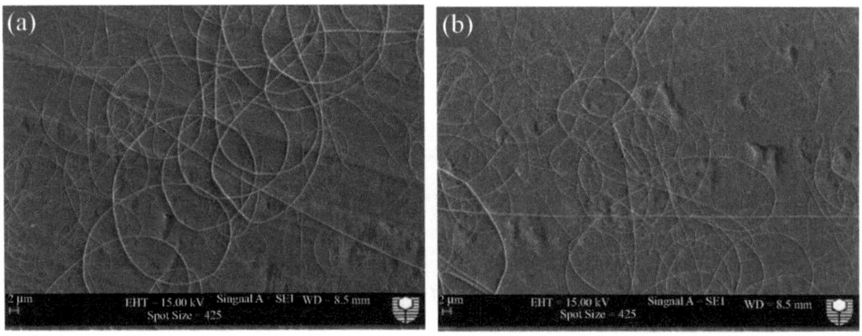

Figure 6.7 SEM micrographs of the confirmation experiment for the optimum combination of factors ($A_{60}B_1C_aD_{18}$) using two different needle sizes: (a) 0.514 mm and 0.819 mm. All scale bars represent 2 μm.

In Fig. 6.8, most of the frequency contribution diagrams for both confirmation experiments lie in the diameter range of 50–100 nm with smallest standard deviations compared to the orthogonal array with nine trials (L9).

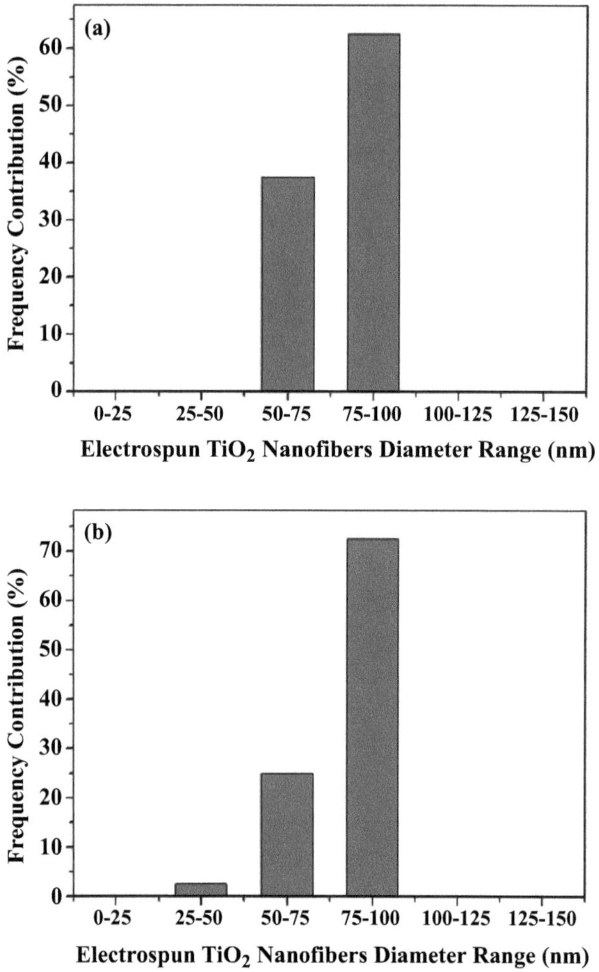

Figure 6.8 Frequency contributions to electrospun TiO_2/PVP nanofiber diameter range for confirmation experiments using needle size of (a) 0.514 mm and (b) 0.819 mm.

6.4 Conclusion

An L_9 orthogonal array along with *S*/*N* ratios and ANOVA in Taguchi DoE method was used to investigate TiP concentration, flow rate, needle-tip-to-collector distance, and applied voltage at three different levels on the nanofiber diameter of electrospun TiO_2/PVP. The small nanofiber diameter with the minimum variance has been found to be controlled mainly by two significant factors, namely, TiP concentration and needle-tip-to-collector distance. The optimum combination of factors with the highest level of TiP concentration and the highest relative needle-tip-to-collector distance (position *a*) was found along with a flow rate of 1 ml/h and an applied voltage setting of 18 kV ($A_{60}B_1C_aD_{18}$). The effects of needle size on the nanofiber diameter and its variation of electrospun TiO_2/PVP were also investigated based on this optimal combination. It was determined that the needle size is a non-significant factor in the nanofiber diameter of electrospun TiO_2/PVP.

Acknowledgments

The authors wish to acknowledge Mrs. E. Miller for the technical assistance with SEM analysis. Moreover, H. Albetran is grateful to the College of Education, University of Dammam, for the financial support in the form of a PhD Scholarship.

References

1. P. Panda and S. Ramakrishna, *J. Mater. Sci.*, **42**, 2189–2193 (2007).
2. J. Y. Park and S. S. Kim, *Met. Mater. Int.*, **15**, 95–99 (2009).
3. C. Wang, Y. Tong, Z. Sun, Y. Xin, E. Yan, and Z. Huang, *Mater. Lett.*, **61**, 5125–5128 (2007).
4. W. Sigmund, J. Yuh, H. Park, V. Maneeratana, G. Pyrgiotakis, A. Daga, J. Taylor, and J. C. Nino, *J. Am. Ceram. Soc.*, **89**, 395–407 (2006).
5. Y. Dong, T. Mosaval, H. J. Haroosh, R. Umer, H. Takagi, K. T. Lau, *J. Polymer. Sci. Part. B: Polymer. Phys.*, **52**, 618–623 (2014).
6. H. Wu, W. Pan, D. Lin, H. Li, *J. Advanced. Ceram.*, **1**, 2–23 (2012).
7. T. Jamnongkan, R. Shirota, S. K. Sukumaran, M. Sugimoto, and K. Koyama, (2013); 10.2417/spepro.005181.

8. J. Y. Park, J. J. Yun, C. H. Hwang, and I. H. Lee, *Mater. Lett.*, **64**, 2692–2695 (2010).
9. M. Samadi, H. A. Shivaee, M. Zanetti, and A. Moshfegh, *J. Molecular. Catalysis. A: Chem.*, **359**, 42–48 (2012).
10. D. A. Hanaor and C. C. Sorrell, *J. Mater. Sci.*, **46**, 855–874 (2011).
11. H. Li, W. Zhang, B. Li, and W. Pan, *J. Am. Ceram. Soc.*, **93**, 2503–2506 (2010).
12. Q. Li, D. J. G. Satur, H. Kim, and H. G. Kim, *Mater. Lett.*, **76**, 169–172 (2012).
13. C. Wessel, R. Ostermann, R. Dersch, and B. M. Smarsly, *J. Phys. Chem. C*, **115**, 362–372 (2011).
14. H. Albetran, H. Haroosh, Y. Dong, V. Prida, B. O'Connor, and I. Low, *Appl. Phys. A*, **116**, 161–169 (2014).
15. Z. M. Huang, Y. Z. Zhang, M. Kotaki, and S. Ramakrishna, *Composites. Sci. Tech.*, **63**, 2223–2253 (2003).
16. S. V. Fridrikh, J. H. Yu, M. P. Brenner, and G. C. Rutledge, *Phys. Review. Lett.*, **90**, 144502–144506 (2003).
17. S. Patra, A. Easteal, and D. Bhattacharyya, *J. Mater. Sci.*, **44**, 647–654 (2009).
18. R. Inai, M. Kotaki, and S. Ramakrishna, *Nanotechnology*, **16**, 208–213 (2005).
19. D. Li and Y. Xia, *Nano. Lett.*, **3**, 555–560 (2003).
20. A. Kumar, R. Jose, K. Fujihara, J. Wang, and S. Ramakrishna, *Chem. Mater.*, **19**, 6536–6542 (2007).
21. C. M. Wu, H. G. Chiou, S. L. Lin, and J. M. Lin, *J. Appl. Polymer. Sci.*, **126**, 89–97 (2012).
22. B. Ksapabutr, T. Chalermkiti, and M. Panapoy, *Chiang. Mai. Uni. J.*, **4**, 115–119 (2005).
23. O. Hardick, B. Stevens, and D. G. Bracewell, *J. Mater. Sci.*, **46**, 3890–3898 (2011).
24. W. K. Son, D. Cho, and W. H. Park, *Nanotechnology*, **17**, 439–443 (2006).
25. J. A. Park, J. Moon, S. J. Lee, S. H. Kim, T. Zyung, and H. Y. Chu, *Thin Solid Films*, **518**, 6642–6645 (2010).
26. W. Nuansing, S. Ninmuang, W. Jarernboon, S. Maensiri, and S. Seraphin, *Mater. Sci. Eng. B*, **131**, 147–155 (2006).
27. S. Chuangchote, J. Jitputti, T. Sagawa, and S. Yoshikawa., *ACS Appl. Mater. Inter.*, **1**, 1140–1143 (2009).

28. Y. Dong, T. Bickford, H. J. Haroosh, K. T. Lau, and H. Takagi, *Appl. Phys. A*, **112**, 747–757 (2013).

29. J. Ghani, I. Choudhury, and H. Hassan, *J. Mater. Process. Technol.*, **145**, 84–92 (2004).

30. Y. Dong and D. Bhattacharyya, *Composites. Part. A: Appl. Sci. Manufacturing*, **39**, 1177–1191 (2008).

31. R. Ranjit, *A Primer on the Taguchi Method*, van Nostrand Reinhold, New York (1990).

Chapter 7

Effect of Pressure on TiO_2 Crystallization Kinetics Using In Situ Sealed Capillary High-Temperature Synchrotron Radiation Diffraction

The phase transformation behavior of TiO_2 sol–gel synthesized nanopowder heated in a sealed quartz capillary from room temperature to 800°C was studied using in situ synchrotron radiation diffraction (SRD). Sealing of the capillary resulted in an increase in capillary gas pressure with temperature. The pressures inside the sealed capillary were calculated using Gay-Lussac's law, and they reached 0.36 MPa at 800°C. The as-synthesized material was entirely amorphous at room temperature, with crystalline anatase first appearing by 200°C (24 wt% absolute), then increasing rapidly in concentration to 89 wt% by 300°C and then increasing more slowly to 97 wt% by 800°C, with there being no indication of the anatase-to-rutile transformation up to 800°C. The best estimate of activation energy for the amorphous-to-anatase transformation from the SRD data was 10(2) kJ/mol, which is much lower than that observed when heating the material under atmospheric pressure in a laboratory X-ray diffraction (XRD) experiment, 38(5) kJ/mol. For the experiment under atmospheric pressure, the anatase crystallization temperature was delayed by ca. 200°C, first appearing after heating

Nanostructured Titanium Dioxide in Photocatalysis
It-Meng Low, Hani Manssor Albetran, Victor Manuel de la Prida Pidal, and Fong Kwong Yam

ISBN 978-981-4877-07-7 (Hardcover), 978-1-003-14853-1 (eBook)
www.jennystanford.com

the sample to 400°C, after which crystalline rutile was first observed after heating to 600°C. The estimated activation energy for the anatase-to-rutile transformation was 120(18) kJ/mol, which agrees with estimates for titania nanofibers heated under atmospheric pressure. Thus, heating the nanopowder material under pressure promoted the amorphous-to-anatase transformation, but retarded the anatase-to-rutile transformation. This behavior is believed to occur in an oxygen-rich environment, and interstitial titanium is also expected to form when the material is heated under high gas pressure. This suggests that atmospheric oxygen appears to accelerate the amorphous-to-anatase transformation, whereas interstitial titanium inhibits TiO_2 structure relaxation, which is required for the anatase-to-rutile transformation.

7.1 Introduction

Titanium dioxide (TiO_2) is a commercially important semiconductor oxide material with two common crystalline phases, namely, anatase and rutile [1]. Anatase is used in photocatalytic applications such as photocatalysts, solar energy cells, antibacterial applications, gas sensors, and catalyst support [2]. Rutile has a high refractive index and is often used in optical devices and as a pigment [3].

TiO_2 photocatalysis has received much attention as a candidate for the photodecomposition of organic pollutants [4, 5]. TiO_2 photoactivity is influenced by various factors, including phase composition (anatase versus rutile), band gap, particle size, and morphology.

Anatase has a higher photoactivity than rutile, whereas their mixtures have been reported to exhibit superior photoactivity [4, 6, 7]. Anatase–rutile mixtures have been synthesized by thermal treatment of amorphous TiO_2, where the initial crystalline TiO_2 phase formed from amorphous TiO_2 is anatase, which transforms to rutile at elevated temperature [8]. The anatase-to-rutile transformation temperature ranges from 400 to 1200°C depending on reaction atmosphere, doping and synthesis method, particle size, and surface area [9–19].

Nanoscale TiO_2 particle morphology can provide a large specific surface area to increase the photodegradation rate since photocatalytic reactions occur mainly on the TiO_2 photocatalyst surface, which is very sensitive to particle size [20–25]. TiO_2 in nanopowder form is used mostly as a catalyst and can be synthesized by many approaches such as inert gas condensation, solution combustion, hydrothermal techniques, solvothermal methods, plasma spray coating, and sol–gel methods [2, 20, 26]. TiO_2 nanopowder fabrication by the sol–gel method is a simple and cost-effective technique for industrial applications [27, 28].

In this study, the phase transformation behavior of TiO_2 nanopowders synthesized by the sol–gel method has been examined with reference to the influence of gas pressure on phase transformation kinetics. The phase transformation effect was examined by in situ SRD by heating the material in a sealed capillary from 25 to 800°C. Over the temperature range examined, the pressure would have varied linearly from 0.10 to 0.36 MPa according to Gay-Lussac's law. In order to demonstrate the influence of pressure, an ex situ XRD experiment was conducted with a suite of samples heated under atmospheric pressure to a range of target temperatures and then returned to room temperature. For safety reasons, the measurements at atmospheric pressure could not be conducted using SRD as safety dictates precluded SRD data measurements with an unsealed capillary. The material under study was characterized for particle size and chemistry by field emission scanning electron microscopy (FESEM) and associated energy dispersive X-ray spectroscopy (EDS), respectively.

7.2 Results and Discussion

7.2.1 Microstructural Imaging

The material particle sizes and morphologies were assessed using FESEM. Figures 7.1(a) and (b) show typical secondary electron images of material at room temperature for (i) TiO_2 nanopowders heated in a sealed capillary to 800°C for the in situ SRD experiment and (ii) bulk material heated to 800°C at atmospheric pressure for the ex situ XRD assessment. The morphology of TiO_2 nanopowder heated under pressure and then cooled to room temperature predominantly comprised anatase (see Sections 3.2 and 3.3), with the size range being 85 ± 31 nm (estimated from 50 particles). By contrast, powders heated under atmospheric pressure consisted of nanograins, which were less regular in shape presumably due to the grains having mainly transformed from anatase to rutile and also due to some grain growth. The grains were significantly larger and uneven surfaces formed for the unpressurized material (118 ± 44 nm, estimated from 50 particles) on heating to 800°C. The variation in particle size from pressurized and unpressurized material provides clear evidence of the effect of atmospheric pressure on TiO_2 crystal formation. The particle size increases with increasing temperature because of anatase-to-rutile transformation in addition to enhanced atomic diffusion, thus indicating that raised pressure (≥0.36 MPa) within the sealed capillary had inhibited grain growth for the pressurized material. For the unpressurized material, at a certain particle size, amorphous TiO_2 transformed into anatase, which subsequently transformed into rutile at elevated temperature. In contrast, the pressurized material was observed to remain as anatase. The much sluggish anatase-to-rutile transformation for the pressurized material appears to result from the relatively high gas pressure inside the sealed capillary at elevated temperature, which provides a greater structural rigidity and thus prevents relaxation in the Ti bonding environment. The latter is essential for anatase-to-rutile phase transformation.

Figure 7.1 FESEM micrographs for (a) TiO_2 nanopowder material sealed in a capillary and heated to 800°C in the in situ SRD experiment; and (b) bulk nanopowder material heated to 800°C under atmospheric pressure. Both samples cooled to room temperature prior to imaging.

The spectra for the as-prepared TiO_2 nanopowder, unpressurized and pressurized materials, are essentially the same. The EDS spectrum in Fig. 7.2 corresponds to the image in Fig. 7.1(a) for material that had been heated under pressure to 800°C. The qualitative EDS spectra show strong Ti and O signatures and also Pt signatures from the sample platinum coating. The intense Ti and O lines were expected from the observation of TiO_2 phases in the material as measured by SRD patterns (Fig. 7.3), and also by laboratory XRD (Fig. 7.4). The quantitative EDS analysis for the as-prepared TiO_2 powder indicated 66.35% O k and 33.65% Ti k, which is approximately coincident with TiO_2 stoichiometry powder and is almost the same for unpressurized and pressurized material.

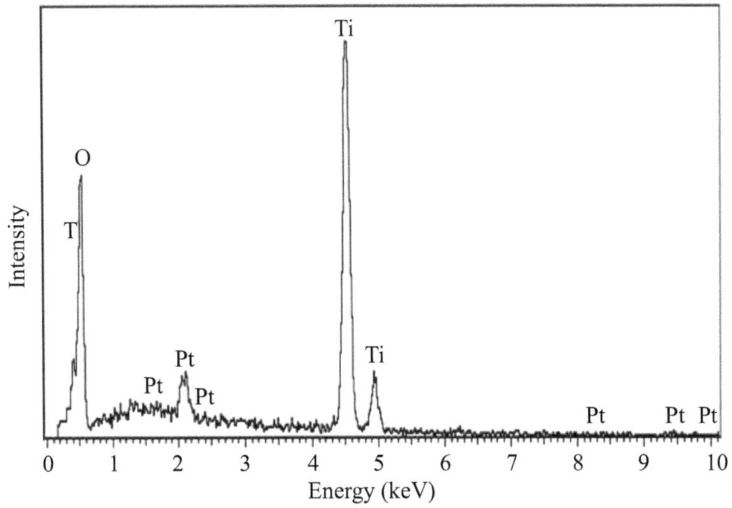

Figure 7.2 EDS spectrum from FESEM/EDS analysis for TiO_2 nanopowder following the in situ SRD measurements for material heated to 800°C in a sealed capillary and then cooled to room temperature [Legend: Ti: titanium, O: oxygen, and Pt: platinum].

7.2.2 SRD Patterns for In Situ Heating of Material Contained in Sealed Capillary

Figure 7.3 shows stacked SRD plot recorded while heating under pressure from ~0.10 MPa at 25°C to ~0.36 MPa at 800°C with the capillary pressure increasing linearly. The anatase was formed here at 200°C, compared to that reported in the literature (400°C) for TiO_2 powder heated in air in ambient condition [19]. Clearly heating under pressure hastens the amorphous-to-anatase transformation. This observation is consistent with that of Pin et al. [31] who studied the effects of oxygen working pressure on the transformation in TiO_2 films grown on a glass substrate using a pulsed laser. In that experiment, the laser was applied to the TiO_2 glass substrate at 400°C and for pressures ranging from ~1.33 to 10 × 10^{-6} MPa. The intensity of the (1 0 1) anatase line was found to increase with an increase in working pressure up to 3.33 × 10^{-6} MPa, and then the line intensity decreased and eventually disappeared with further increase in working pressure. Thus, anatase formation at a much reduced temperature in this study is attributed to an oxygen-richer environment inside the sealed capillary because the oxygen partial pressure increases with increasing temperature together with the desorption of propanol, which is 27% rich in oxygen. However, the anatase-to-rutile transformation was inhibited as the gas pressure increased linearly inside the sealed capillary from ~0.10 MPa at 25°C to ~0.36 MPa at 800°C. This higher gas pressure at elevated temperature caused interstitial Ti formation and enhanced structural rigidity, both of which did not allow relaxation of the Ti bonding, which is required for anatase-to-rutile transition.

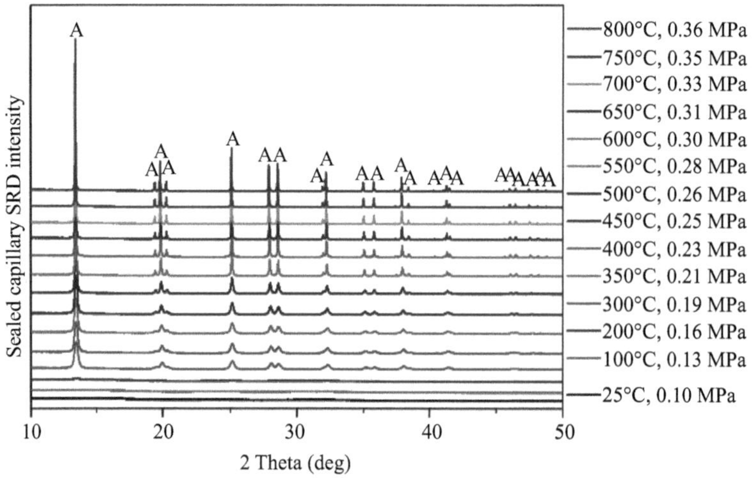

Figure 7.3 Stacked in situ high-temperature SRD plots for TiO_2 nanopowders heated in a sealed capillary from 25 to 800°C [A: anatase]. Wavelength = 0.0825 nm.

7.2.3 Use of Ex Situ XRD at Atmospheric Pressure to Determine the Influence of Capillary Pressure in SRD Experiment

Figure 7.4 shows the stacked ex situ XRD plot after heating a suite of samples separately from 25 to 800°C at increments of 100°C and at 2°C/min and then cooling to room temperature. The TiO_2 nanopowders at atmospheric pressure crystallized partially to initially form anatase and rutile by 400°C and 600°C, respectively. This finding is consistent with literature data reporting crystallization-onset temperatures for anatase and rutile ranging between ~400 and 600°C, and 400 and 1200°C, respectively, which values are consistent with the literature on TiO_2 transformations [9–19]. Therefore, the initial anatase crystallization temperature under pressure was reduced by 200°C compared with material heated at atmospheric pressure, whereas the anatase-to-rutile transformation had not commenced by 800°C for the pressurized

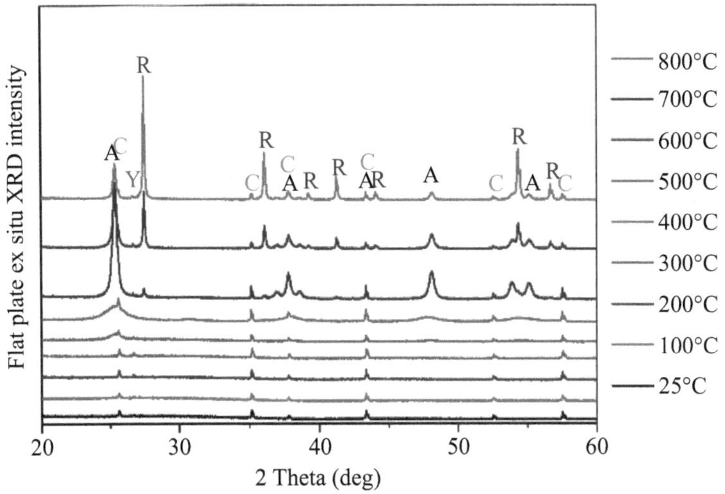

Figure 7.4 Stacked ex situ XRD plots for bulk TiO_2 nanopowder after heating in air under atmospheric pressure at 2°C/min to the target temperatures 100–800°C in increments of 100°C, and then cooling to room temperature [Legend: A: anatase, R: rutile, C: corundum (Al_2O_3), Y: unknown phase]. Samples pressed into "infinitely thick," flat plate format for Bragg–Brentano diffractometer at a wavelength of 0.15419 nm.

material. For material heated under pressure, the amorphous-to-anatase transformation was accelerated by the increase in oxygen partial pressure within the sealed capillary compared with the relatively low oxygen partial pressure for the material heated under atmospheric pressure. Thus, an increase in oxygen partial pressure lowered the anatase crystallization temperature.

In support of Figs. 7.3 and 7.4, Fig. 7.5 compares plots of (i) the in situ high-temperature SRD pattern recorded at 800°C under increased pressure and (ii) the ex situ room-temperature XRD pattern after cooling from 800°C under atmospheric pressure. Because the patterns were recorded at different wavelengths, the instrument 2θ values have been converted to interplanar d-spacings. The figure demonstrates the dramatic differences in the patterns due to the substantial conversion of anatase-to-rutile when the material is heated under atmospheric pressure. The XRD pattern shows a peak for an unidentified phase at 26.64°, which is likely to be due to a contaminant from micronizing. The contaminant phase was ignored in the Rietveld calculations.

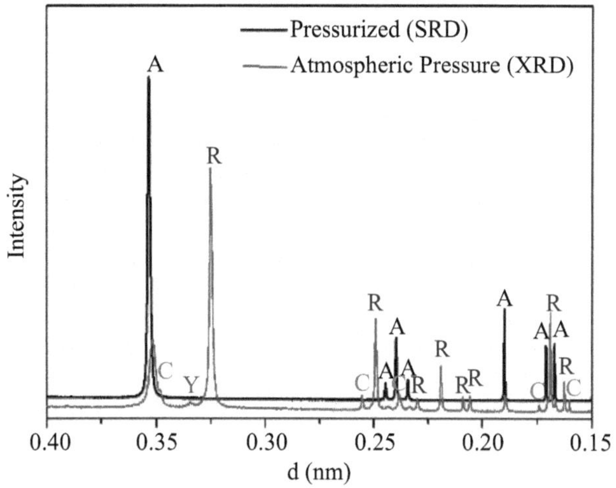

Figure 7.5 Comparison of (i) SRD pattern at 800°C measured in a sealed capillary during in situ heating; and (ii) the corresponding ex situ XRD pattern after heating the TiO_2 nanopowder to 800°C under atmospheric pressure and then cooling to room temperature [Legend: A: anatase, R: rutile, C: corundum (α-Al_2O_3), Y: unknown phase]. Ex situ pattern measured by Bragg–Brentano diffractometer with sample pressed into "infinitely thick," flat plate specimen.

Figures 7.6(a) and (b) show the typical in situ SRD and ex situ XRD Rietveld residual difference plots for the 800°C data sets. In general, the quality of the Rietveld fits is reasonable according to a visual assessment of the difference plots and: (i) for the SRD data, R_{exp} = 3.69%, R_{wp} = 7.57%, R_B (anatase) = 6.52%; and (ii) for the XRD data, R_{exp} = 5.53%, R_{wp} = 8.33%, R_B (anatase) = 0.76%, R_B (rutile) = 4.53%, and R_B (corundum) = 1.86%. The goodness-of-fit (χ^2) values were 2.05 for the SRD data, and 1.51 for the XRD pattern.

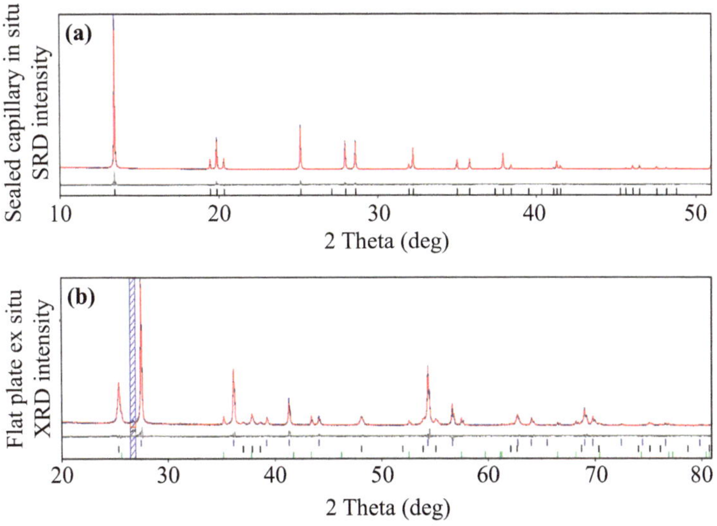

Figure 7.6 Rietveld difference plots for TiO_2 nanopowders: (a) the SRD pattern at 800°C measured during in situ heating; and (b) ex situ XRD pattern heating the TiO_2 nanopowder to 800°C and then cooling to room temperature. Measured patterns are indicated by black crosses and calculated patterns by solid red lines. Gray residual plot shows the difference between the calculated and the measured patterns. Blue region indicates the excluded region encompassing the unknown phase peak. Peak positions for anatase and rutile are indicated by the black and blue bars, respectively. The plots were computed with TOPAS (Version 4.2).

The relative concentration of crystalline TiO_2 was combined with the amorphous TiO_2 estimates to provide absolute amorphous TiO_2 and crystalline TiO_2 concentrations [29, 30]. Figure 7.7 shows the variation in absolute phase concentrations (amorphous titania, crystalline anatase, and crystalline rutile) as the material is heated under pressure compared with the phase development when the material is heated at atmospheric pressure. The plot demonstrates in terms of phase concentrations how the application of pressure accelerates the amorphous-to-anatase transformation but delays the anatase-to-rutile transformation.

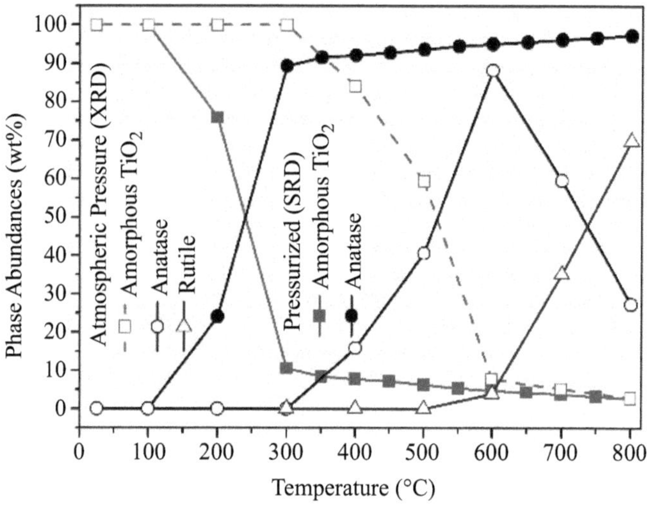

Figure 7.7 Phase abundances of amorphous titania, crystalline anatase, and crystalline rutile for pressurized material heated in a sealed capillary and a suite of TiO_2 nanopowders heated under atmospheric pressure.

The following features in Fig. 7.7 are noted:

- Under atmospheric pressure, anatase is first observed at 400°C (16 wt%) and then reaches a maximum concentration by 600°C (88 wt%) before starting its transformation to rutile. This TiO_2 transformation behavior is approximately consistent with data from the literature with starting pure titania powders used were titanium tetraisopropoxide [19]. The crystallization temperatures of anatase and rutile were 415°C and 680°C, respectively.
- By contrast, anatase is first seen by 200°C (24 wt%) when the material is pressurized, which is a reduction of ca. 200°C compared with heating under normal pressure. The anatase concentration then rapidly climbs to 89 wt% within a further

100°C, and then climbs more slowly, increasing linearly with temperature to 97 wt% by 800°C.

- From 600°C under atmospheric pressure, the anatase (88 wt%) transforms progressively to rutile by 800°C (39 wt% anatase by weight).
- The transformation of anatase to rutile observed under atmospheric pressure does not occur by 800°C when the material is pressurized.

Increasing the concentration of atmospheric oxygen is known to accelerate the amorphous-to-anatase transformation [8, 30], whereas the formation of interstitial titanium in anatase may inhibit the anatase-to-rutile transformation [1, 9]. On this basis, the increase in oxygen partial pressure in the sealed capillary accelerated the transformation from amorphous TiO_2 to crystalline anatase. Also titanium interstitials that were likely to have developed at elevated temperatures and under the higher partial pressure inside the sealed capillary would have inhibited the anatase-to-rutile transformation through structural constraint due to an increase in the TiO_2 strain energy [1, 9].

7.2.4 Crystallization Kinetics Modelling

Table 7.1 shows the average activation energies calculated using the Arrhenius equation and that of Matusita et al. [29, 30] for an amorphous-to-anatase transformation in pressurized TiO_2 material and for amorphous-to-anatase and anatase-to-rutile transformations for material heated at atmospheric pressure. Using the phase levels reported in Fig. 7.7, the estimated activation energy for the amorphous-to-anatase transformation when the material is heated under pressure in a sealed capillary is 10(2) kJ/mol. In contrast, the estimated activation energy under atmospheric pressure for the amorphous-to-anatase transformation is 38(5) kJ/mol, and 120(18) kJ/mol for the anatase-to-rutile transformation. Thus, increasing the oxygen partial pressure apparently lowered the activation energy substantially for the amorphous-to-anatase transformation.

Table 7.1 Phase transformation activation energies (kJ/mol) for pure titania nanopowders from non-isothermal SRD data analysis using the Arrhenius and Matusita et al. equations

TiO_2 Material	Technique	Amorphous-to-Anatase	Anatase-to-Rutile	Environment
Nanopowder (this study)	In situ SRD	10(2)	—	Heated under pressure using sealed capillary
	Ex situ XRD	37(5)	120(18)	Heated in air under atmospheric pressure
Nanofibers [29, 30]	In situ SRD	45(9)	97(7)	Heated in argon under atmospheric pressure
		69(17)	129(5)	Heated in air under atmospheric pressure

By contrast, the anatase-to-rutile activation energy value under atmospheric pressure is comparable with literature values for titania nanofibers under atmospheric pressure with air and argon atmospheres. It is noted from Table 7.1 that the amorphous-to-anatase activation energies for nanopowders are lower than that for nanofibers. For the anatase-to-rutile transformation energy, the value for nanopowder material heated in air at atmospheric pressure agreed closely with earlier estimates for nanofiber material heated in air or/and argon at atmospheric pressure.

7.3 Conclusion

In situ and ex situ heating experiments were conducted with sol–gel synthesized TiO_2 nanopowders using SRD (by heating in a sealed pressurized capillary) and XRD (heating at atmospheric pressure) analysis, respectively, to study the effect of pressure on phase transformation behavior. The material was amorphous as synthesized, with crystalline anatase first formed by 200°C in the sealed capillary, whereas crystalline anatase first formed by 400°C under atmospheric pressure. Under pressure, rutile has not been seen by 800°C, whereas the phase was first observed by 600°C at atmospheric pressure, which is consistent with the literature. Thus, increasing the oxygen partial pressure within the sealed capillary accelerates the amorphous-to-anatase transformation because of oxygen-rich environment but inhibits the anatase-to-rutile transformation because of the interstitial titanium formation, which inhibits TiO_2 structure relaxation.

The study has pointed to the interesting possibility of suppressing the amorphous-to-anatase transformation by reducing the oxygen partial pressure below that for air at atmospheric pressure.

Acknowledgments

This work was funded by the Australian Synchrotron (AS132/PD/6516). The authors thank Dr. J. Kimpton at the Australian Synchrotron for advising on instrumentation at the Powder Diffraction Beamline. The authors also thank E/Prof. B. H. O'Connor for guidance on the Rietveld work, Ms. E. Miller and Ms. V. Avery of John de Laeter Center, Curtin University, for kind assistance with FESEM, and XRD data analysis, respectively.

References

1. D. A. Hanaor and C. C. Sorrell, "Review of the anatase to rutile phase transformation," *J. Mater. Sci.*, **46**, 855–874 (2011).
2. Y. Ren, C. Sun, K. Li, L. Wang, and M. Song, "Preparation of TiO_2–reduced graphene oxide–Pd nanocomposites for phenol photocatalytic degradation," *Ceram. Int.*, **42**, 1339–1344 (2016).
3. J. H. Braun, "Titanium dioxide: A review," *J. Coat. Technol.*, **69**, 59–72 (1997).
4. A. Natoli, A. Cabeza, A. G. Torre, M. A. G. Aranda, and I. Santacruz, "Colloidal processing of macroporous TiO_2 materials for photocatalytic water treatment," *J. Am. Ceram. Soc.*, **95**, 502–508 (2012).
5. D. Wu and M. Long, "Realizing visible-light-induced self-cleaning property of cotton through coating N-TiO_2 film and loading AgI particles," *ACS Appl. Mater. Interfaces*, **3**, 4770–4774 (2011).

6. H. Li, W. Zhang, and W. Pan, "Enhanced photocatalytic activity of electrospun TiO_2 nanofibers with optimal anatase/rutile ratio," *J. Am. Ceram. Soc.*, **94**, 3184–3187 (2011).
7. S. Kment, P. Klusonc, V. Stranak, P. Virostko, J. Krysa, M. Cada, J. Pracharova, M. Kohout, M. Morozova, P. Adamek, and Z. Hubicka, "Photo-induced electrochemical functionality of the TiO_2 nanoscale films," *Electrochim. Acta.*, **54**, 3352–3359 (2009).
8. H. Albetran, B. H. O'Connor, and I. M. Low, "Effect of calcination on band gaps for electrospun titania nanofibers heated in air–argon mixtures," *Mater. Des.*, **92**, 480–485 (2016).
9. R. D. Shannon and J. A. Pask, "Kinetics of the anatase–rutile transformation," *J. Am. Ceram. Soc.*, **48**, 391–398 (1965).
10. J. A. Gamboa and D. M. Pasquevich, "Effect of chlorine atmosphere on the anatase–rutile transformation," *J. Am. Ceram. Soc.*, **75**, 2934–2938 (1992).
11. H. Albetran, H. Haroosh, Y. Dong, V. M. Prida, B. H. O'Connor, and I. M. Low, "Phase transformations and crystallization kinetics in electrospun TiO_2 nanofibers in air and argon atmospheres," *Appl. Phys. A*, **116**, 161–169 (2014).
12. I. M. Low, H. Albetran, V. M. Prida, V. Vega, P. Manurung, and M. Ionescu, "A comparative study on crystallization behavior, phase stability, and binding energy in pure and Cr-doped TiO_2 nanotubes," *J. Mater. Res.*, **28**, 304–312 (2013).
13. R. A. Eppler, "Effect of antimony oxide on the anatase–rutile transformation in titanium dioxide," *J. Am. Ceram. Soc.*, **70**, 64–66 (1987).
14. I. M. Low, B. Curtain, M. Philipps, Z. Q. Liu, and M. Ionescu, "High temperature diffraction study of in situ crystallization of nanostructured TiO_2 photocatalysts," *J. Aus. Ceram. Soc.*, **48**, 198–204 (2012).
15. H. Zhang and J. F. Banfield, "Understanding polymorphic phase transformation behavior during growth of nanocrystalline aggregates: Insights from TiO_2," *J. Phys. Chem. B*, **104**, 3481–3487 (2000).
16. F. Gennari and D. Pasquevich, "Kinetics of the anatase–rutile transformation in TiO_2 in the presence of Fe_2O_3," *J. Mater. Sci.*, **33**, 1571–1578 (1998).
17. J. Arbiol, J. Cerda, G. Dezanneau, A. Cirera, F. Peiro, A. Cornet, and J. R. Morante, "Effects of Nb doping on the TiO_2 anatase-to-rutile phase transition," *J. Appl. Phys.*, **92**, 853–861 (2002).

18. R. Plugaru, A. Cremades, and J. Piqueras, "The effect of annealing in different atmospheres on the luminescence of polycrystalline TiO_2," *J. Phys: Condens. Matter.*, **16**, 261–268 (2004).

19. K. Okada, N. Yamamoto, Y. Kameshima, A. Yasumori, and K. J. D. MacKenzie, "Effect of silica additive on the anatase-to-rutile phase transition," *J. Am. Ceram. Soc.*, **84**, 1591–1596 (2001).

20. K. Vajda, K. Saszet, E. Z. Kedves, Z. Kasa, V. Danciu, L. Baia, K. Magyari, K. Hernadi, G. Kovacs, G. Kovacs, and Z. Pap, "Shape-controlled agglomeration of TiO_2 nanoparticles. new insights on polycrystallinity vs. single crystals in photocatalysis," *Ceram. Int.*, **42**, 3077–3087 (2016).

21. M. Takeuchi, H. Yamashita, M. Matsuoka, M. Anpo, T. Hirao, N. Itoh, and N. Iwamoto, "Photocatalytic decomposition of NO under visible light irradiation on the Cr-ion-implanted TiO_2 thin film photocatalyst," *Catal. Lett.*, **67**, 135–137 (2000).

22. I. H. Chowdhury, S. Ghosh, and M. K. Naskar, "Aqueous-based synthesis of mesoporous TiO_2 and Ag–TiO_2 nanopowders for efficient photodegradation of methylene blue," *Ceram. Int.*, **42**, 2488–2496 (2016).

23. J. E. Mathis, J. J Lieffers, C. Mitra, F. A. Reboredo, Z. Bi, C. A. Bridges, M. K. Kidder, and M. P. Paranthaman, "Increased photocatalytic activity of TiO_2 mesoporous microspheres from codoping with transition metals and nitrogen," *Ceram. Int.*, **42**, 3556–3562 (2016).

24. G. Lu, A. Linsebigler, and J. T. Yates Jr, "Photooxidation of CH_3Cl on TiO_2 (110): A mechanism not involving H_2O," *J. Phys. Chem.*, **99**, 7626–7631 (1995).

25. R. K. Madhusudan, C. Gopal Reddy, and S. Manorama, "Preparation, characterization, and spectral studies on nanocrystalline anatase TiO_2," *J. Solid State Chem.*, **158**, 180–186 (2001).

26. Q. Wu and R. Krol, "Selective photoreduction of nitric oxide to nitrogen by nanostructured TiO_2 photocatalysts: Role of oxygen vacancies and iron dopant," *J. Am. Chem. Soc.*, **134**, 9369–9375 (2012).

27. M. Crisan, N. Dragan, D. Crisan, A. Ianculescu, I. Nitoi, P. Oancea, L. Todan, C. Stan, and N. Stanica, "The effects of Fe, Co and Ni dopants on TiO_2 structure of sol–gel nanopowders used as photocatalysts for environmental protection: A comparative study," *Ceram. Int.*, **42**, 3088–3095 (2016).

28. M. Barakat, G. Hayes, and S. I. Shah, "Effect of cobalt doping on the phase transformation of TiO_2 nanoparticles," *J. Nanosci. Nanotechnol.*, **5**, 759–765 (2005).

29. H. Albetran, V. M. Prida, B. H. O'Connor, and I. M. Low, "Effect of vanadium ion implantation on the crystallization kinetics and phase transformation of electrospun TiO_2 nanofibers," *Appl. Phys. A*, **120**, 623–634 (2015).

30. H. Albetran, B. H. O'Connor, and I. M. Low, "Activation energies for phase transformations in electrospun titania nanofibers: Comparing the influence of argon and air atmospheres," *Appl. Phys. A*, **122**, 367 (2016).

31. M. W. Pin, G. H. Lee, W. J. Lee, E. H. Choi, and Y.Z. Yoo, "Effects of oxygen working pressure on structural and optical properties of TiO_2 films grown on glass substrate," *Jpn. J. Appl. Phys.*, **51**, 1–12 (2012).

Chapter 8

Characterization of Chemical-Bath-Deposited TiO_2 Thin Films

In this study, a series of titania thin films was prepared by chemical bath deposition (CBD) of $TiCl_3$ on indium tin oxide (ITO) glass at room temperature, followed by calcinations at 500°C for 4 h. The effect of cyclic deposition on phase composition, microstructure, and electrical resistivity of TiO_2 thin films was characterized using X-ray diffraction (XRD), scanning electron microscopy (SEM), and four-point probe, respectively. Results showed that TiO_2 films produced by single deposition cycle were amorphous. In contrast, those produced by five and six deposition cycles were partly amorphous and partly crystalline with the formation of rutile. Both the film thickness and electrical resistivity increased with an increase in the number of deposition cycles.

8.1 Introduction

Thin films are important components of various devices widely used in electronic, protective, membrane, and sensor applications. The most common preparation method of thin films is chemical deposition since it offers several advantages such as low cost and the ability to produce samples with large surface area [1]. Some authors [1, 2] have reported various aspects of chemically deposited

Nanostructured Titanium Dioxide in Photocatalysis
It-Meng Low, Hani Manssor Albetran, Victor Manuel de la Prida Pidal, and Fong Kwong Yam

ISBN 978-981-4877-07-7 (Hardcover), 978-1-003-14853-1 (eBook)
www.jennystanford.com

thin films, while others [3, 4] have studied the effects of varying the growth parameters such as deposition rates, bath compositions, and bath temperature on the microstructures and properties of thin films.

TiO_2 exists in three main phases: anatase, brookite, and rutile where the latter is a stable phase. This oxide is an attractive material due to its wide applications, such as in paints, sunscreens, and as photocatalyst under ultraviolet light [5–7]. Hitherto, the synthesis of TiO_2 thin films or bulk has led to the emergence of nanoscience and nanotechnology by virtue of their superior physical and chemical properties [8, 9].

In this study, we have used chemical bath deposition to synthesize a series of titania thin films on ITO glass and investigated the effect of cyclic deposition on phase composition, microstructure, and electrical resistivity of TiO_2 thin films.

8.2 Results and Discussion

8.2.1 XRD Analysis

It was found that the color of TiO_2 thin films is clear or transparent and adheres well to the ITO glass substrate. Visually the films are homogenous with smooth surface appearance. The XRD spectra were collected for samples A, D, and F. The XRD data were also collected on ITO glass as a control. The diffraction pattern of sample F is shown in Fig. 8.1, which depicts the existence of rutile only in the thin film of TiO_2. In all samples, only rutile was detected and no peaks due to anatase (101) were detected probably due to their transformation to rutile during sintering at 500°C for 4 h.

It should be mentioned that the normal XRD is not ideal for characterizing the phase compositions of thin films due to relatively high penetrating (≥30 μm) depth of X-rays in symmetric mode. Since the TiO_2 thin films produced in this study are less than 5 μm thick, the X-rays will penetrate into the ITO substrate and produce an undesirable strong background noise due to the amorphous nature of ITO. Hence, it will be more accurate to use grazing-incidence XRD to probe their near-surface phase compositions of TiO_2 thin films. The critical angle (α_c) for grazing-incidence diffraction can be approximated by [10]:

$$\alpha_c = 1.6 \times 10^{-3} \rho\lambda \tag{8.1}$$

where α_c is in radians, ρ is the density of TiO_2 (rutile = 4.25 g/cm^3), and λ is wavelength in Å. Hence, the calculated value of α_c for TiO_2 thin films is 0.010472 or 0.6°. Above the critical angle, the penetration depth (d) can be calculated from [11]:

$$d = 2\alpha/\mu \tag{8.2}$$

where μ is the linier absorption coefficient of matter. Since the value of μ for rutile is 907.0775/cm [12] and the normal XRD uses an incidence angle of 8°, the penetration depth of X-rays will be ~30.8 μm, which is deeper than the thickness of the thin films. As a result, the X-rays will also detect the ITO glass and give rise to undesirable large amorphous background.

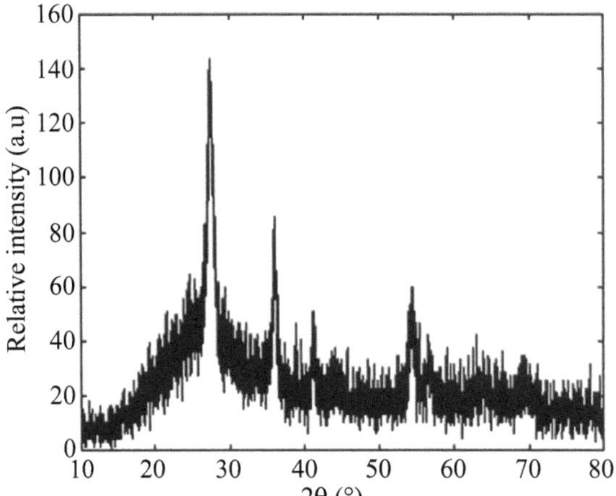

Figure 8.1 The XRD plot for sample F with six cycles of deposition.

By using the Scherrer equation [13], one can determine the crystallite size of this sample. By taking the width of diffraction peak into consideration, the crystallite size of TiO_2 in sample F was determined to be ~10 nm. This value is much less than the particle size of ~200 nm indicated by the SEM image [Fig. 8.2(d)]. Since particles or grains are made up of small crystallites, this suggests that there are about 20 crystallites in each TiO_2 particle on sample F.

The PDF file number 21-1276 for rutile from the database was used to identify the phases formed [14]. According to Lokhande [15], there is no difference between ITO glass and TiO_2 having low grade of crystallinity. In fact, it is not necessary for TiO_2 to have a high grade of crystallinity when it is used in applications such as a photodetector [16].

8.2.2 Microstructure Analysis

The SEM images of TiO_2 thin films prepared by multiple depositions are shown in Fig. 8.2. These images clearly show distinct contrast in microstructures in films prepared between one and six deposition cycles. This indicates that the number of deposition cycles can lead to a significant change in the microstructures of TiO_2 films.

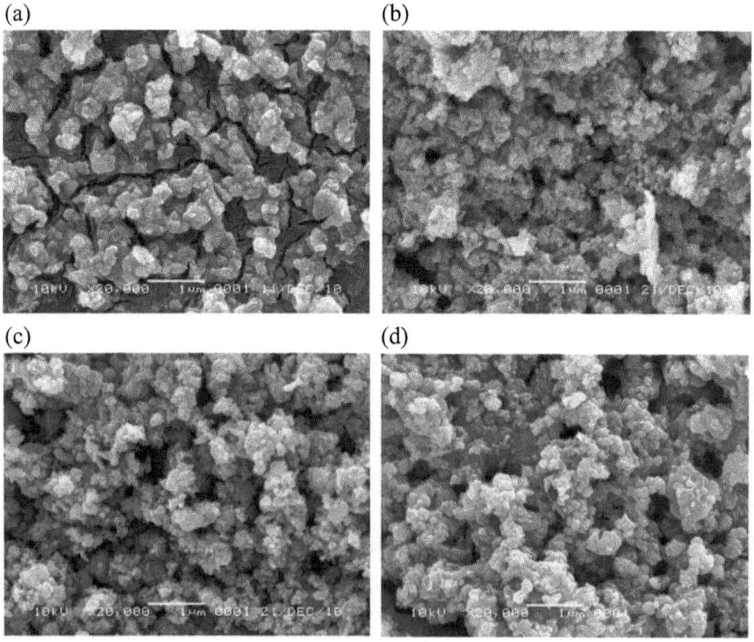

Figure 8.2 SEM images of samples with various processing conditions: (a) sample A (1 cycle @ 6 h), (b) sample B (2 cycles @ 3 h), (c) sample D (4 cycles @ 1.5 h), and (d) sample F (6 cycles @ 1 h).

The microstructures of the thin films shown in Fig. 8.2 are for samples with one, two, four, and six deposition cycles. It is interesting to note that the microstructures for thin films prepared from one to four cycles are quite similar although cracking can be seen in sample A with one deposition cycle. In films prepared from four and six deposition cycles, the cracks had disappeared and the amount of pores also decreased. It appears that multiple depositions had helped to fill the pores within the TiO_2 layer, thus making the surface denser. If the film remains attached to the substrate and does not crack during sintering, shrinkage in the plane of the substrate can be inhibited. The stresses that arise during the sintering of constrained films are analogous to those in a sandwich seal caused by mismatch in the thermal expansion coefficients in the layers.

It is interesting to note that the films have pores distributed all over the surfaces. These pores can play the role of enhancing dye sensitization when the TiO_2 films are used for solar cell application [17] as the dye can fill the pores. Dye can also absorb the incident ray and inject electrons to the TiO_2 semiconductor.

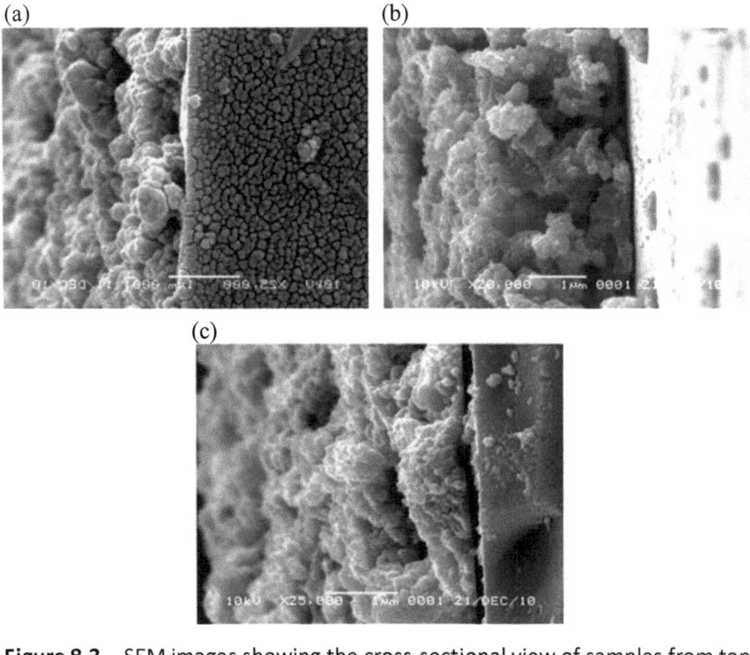

Figure 8.3 SEM images showing the cross-sectional view of samples from top surface to the ITO glass substrate where the TiO_2 layer is on the left and the ITO glass is on the right. (a) Sample A, (b) Sample B, and (c) Sample E.

Furthermore, samples were also observed using SEM in cross-sectional view from the top surface of the film to the ITO glass in order to estimate the film thickness. Microstructures of samples A, B, and E are shown in Fig. 8.3. Using the scale bar on each image, one can estimate the film thickness deposited on the substrate. The variation in film thickness as a function of deposition cycles is depicted in Fig. 8.4. By taking into account the two estimated standard deviations, the thickness of the films can be observed to increase with the number of deposition cycles. This is self-evident because when repeated depositions were done, the material deposited on the substrate also increased. From this figure, one can tailor-design the thickness of the film based on the number of deposition cycles.

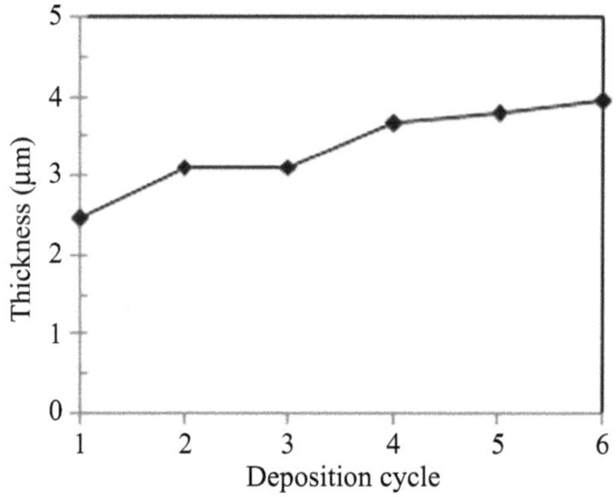

Figure 8.4 Variations in film thickness as a function of deposition cycle.

8.2.3 Electrical Resistivity

The electrical resistivity or conductivity of TiO_2 films was measured using four-point probes. By knowing the resistivity, one can infer whether the film produced is a conductor or semiconductor. The application of TiO_2 as a photoconductor must have properties of a semiconductor because a conductor has no band-gap energy. If the incident ray overcomes the band-gap energy, electrical conduction will occur. The variation in electrical resistivity as a function of deposition cycles is depicted in Fig. 8.5.

The resistivities of conductors and semiconductors normally range from 10^{-3} to 10^{8} Ωcm [18]. Based on this value, the resistivity of TiO_2 thin films synthesized in this study can be classified as a semiconductor. Since the resistivity increases with deposition cycles, this indicates that multiple depositions serve to increase the resistivity by virtue of more pores and materials being deposited in the film as the thickness increases. When the deposited material increases, the resistivity of a layer increases with each deposition cycle. It has been reported that the resistivity value of TiO_2 thin films prepared by DC magnetron sputtering ranged from 10^7 to 10^{10} Ωcm [19]. Thus, the resistivity results obtained in this work are comparable to results reported in the literature.

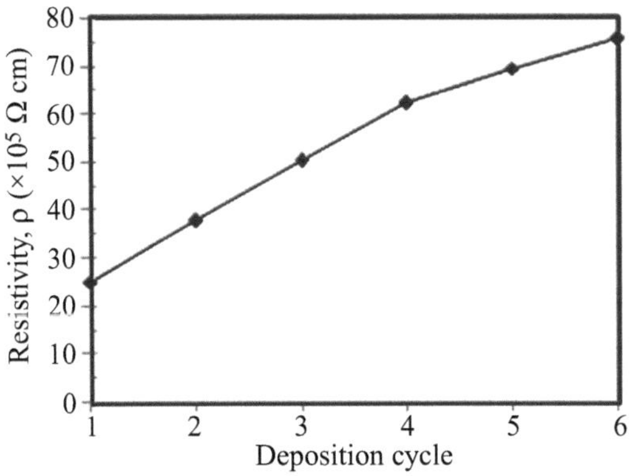

Figure 8.5 Variations in electrical resistivity of TiO_2 films as a function of deposition cycle.

8.3 Conclusion

Thin films of TiO_2 produced by multiple depositions were more crystalline with rutile being the only phase formed. In contrast, TiO_2 films produced by a single deposition cycle were amorphous. Both the film thickness and electrical resistivity of TiO_2 increased with an increase in the number of deposition cycles.

Acknowledgments

The authors are particularly grateful to Prof. P. Manurung, Dr. Y. Putri, and Dr. W. Simanjuntak of the University of Lampung for conducting this work using various analytical facilities.

References

1. J. Y. Choi, K. J. Kim, J. B. Yoo, and D. Kim, *Solar Energy*, **64** (1998), 41–45.
2. B. R. Ortega and D. Lincot, *J. Electrochem. Soc.*, **140** (1993), 3464–3468.
3. F. C. Eze and C. E. Okeke, *Mater. Chem. Phys.*, **47** (1997), 31–33.
4. P. J. Sabestian and H. Hu, *Adv. Mater. Opt. Electron.*, **4** (1994), 407–409.
5. A. Fujishima and K. Honda, *J. Nature*, **238** (1972), 37–38.
6. A. Fujishima, T. N. Rao, and D. A. Tryk, *J. Photochem. Photobiol. C*, **1** (2000), 1–21.
7. D. A. Tryk, A. Fujishima, and K. Honda, *Acta*, **45** (15–16) (2000), 2363–2267.
8. M. Tejos, G. E. Buono-Cuore, F. R. Diaz, M. A. del Valle, and J. J. Palomares, *J. Chil. Chem. Soc.*, **49** (2004), 297–301.
9. D. Reyes-Coronade, G. Rodriguez-Gattorno, M. E. Espinosa-Pesqueira, C. Cab, R. de Ros, and G. Oskam, *Nanotechnology*, **19** (2008), 1–6.
10. P. Barner, *Trans. J. Br. Ceram. Soc.*, **91** (1992), 26–29.
11. G. H. Vineyard, *Phys. Rev. B*, **26** (1982), 4146–4158.
12. H. A. Liebhfsky, H. G. Pfeiffer, E. H. Winslow, and P. D. Zemany, *X-Ray, Electrons and Analytical Chemistry*. Wiley-Interscience, New York (1972), 525–528.
13. B. D. Cullity, *Elements of X-Ray Diffraction*. Addison-Wesley Publishing Company (1978), 105.
14. Natl. Bur. Stand (US), Monogr. 25.7, 83 (1969). PCPDFWIN Version 1.30. 1997.
15. C. D. Lokhande, Eun-Ho, K. Deong, Shimjaoo, *J. Mater. Sci.*, **39** (2004), 2915–2918.
16. Y. C. Lee, Z. Hasan, F. K. Yam, K. Ibrahim, M. E. Kordesch, W. Halven, and P. C. Colter, *J. Solid State Commun.*, **133** (2005), 283–287.
17. G. Phani, G. Tulloch, I. Vittorio, and I. Skyrabbin, *Renewable Energy*, **22**(1) (2001), 303–309.
18. S. M. Sze, *Semiconductor Devices: Physics and Technology*. John Wiley & Sons (1985), 1.
19. M. Stamate, I. Vascan, I. Lazar, G. Lazar, I. Caraman, and M. Caraman, *J. Opto. Adv. Mater.*, **7** (2005), 771–774.

Chapter 9

Influence of Electrolyte and Temperature on Anodic Nanotubes

Here we report about the influence of temperature and acidic electrolyte media on the growth of self-aligned TiO_2 nanotube arrays, synthesized by using a well-known electrochemical anodization technique. We have focused on the nanotube size distribution and the formation of stable wall thickness on the large surface area of titania nanotube arrays obtained during the TiO_2 anodic growth under different ambient conditions. We study the growth of TiO_2 nanotube arrays by varying the anodization temperatures between 20 and 2°C, and also using several acidic electrolyte media, with different chemical concentrations and variation of the anodization voltage.

9.1 Introduction

The synthesis of self-organized nanostructured functional oxides with controlled structure and morphology has attracted great interest among the scientific community due to their potential applications in various technological fields, such as nano-electronics, magnetism, optic and information storage devices, which are motivated by the broad range of enhanced mechanical, optical, chemical, magnetic, and electronic properties exhibited for these

Nanostructured Titanium Dioxide in Photocatalysis
It-Meng Low, Hani Manssor Albetran, Victor Manuel de la Prida Pidal, and Fong Kwong Yam

ISBN 978-981-4877-07-7 (Hardcover), 978-1-003-14853-1 (eBook)
www.jennystanford.com

nanostructured materials with respect to those from their bulk analogues. Actually, a lot of effort has been put to obtain self-ordered nanostructures of TiO_2 prepared by sol–gel coating, nanoimprint, or electrochemical processes [1].

Titanium dioxide nanotube array is a well-known nanostructured semiconductor oxide with improved functional photocatalytic and sensing properties [2, 3]. Considering these properties, TiO_2 nanotube arrays attract wide scientific interest in view of their applications in self-cleaned gas sensors, photocatalytic materials, dye-sensitized solar cells, and water photoelectrolysis [4, 5]. Furthermore, its biocompatible nature makes it an excellent choice to be used in biotechnological applications [6, 7]. The physical and chemical properties of these nanodimensional structures are strongly dependent on their geometrical features such as tube diameter, tube length, wall thickness, etc., which also determine the efficiency of the devices based on these nanostructured functional materials. Potentiostatic anodization of titanium in fluoride-containing electrolytes is one of the most promising techniques for the synthesis of vertically oriented and parallel self-aligned TiO_2 nanotube arrays. A required specific nanoarchitecture of the oxide can be well controlled by varying the settings in the synthesis procedure [3]. In the present study, the structural and morphological features of self-aligned titanium oxide nanotube arrays grown by electrochemical anodization in different electrolytic media with variation in the potentiostatic anodization voltage and temperature of electrolyte have been investigated.

9.2 Results and Discussion

High-purity titanium foils (99.6%, 0.50 mm thickness) were mechanically polished with an automatic polisher in order to obtain an appropriate brittle and flat surface. Hereafter, Ti samples were ultrasonically degreased in isopropanol, ethanol, and acetone and rinsed in plenty of de-ionized (DI) water. Samples were mounted like an anodic electrode in a home-made electrochemical cell with a two-electrode configuration, serving as cathode a Pt grid, as shown in Fig. 9.1. A constant bias voltage between anode and cathode was applied by a power supply, and current transients were recorded with digital multimeter interfaced with a computer. Different compositions of the electrolytic media were employed together with the variation of the parameters of the anodization conditions. For TiO_2 nanotubes grown in acidic electrolytes, an aqueous dissolution of HF (0.5%), or a mixture of H_2SO_4 (1 M) + 0.15% HF, anodized at 20 V was employed. However, for TiO_2 nanotubes grown under neutral electrolytes, a mixture of 0.3% NH_4F + 1.8% H_2O + ethylene glycol, at the anodization voltage of 60 V was employed. Some of the anodic treatments were performed with varying the temperature of the electrolyte between room temperature and 2°C [8, 9].

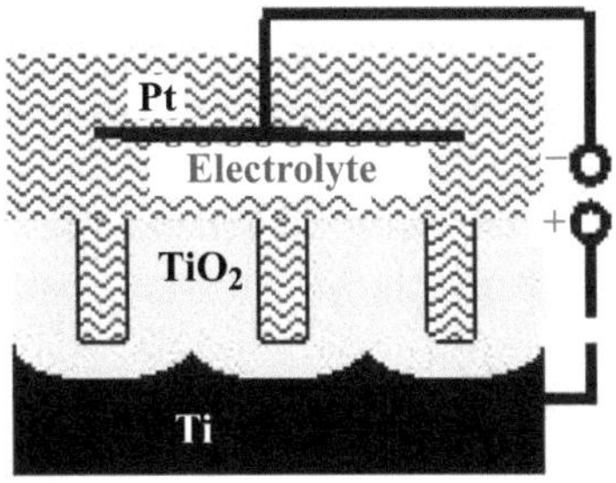

Figure 9.1 Schematic of the home-made electrochemical anodization cell for the growth of TiO_2 nanotubes.

9.2.1 Influence of Electrolyte Composition on TiO_2 Nanotubes Formation

Figure 9.2 shows SEM images of the TiO_2 nanotube arrays prepared in different electrolytes, which consist of an aqueous acidic dissolution of HF (0.15 wt.%) plus H_2SO_4 (1 M), and a neutral electrolyte formed by a mixture of 0.3% NH_4F + 1.8% H_2O + ethylene glycol, with the above-mentioned anodization conditions. In aqueous acidic electrolytes, a local thinning of the oxide barrier layer happens, and random growth of TiO_2 nanotube starts, as can be seen in Fig. 9.2(a). At this stage, the nanotube formation takes place and the thickness does not change at the oxide barrier layer. However, in the case of TiO_2 nanotubes anodized in neutral organic electrolytes, such as in the case of sample shown in Fig. 9.2(b), the electrochemical process leads to a less passivation of the Ti substrate. As a result, the growth of nanotubes occurs without the formation of random pores due to a less aggressive nature of the non-acidic organic electrolyte [11].

The as-grown TiO_2 nanotube arrays made in acidic electrolyte vary significantly in length (from 110 up to 300 nm) and average tube diameter (from 40 to 80 nm). However, in the organic electrolyte, longer nanotube lengths and better self-organized TiO_2 nanotubular layers were obtained. The length of the nanotube array was 30 microns, and the tube diameter was 80 nm for this anodic sample. Those differences can be explained considering the different nature of the employed media. Neutral and viscous electrolytes allow for a better control in the anodic conditions, creating two different environments along the tube, one inert at the top of it and the other chemically reactive at the bottom. Thus, the inhibition of the oxide dissolution at the top side of the nanotube allows the formation of longer TiO_2 nanotubes [11]. However, in acidic electrolytes, the growth rate of oxide layer and the dissolution reaction rate reach the equilibrium state, and after that the length of nanotubes is kept constant during the rest of the anodization process [10].

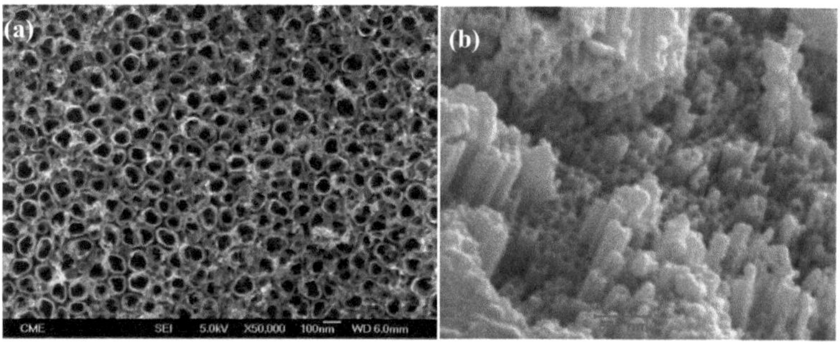

Figure 9.2 SEM images of TiO_2 nanotube arrays for a sample anodized in (a) an aqueous acidic dissolution of HF (0.15 wt%) plus H_2SO_4 (1 M) and (b) a neutral electrolyte of 0.3% NH_4F + 1.8% H_2O + ethylene glycol.

9.2.2 Temperature Dependence on Anodic Synthesis of TiO_2 Nanotubes

The titanium samples anodized under HF acidic solutions yield a regular self-aligned and randomly disordered nanotube arrays over large surface areas of the sample substrate. More than hundreds of squared microns of these arrays can be seen from the top-view SEM image for the sample anodized at 20 V during 90 min in HF (0.5 wt%)

acidic electrolyte at room temperature (RT), shown in Fig. 9.3(a); and also in the case of sample anodized at 20 V during 18 h in a mixture of HF (0.15 wt%) + H_2SO_4 (1 M), but at 4°C, which is shown in Fig. 9.3(b), respectively [8]. The self-aligned tubular structure exhibited in the SEM image of Fig. 9.3(a) has an average inner hole diameter of 65 ± 10 nm and a nearly uniform wall thickness above 25 ± 10 nm, similar to the values reported elsewhere [2, 8, 12]. The resulting average outer diameter of the titania nanotubes ranges between 90 and 110 nm. Meanwhile, for the sample anodized in a mixture of HF (0.15 wt%) and H_2SO_4 (1 M) acidic electrolytes at 4°C, as shown in Fig. 9.3(b), it consists of an averaged inner tube diameter of about 55 ± 10 nm and wall thickness of the nanotubes of about 40 ± 10 nm. Therefore, the fluoride concentration does not influence, so much,

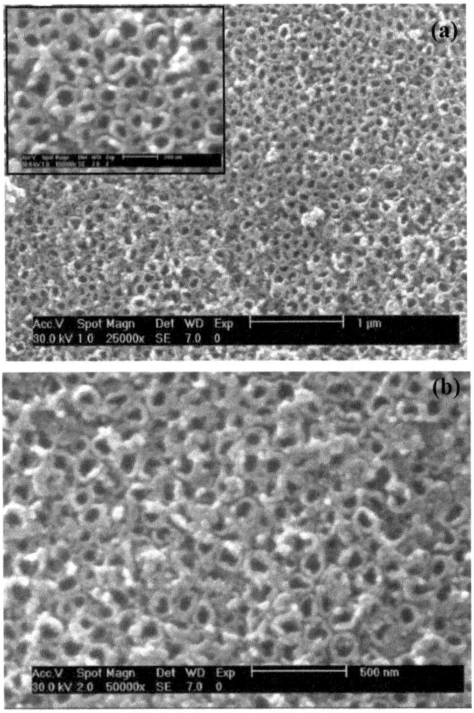

Figure 9.3 Top-view surface SEM images of TiO_2 nanotube arrays grown by electrochemical anodization processes: (a) for a sample anodized at 20 V during 90 min in HF (0.5 wt%) acid at RT and (b) for a sample anodized at 20 V during 18 h in a mixture of HF (0.15 wt%) plus H_2SO_4 (1 M), but at 4°C. The scale bar shown in the inset of Fig. 9.3(a) corresponds to 100 nm.

the TiO_2 nanotube diameter, but it appears that the temperature of the anodic process does [13, 14]. The different anodizing conditions employed lead to different stages of the surface structures of nanotubular TiO_2, where highly self-aligned nanotube arrays are formed at low anodic potentials as a result of the electrochemical etching, as can be seen from the top-view SEM images. Low-temperature anodization appears to inhibit the formation of nanotubular structure. Meanwhile, the straight nanotubes obtained at RT anodic processes have a nanotube size ranging from 60 to 100 nm and a wall thickness of 25–40 nm [8].

9.2.3 Optical Properties of Anodic TiO_2 Nanotubes

The semiconducting and photoelectrochemical properties exhibited by the amorphous structure of the nanoporous anodic TiO_2 are very promising for their application in the designing of photo-catalyst and solar-cells-based devices [15] such as in the construction of hydrogen storage and sensor devices [16–19]. Figure 9.4 shows the photoluminescence (PL) spectrum corresponding to a nanoporous sample formed by self-aligned TiO_2 nanotubes. In the PL spectrum of a self-aligned titania nanotube array, a broad emission band of around 0.8 eV can be clearly observed in the wavelength range of 580–780 nm. The Raman spectra shown in Fig. 9.5 correspond to a nanotubular TiO_2 membrane synthesized after following the anodic conditions as reported elsewhere [14]; the spectrum has been collected at three different wavelength ranges, for a data storage time of 240 s. Si was used as the reference spectrum, and the appearance of spurious signals observed in the TiO_2 nanotube membrane spectrum is actually under analysis. It can be derived from the experimental results that the semiconducting optimal conditions exhibited by the nanoporous structure of the self-assembled titania nanotube templates are a consequence of the impurities and defects present in the titania nanotube walls, which exhibit different densities of donor concentrations from the bottom of the nanopores to the surface along the nanotube walls. The semiconductor behavior exhibited by these nanotubular oxides can be well explained considering a high concentration of electronic defects in the wall of nanotube oxide, which are in electronic equilibrium with the underlying compact oxide at the bottom. In this structure, the surface chemistry of nanotubes introduces considerable changes in the semiconductor behavior of the underlying oxide, which in turn is manifested in surface-related effects, giving rise to modifications of the redox reactions and a shift in the flat band potential respect to the base oxide, and trapping of photoelectrons [20].

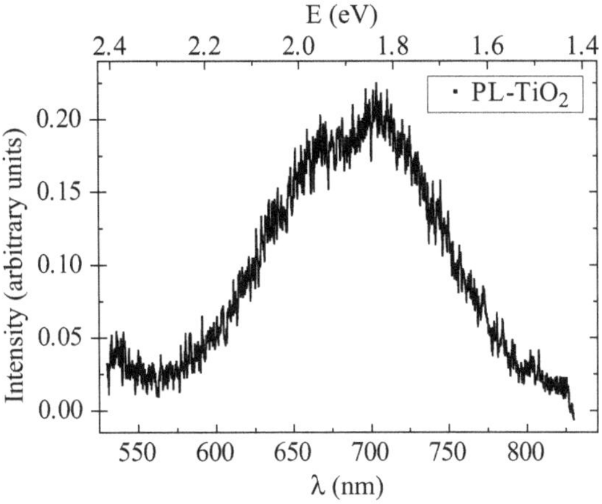

Figure 9.4 Photoluminescence spectrum of a self-aligned TiO_2 nanotube array. Adapted from Ref. [20].

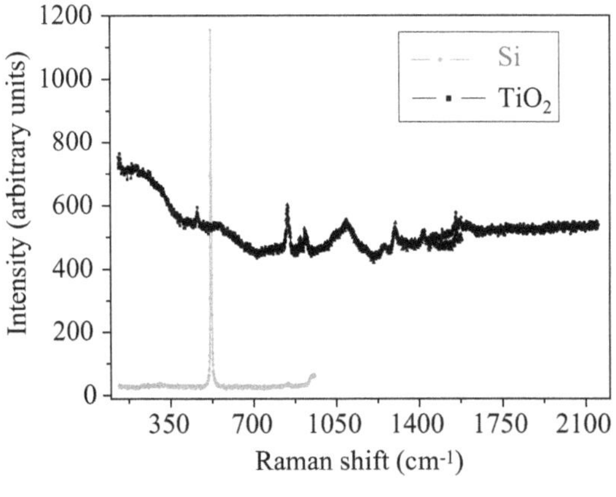

Figure 9.5 Raman spectra of a membrane of TiO_2 nanotubes anodized in HF acidic electrolyte. Si is used as the reference spectrum. Adapted from Ref. [20].

9.3 Conclusion

Self-aligned nanotubular Ti oxide layers have been grown by the anodization of titanium foils in different fluoride-containing electrolytes. These electrolytes allow synthesizing TiO_2 nanotube arrays with thickness in the range of 150 nm up to almost 30 microns, by using a range of potentials between 12 and 60 V and different anodization times. The results show an improvement in the self-alignment of the nanotube arrays and an increase of about 10,000% in the length of nanotubes obtained by anodization with electrolytes containing NH_4F in ethylene glycol, with respect to the ones obtained with the aqueous acidic electrolytes. Ti anodization in HF solutions and its mixtures with sulfuric acid, performed at lower temperature than RT, appears to inhibit the nanotubular structure formation. The straight nanotubes obtained at RT have a nanopore size ranging between 60 and 100 nm, 25 and 40 nm wall thickness, and about 300 nm length. The semiconducting behavior exhibited by the nanoporous structure of the self-assembled titania nanotube templates is a consequence of the impurities and defects present in the titania nanotube walls.

References

1. S. Shena, J. Chena, M. Wanga, X. Shengb, X. Chena, X. Fengb, and S. S. Maoc, *Prog. Mater. Sci.*, **98** (2018), 299–385.
2. D. Gong, C. A. Grimes, O. K. Varghese, W. Hu, R. S. Singh, Z. Chen, and E. C. Dickey, *J. Mater. Res.*, **16** (2001), 3331.
3. O. K. Varghese, X. Yang, J. Kendig, M. Paulose, K. Zeng, C. Palmer, K. G. Ong, and C. A. Grimes, *Sensor. Lett.*, **4** (2006), 120.
4. A. G. Muñoz, *Electrochim. Acta*, **52** (2007), 4167.
5. C. A. Grimes, *J. Mater. Chem.*, **17** (2007), 1451.
6. S. C. Roy, M. Paulose, and C. A. Grimes, *Biomaterials*, **28** (2007), 4667.
7. Y. Fu and A. Mo, *Nanoscale Res. Lett.*, **13** (2018) 187.
8. V. M. Prida, E. Manova, V. Vega, M. Hernandez-Velez, P. Aranda, K. R. Pirota, M. Vázquez, and E. Ruiz-Hitzky, *J. Magn. Magn. Mater.*, **316** (2007), 110–113.

9. V. Vega, M. A. Cerdeira, V. M. Prida, D. Alberts, N. Bordel, R. Pereiro, F. Mera, S. García, M. Hernández-Vélez, and M. Vázquez, *J. Non-Crystalline Solids*, **354** (2008), 5233–5235.

10. J. Bai, B. Zhou, L. Li, Y. Liu, Q. Zheng, J. Schao, X. Zhu, W. Cai, J. Liau, and L. Zou. *J. Mater. Sci.*, **43** (2008), 1880.

11. X. Feng, J. M. Macak, and P. Schumuki, *Chem. Mater.*, **10** (2007), 1534.

12. R. Beranek, H. Tsuchiya, T. Sugishima, J. M. Macak, L. Traveira, S. Fujimoto, H. Kisch, and P. Schmuki, *Appl. Phys. Lett.*, **87** (2005), 243114.

13. Q. Cai, M. Paulose, O. K. Varghese, and C. A. Grimes, *J. Mater. Res.*, **20** (2005), 230.

14. V. M. Prida, K. R. Pirota, D. Navas, A. Asenjo, M. Hernández-Vélez, and M. Vázquez, *J. Nanosci. Nanotechnol.*, **7** (2007), 272.

15. L.-L. Li, Y.-J. Chen, H.-P. Wu, N. S. Wang, and E. Wei-Guang Diau, *Energy Environ. Sci.*, **4** (2011), 3420.

16. B. Tan, Y. Zhang, and M. Long, *Nanoscale Res. Lett.*, **9** (2014), 190.

17. S. Ali, H. Granbohm, J. Lahtinen, and S.-P. Hannula, *Nanoscale Res. Lett.*, **13** (2018), 179.

18. J. Liang, C. Hao, K. Yu, and Y. Li, *Nanomater. Nanotechnol.*, **6** (2016), 1–5.

19. N. K. Shrestha, J. M. Macak, F. Schmidt-Stein, R. Hahn, C. T. Mierke, B. Fabry, and P. Schmuki, *Angew. Chem. Int. Ed.*, **48** (2009), 969 –972.

20. V. Vega, V. M. Prida, M. Hernández-Vélez, E. Manova, P. Aranda, E. Ruiz-Hitzky, and M. Vázquez, *Nanoscale Res. Lett.*, **2** (2007), 355–363.

Part IV

Materials Properties and Applications

Chapter 10

Phase Transformations and Crystallization Kinetics of Electrospun TiO_2 Nanofibers in Air and Argon Atmospheres

The effects of atmospheric air and argon environments on thermal-induced phase transformations in electrospun TiO_2 nanofibers have been investigated in situ using synchrotron radiation diffraction (SRD). Diffraction results showed that the as-synthesized TiO_2 nanofibers were initially amorphous but crystallized to form anatase and rutile after thermal annealing in air or argon at elevated temperatures. The crystallization temperature of anatase was delayed by 100°C in argon relative to air, and the transformation of anatase to rutile occurs faster in argon atmosphere than in air due to the formation of oxygen vacancies. Nonlinear strains formed in both polymorphs and the substantial elevation of rutile thermal expansion pointed to strain anisotropy in the rutile phase and the concomitant fiber breakage.

10.1 Introduction

Titanium dioxide (TiO_2) is one of the most widely studied materials due to its high photoactivity, photodurability, mechanical robustness,

Nanostructured Titanium Dioxide in Photocatalysis
It-Meng Low, Hani Manssor Albetran, Victor Manuel de la Prida Pidal, and Fong Kwong Yam

ISBN 978-981-4877-07-7 (Hardcover), 978-1-003-14853-1 (eBook)
www.jennystanford.com

low cost, and chemical and biological inertness [1–6]. Titanium dioxide has three polymorphs: anatase, brookite, and rutile [7–10]. Anatase and rutile are the two main polymorphs that exhibit different properties and thus different photocatalytic performances. They both have tetragonal crystal structures, but different space groups and atoms per unit cell (Z): space group $I4/amd$ and $Z = 4$ for anatase and $P4_2/mnm$ and $Z = 2$ for rutile [9, 11]. At room temperature, the lattice parameters of anatase are $a = b = 0.3785$ nm and $c = 0.9514$ nm, while for rutile the parameters are $a = b = 0.4594$ nm and $c = 0.29589$ nm [12]. These differences in the interatomic spacing and crystal structures for anatase and rutile can result in different densities and electronic band structures. Anatase is considered more photochemically active than rutile by virtue of its lower rates of electronic recombination and higher surface absorptive capacity [13]. However, recent work has indicated that a mixed-phase TiO_2 tends to make a superior photocatalyst than a single-phase TiO_2. For example, a mixed-phase sample of 70% anatase and 30% rutile has been found as the best photocatalyst for wastewater treatment [14]. Similarly, commercial TiO_2 photocatalyst (P-25) that contains 80 wt% anatase and 20 wt% rutile has been observed to exhibit excellent photocatalytic reactions [12].

The optimum composition in terms of the anatase and rutile levels for maximum photocatalytic performance can be achieved by thermal annealing. In general, anatase forms amorphous TiO_2 at low temperatures, which subsequently transforms to rutile at higher temperatures [15, 16]. The transformation is affected by various conditions, which include the synthesis method, the amount of impurities, temperature, calcination time, and atmosphere type [15, 17]. Reaction atmospheres have been observed to have a significant influence on the thermal transformation of TiO_2 phases because oxygen defect levels and interstitial titanium ion levels are influenced by the type of atmosphere [15, 18]. The effects of air, vacuum, argon, argon–chlorine mixture, hydrogen, steam, and nitrogen on the anatase-to-rutile phase transformation have been reported [18]. For TiO_2 films, the amorphous-to-anatase and anatase-to-rutile phase transformations were observed to occur at a lower temperature when annealed in a hydrogen atmosphere than in air or vacuum [19]. The rate of anatase-to-rutile phase transformation was found to decrease in vacuum but increase in hydrogen atmosphere due to the formation of either titanium interstitials or oxygen vacancies. Since the phase transition involves an overall shrinkage of the oxygen structure and a cooperative movement of ions, the formation of oxygen vacancies would accelerate the phase transition and vice versa for the formation of titanium interstitials [15].

Nanostructured TiO_2 with elongated morphology exhibits high surface-area-to-volume ratio, which is of great significance for increasing the decomposition rate of organic pollutants because photocatalytic reactions take place rapidly and drastically on the surface of the catalyst [14, 20]. Hitherto, various techniques have been used to synthesize nanostructured TiO_2 with various morphologies (e.g., nanowires, nanorods, nanotubes, and nanofibers). These include microemulsion, sol–gel, hydrothermal, and vapor deposition [8, 21]. Elongated nanostructures are of particular importance, as long and thin nanotubes or nanofibers can provide a high specific surface area for photocatalysis [9]. Furthermore, a variety of techniques such as self-assembly, evaporation, anodization, and electrospinning have been developed for fabricating one-dimensional nano-TiO_2 [1, 22, 23]. Among the fabrication methods, electrospinning is a simple and cost-effective technique applicable at industrial levels for fabricating one-dimensional nanofibers [22–25]. In this process, a solution of polymer and TiO_2 precursor is ejected through a needle in a high electric field whereby composite nanofibers of polymer and amorphous TiO_2 are formed [25–29]. To the best of our knowledge, the effect of atmospheres on the in situ crystallization of anatase and rutile in electrospun TiO_2 nanofibers has not been previously reported.

In the present work, the effect of air and argon atmospheres on the transformation of anatase to rutile and the in situ crystallization kinetics of TiO_2 nanofibers were investigated using SRD over the temperature range of 25–900°C. TiO_2 nanofibers were synthesized using electrospinning, and their morphology, structure, and thermal expansion were characterized by scanning electron microscopy (SEM) and associated energy dispersive spectroscopy (EDS), and high-temperature SRD.

10.2 Results and Discussion

10.2.1 Microstructures of Electrospun TiO_2 Nanofibers

Figure 10.1 illustrates the SEM images of electrospun nanofibers using secondary and backscattered electrons, before and after thermal annealing. In the former, the nanofibers have smooth surfaces with a random distribution and annealing. The average diameter of electrospun nanofibers was 614 ± 190 nm, with the fibers being well dispersed within the PVP matrix, as shown in Figs. 10.1(a) and (b).

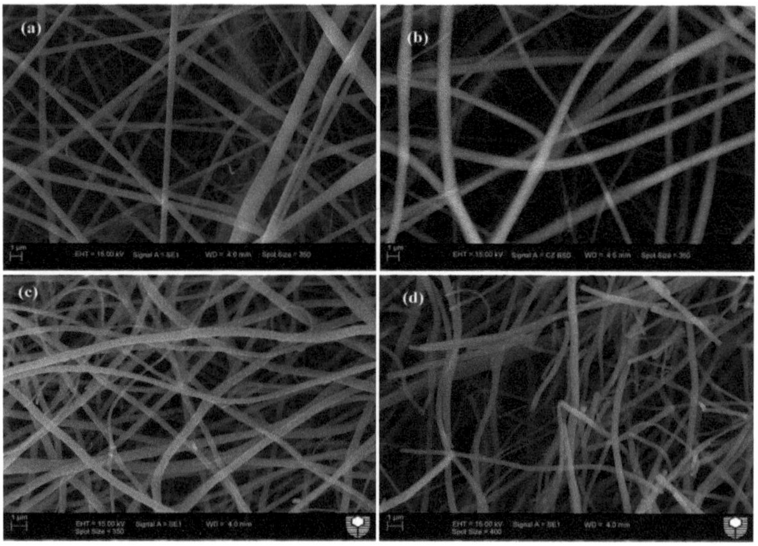

Figure 10.1 SEM micrographs of as-spun TiO_2 nanofibers imaged with (a) secondary electrons and (b) backscattered electrons; and with secondary electrons after cooling from 900°C in (c) air and (d) argon atmospheres.

However, after thermal annealing at 900°C, the average diameters of the nanofibers shrank by approximately 42% in air [Fig. 10.1(c)] and 47% in argon [Fig. 10.1(d)] compared to as-spun nanofibers, due to the loss of the organic binder during heating. The average diameters of nanofibers were 358 ± 195 nm in air and 328 ± 113 nm in argon. In argon, the nanofibers were partially broken, while in air the extent of fiber breakage was less. A probable reason for the fiber breakage could be due to damage arising from sample handling or during the SEM examination. A more likely explanation is the thermal expansion anisotropy of rutile and the associated strains induced at elevated temperatures (see Table 10.2, and the thermal expansion section).

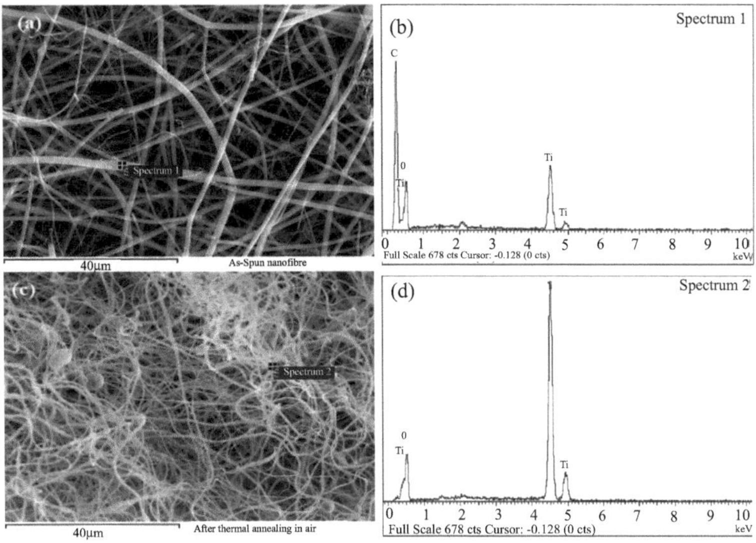

Figure 10.2 (a) SEM micrograph of the as-received electrospun TiO_2 nanofibers and (b) the corresponding EDS spectrum (#1); and (c) SEM micrograph for electrospun TiO_2 nanofibers after cooling in air from 900°C, and (d) the corresponding EDS spectrum (#2).

The EDS spectrum of the as-spun TiO_2/PVP nanofibers in Fig. 10.2(b) showed strong signatures for Ti, O, and C, with the C content being due to the PVP polymer binder. The EDS spectrum of the nanofiber sample after thermal annealing at 900°C [Fig. 10.2(d)] was similar to that for the original material, other than showing the absence of the C peak due to the complete loss of the organic matrix. The same observation was made when the nanofibers were heated in argon.

10.2.2 Effect of Environmental Atmosphere on Phase Transitions during Thermal Annealing

Figure 10.3 shows the effect of atmosphere on the in situ crystallization kinetics of the as-synthesized TiO_2 nanofibers over the temperature range of 25–900°C, as revealed by SRD. The SRD patterns showed Pt peaks due to the Pt sample mount, which were ignored in the calculations. The TiO_2/PVP nanofibers were initially amorphous as shown by the pronounced amorphous hump for the 25°C pattern and the absence of diffraction peaks. By 200°C, the amorphous feature had largely disappeared and had gone completely by 400°C. In air [Fig. 10.3(a)], anatase and rutile were first observed at 600°C and 700°C, respectively. However, in argon, both anatase and rutile were first observed at 700°C [Fig. 10.3(b)], so that the initial crystallization of anatase was delayed by 100°C in argon compared with air. The reason for this delay in the formation of anatase in argon atmosphere may be attributed to the lack of oxygen. It appears that atmospheric oxygen accelerates the transformation from amorphous TiO_2 into anatase. Moreover, the partial pressure of oxygen also influences the rate of anatase-to-rutile transformation, becoming faster in argon atmosphere, which has a much lower oxygen partial pressure [15]. Defects of oxygen vacancies created in argon atmosphere by the low partial pressure of oxygen accelerate the anatase-to-rutile transformation, which involves the breaking of anatase Ti–O bonds and reforming of rutile Ti–O bonds [12, 18]. Hence, the rate of transformation from anatase to rutile was faster in argon than in air atmosphere by virtue of oxygen vacancies (see Table 10.1).

Table 10.1 Effect of atmospheres on the fraction of anatase-to-rutile transformation as a function of temperature

	Air (x)		Argon (x)	
Temperature (°C)	**Spurr Equation**	**Rietveld method**	**Spurr Equation**	**Rietveld Method**
700	1.5	4.98(10)	9.2	15.41(28)
800	19.2	24.02(32)	46.5	50.99(28)
900	63.04	64.95(60)	84.5	85.98(49)

The fraction of anatase-to-rutile transformation can be estimated using the Spurr equation [33]:

$$x = \frac{1}{[1 + 0.8(I_A + I_R)]} \qquad (10.1)$$

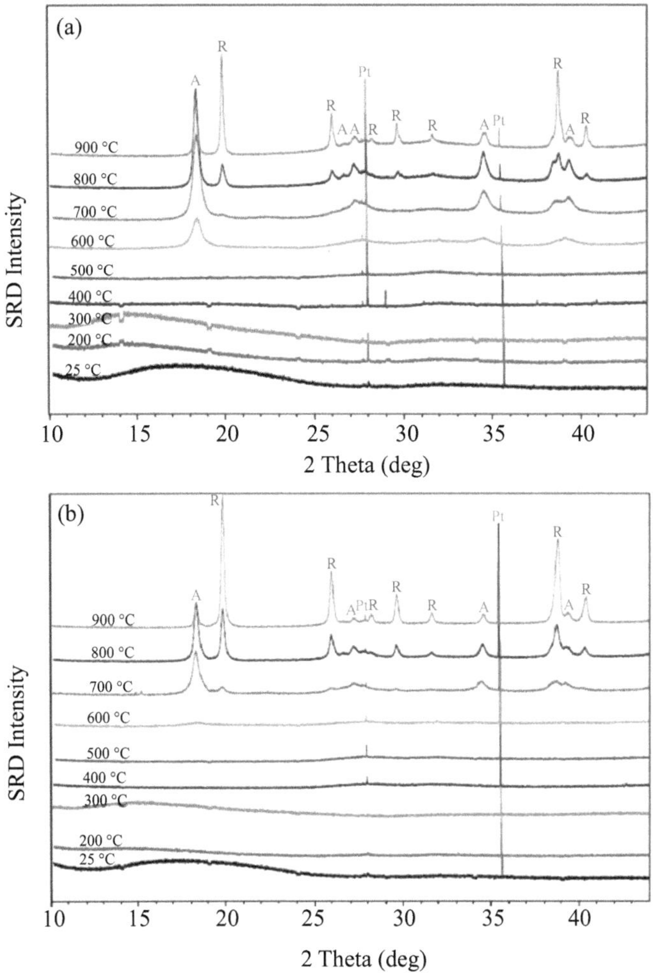

Figure 10.3 In situ SRD plots showing the effect of temperature on the crystallization behavior of the as-synthesized TiO_2 nanofibers when heated at 25–900°C (a) in air and (b) in argon [Legend: anatase (A), rutile (R), and platinum (Pt)].

where I_A and I_R are the X-ray integrated intensities of anatase (101) reflection and rutile (110) reflection, respectively. Table 10.1 shows the fraction of anatase-to-rutile transformation as a function of temperature in air and in argon as determined using the Spurr equation and Rietveld method. It is evident that the weight fraction of rutile increased dramatically as the temperature increased by virtue of the transformation of metastable anatase to rutile at elevated temperatures. The rate of anatase-to-rutile transformation was also observed to increase faster in argon than in air. This enhanced transformation rate may be attributed to the formation of oxygen vacancies in TiO_2 annealed in argon [12]. The formation of oxygen vacancies would accelerate the phase transition since it involves an overall shrinkage of the oxygen structure and a cooperative movement of ions [15].

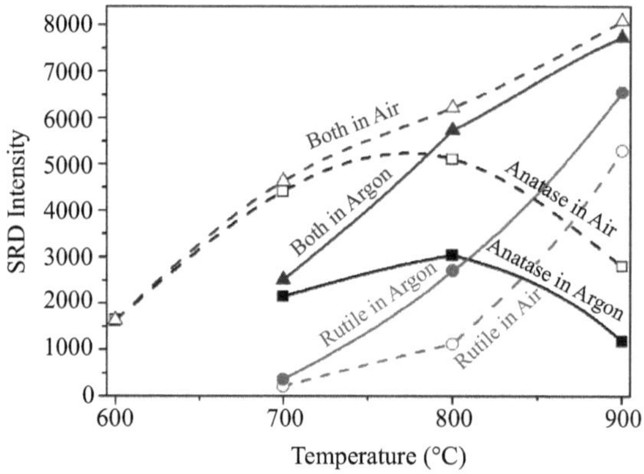

Figure 10.4 The SRD intensities of the anatase (101) and rutile (110) reflections in the range of 600–900°C for TiO_2 nanofibers heated in air and argon. The vertical axis is proportional to the phase concentration for both anatase and rutile, and the sum of the anatase and rutile intensities is proportional to total crystalline titania.

Figure 10.4 shows the intensities of anatase (101) and rutile (110) peaks over the temperature range of 600–900°C. Because the most intense peaks for both crystalline phases have virtually the same reference intensity ratio, the intensity–temperature plots should be seen as "relative concentration–temperature" plots, with the sum of the anatase and rutile peaks being proportional to the total concentration of crystalline TiO_2. The formation of crystalline TiO_2 was first observed at 600°C and increased steadily up to 900°C, when heating in both air and argon. It is evident that the amorphous TiO_2 was not completely converted to crystalline TiO_2 by 900°C. Figure 10.4 also shows the progressive growth of rutile after the formation, whereas the anatase concentration steadily reduces beyond approximately 800°C due to transformation to rutile. Overall,

Fig. 10.4 is consistent with the transformation of anatase to rutile becoming more likely with increases in temperature. This discussion is extended below when the results on relative amounts of anatase and rutile with temperature in Fig. 10.6 are considered.

Figure 10.5 shows the typical SRD plots from Rietveld analysis for data measured in air and argon at 800°C. In general, the quality or goodness of fit is gauged by the values of weighted pattern R-factor (R_{wp}), the expected R-factor (R_{exp}), and the derived Bragg R-factor (R_B) [30]. In this work, values obtained for air and argon atmospheres were R_{wp} = 13.0 and 10.0, R_{exp} = 3.7 and 4.6, R_B (anatase) = 4.6 and 4.9, and R_B (rutile) = 7.3 and 4.2, respectively.

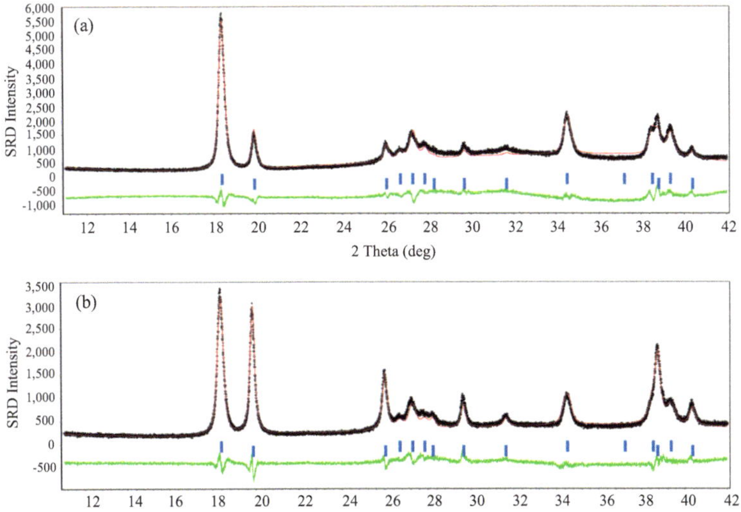

Figure 10.5 SRD Rietveld difference plots for 800°C SRD data (a) in air and (b) under argon. Measured patterns are indicated by black crosses and calculated patterns by solid red lines. The green residual plot shows the difference between the calculated and the measured patterns. The peak positions for anatase and rutile are indicated by the top and bottom blue bars, respectively.

Figure 10.6 shows the variation with temperature of the relative amounts of anatase and rutile from Rietveld analysis, which supplements the information shown in Fig. 10.4. The conversion of crystalline anatase to rutile in both air and argon atmospheres is clearly indicated. The results show that the conversion is less rapid in air than in argon. At 600°C, anatase formed in air but not yet in argon. In argon at 700°C, anatase comprised 85% of the crystalline TiO_2, and 95% of the total in air. By 800°C, the abundance of anatase in crystalline TiO_2 falls to 76% in air and 50% in argon, and then by 900°C to 35% in air and to 14% in argon. In addition, the phase content of rutile increased dramatically as the temperature increased by virtue of the transformation of metastable anatase to rutile. Rutile formed at 700°C in both atmospheres but with different abundances. It increased from about 15% at 700°C to well over 85% at 900°C in argon atmosphere. However, in air, it increased from 5% up to only 65% for the same temperature range.

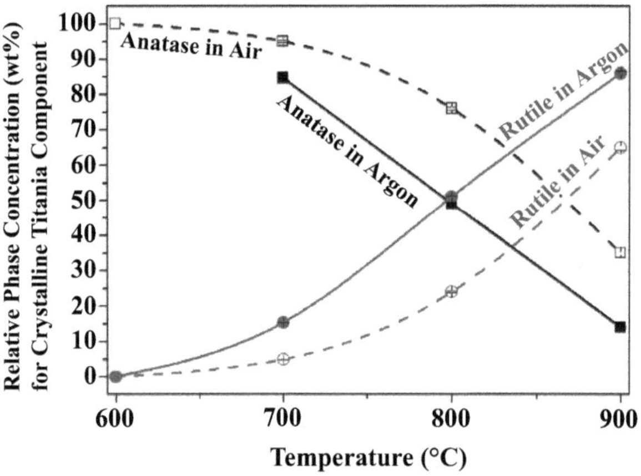

Figure 10.6 Relative phase abundances of anatase and rutile for the crystalline TiO_2 component in the temperature range 600–900°C for TiO_2 nanofibers heated in air and argon atmospheres. The phase levels represent the percentages of crystalline TiO_2 at each temperature.

Figure 10.7 shows the effect of temperature on the growth of anatase and rutile crystallites according to atmosphere. The average crystallite size for anatase was ~13 nm at 700°C in air and increased linearly for both atmospheres, to ~30 nm at 900°C. The significantly larger rutile crystallite sizes were ~18 nm at 700°C in both air and argon, but grew to ~37 nm in air and ~40 nm in argon at 900°C. Therefore, the atmosphere had an insignificant effect on crystallite growth. The crystallite growth with temperature behavior is attributed to the conversion of amorphous material to crystalline TiO_2 through an atomic diffusion-controlled nucleation and growth process. The gradients of the linear regressions for the plots in

Fig. 10.5 amount to approximately 1.5–2.0% volumetric growth per degree increase in temperature. This diffusion growth is orders of magnitude higher than that due to lattice thermal expansion (see Fig. 10.9).

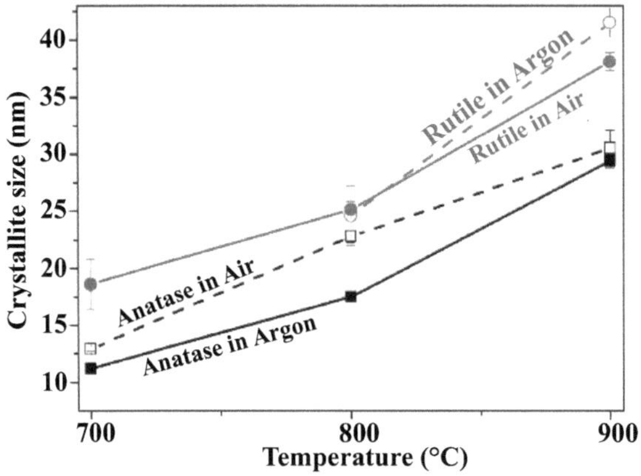

Figure 10.7 Williamson–Hall plot estimates of the effect of temperature on the average crystallite size of (a) anatase and (b) rutile in TiO_2 nanofibers according to temperature.

The effect of temperature on the average lattice strain of both anatase and rutile for both atmospheres is shown in Fig. 10.8. It can be seen that the initial strain present in both phases is progressively relieved as the temperature increases [32]. The observed strain relief is consistent with crystallite–crystalline interactions being negligible up to the maximum temperature considered in the study, at which point there is still substantial amorphous TiO_2 present. The higher strains observed for the anatase phase in air are consistent with the finding of Nicula et al. [34] who reported that the anatase phase exhibits larger strain values than rutile in titania powders over 400–850°C.

The variations in unit-cell parameters and cell volumes for anatase and rutile with temperature are shown in Fig. 10.9(a,b) and Fig. 10.10, and the corresponding linear and volumetric thermal expansion coefficients (TECs) are provided in Table 10.2. The unit-cell parameters and cell volume for both phases in air and argon atmospheres increased linearly with temperature, as expected. The

SRD thermal expansion data of Hummer et al. [31] measured with titania powders over the temperature range of 25–300°C, which are included in Table 10.2, were extrapolated to the temperatures considered in this study. It is seen that the TECs for anatase in both air and argon compare closely with values reported for crystalline TiO_2 powder. By contrast, the rutile expansion coefficients are higher in argon, and higher again in air. There is also clear evidence for thermal expansion anisotropy in the rutile phase, which points to the development of a more pronounced strain in rutile compared with anatase, and with the effect being more pronounced in argon, which may explain the rougher appearance and more breakage of the nanofibers at high temperatures compared with heating in air. In addition, for both anatase and rutile, the TECs were higher in air than in argon atmospheres due to the formation of oxygen vacancies in argon atmosphere. However, the linear expansion coefficient ratio of α_a to α_c was ~9% higher in argon than in air, which could be the main reason for the embrittlement of nanofibers. Furthermore, it should be noted that literature equations of anatase and rutile parameters have hitherto been determined at relatively low temperatures, from room temperature to just over 300°C using anatase and rutile TiO_2 powders, while the present study examined TiO_2 nanofibers at higher temperatures.

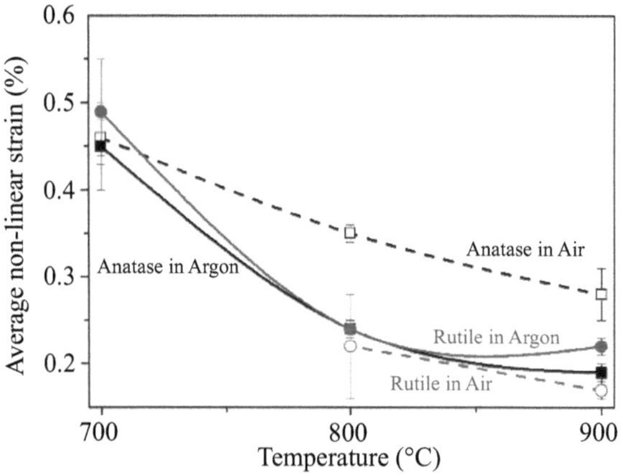

Figure 10.8 Williamson–Hall plot estimates of mean strain for anatase and rutile crystallites in the range of 600–900°C for TiO_2 nanofibers heated in air and argon atmospheres.

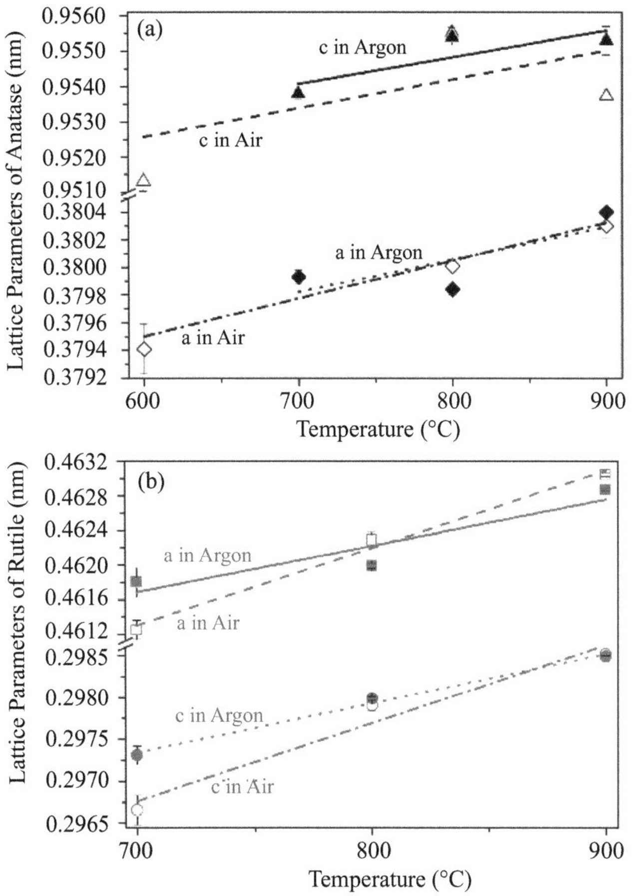

Figure 10.9 Variation with temperature of the lattice parameters for anatase (a) and rutile (b) in air and argon atmospheres.

Table 10.2 Linear (α_a, α_c) and volumetric (β) thermal expansion coefficients (TECs) calculated from Figs. 10.9 and 10.10 at 800°C

	TEC×10^{-5} (°C^{-1})		α_a	α_c	β	α_c/α_a
Atmosphere (this study)	Air	Anatase	0.72	0.86	2.29	1.19
		Rutile	1.94	3.12	6.99	1.61
	Argon	Anatase	0.62	0.79	1.99	1.27
		Rutile	1.16	1.98	4.24	1.70
Hummer et al. [31]	Powdered TiO_2 sample in air	Anatase	0.53	1.04	2.17	1.96
		Rutile	0.87	1.01	2.83	1.16

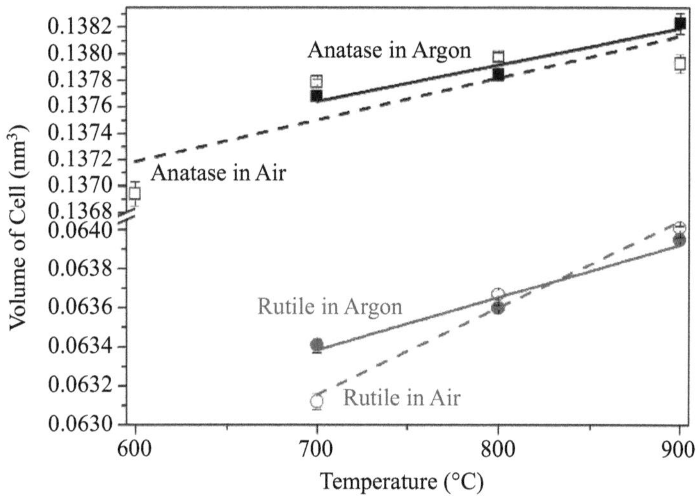

Figure 10.10 Variation with temperature of the cell volume for anatase and rutile in air and argon atmospheres.

10.3 Conclusion

The effect of air and argon on the dynamic crystallization behavior of electrospun TiO_2 nanofibers was investigated in situ using SRD data measured in the temperature range of 25–900°C. The electrospun TiO_2 nanofibers were initially amorphous but crystallized progressively to anatase and rutile at elevated temperatures. In air, anatase was first observed at 600°C and rutile at 700°C, but in argon both appeared at 700°C. The level of crystalline TiO_2 (anatase and rutile) increased monotonically with increase in temperature, with some amorphous TiO_2 still being observed at 900°C. The rates of anatase-to-rutile transformation were faster in argon than in air by virtue of oxygen vacancies in the former. The nonlinear strain in both polymorphs reduced with temperature, with the strain of anatase in air being elevated. The substantial elevation of rutile thermal expansion pointed to strain anisotropy in the rutile phase and the concomitant fiber breakage.

Acknowledgments

The authors gratefully acknowledge the Australian Synchrotron (AS122/PDFI/5075) for work supporting. The authors would like to thank E/Prof. B. H. O'Connor for guidance on the Rietveld work and Ms. E. Miller and Dr. N. Chen-Tan for assistance with the SEM work. We also thank Ms. K. Haynes, Mr. A. Viereckl, and Mr. J. Xiao Hua for laboratory assistance.

References

1. I. M. Low, H. Albetran, V. M. Prida, V. Vega, P. Manurung, and M. Ionescu, *J. Mater. Res.*, **28**, 304 (2013).
2. P. Manurung, Y. Putri, W. Simanjuntak, and I. M. Low, *Ceram. Int.*, **39**, 255 (2013).
3. P. Zhang, C. Shao, X. Li, M. Zhang, X. Zhang, Y. Sun, and Y. Liu, *J. Hazard. Mater.*, **237–238**, 331 (2012).
4. C. Yu, L. Wei, X. Li, J. Chen, Q. Fan, and J. Yu, *Mater. Sci. Eng. B*, **178**, 344 (2013).
5. X. Pan, Y. Zhao, S. Liu, C. L. Korzeniewski, S. Wang, and Z. Fan, *ACS Appl. Mater. Interfaces*, **4**, 3944 (2012).
6. V. J. Babu, A. S. Nair, Z. Peining, and S. Ramakrishna, *Mater. Lett.*, **65**, 3064 (2011).
7. G. Liu, L. Wang, H. G. Yang, H. M. Cheng, and G. Q. Lu, *J. Mater. Chem.*, **20**, 831 (2010).
8. S. Chuangchote, J. Jitputti, T. Sagawa, and S. Yoshikawa, *ACS Appl. Mater. Interfaces*, **1**, 1140 (2009).
9. D. W. Kim, N. Enomoto, Z. E. Nakagawa, and K. Kawamura, *J. Am. Ceram. Soc.*, **79**, 1095 (1996).
10. R. Liu, L. S. Qiang, W. D. Yang, and H. Y. Liu, *Mater. Res. Bull.*, **48**, 1458 (2013).
11. W. Li, C. Ni, H. Lin, C. P. Huang, and S. I. Shah, *J. Appl. Phys.*, **96**, 6663 (2004).
12. D. A. Hanaor and C. C. Sorrell, *J. Mater. Sci.*, **46**, 855 (2011).
13. S. Shang, X. Jiao, and D. Chen, *ACS Appl. Mater. Interfaces*, **4**, 860 (2012).
14. H. Li, W. Zhang, and W. Pan, *J. Am. Ceram. Soc.*, **94**, 3184 (2011).

15. R. D. Shannon and J. A. Pask, *J. Am. Ceram. Soc.*, **48**, 391 (1965).
16. R. A. Eppler, *J. Am. Ceram. Soc.*, **70**, 64 (1987).
17. Y. Iida and S. Ozaki, *J. Am. Ceram. Soc.*, **44**, 120 (1961).
18. J. A. Gamboa and D. M. Pasquevich, *J. Am. Ceram. Soc.*, **75**, 2934 (1992).
19. J. H. Huang and M. S. Wong, *Thin Solid Films*, **520**, 1379 (2011).
20. J. S. Lee, T. J. Ha, M. H. Hong, and H. H. Park, *Thin Solid Films*, **529**, 98 (2013).
21. I. M. Low, B. Curtain, M. Philipps, Q. Z. Liu, and M. Ionescu, *J. Aust. Ceram. Soc.*, **48**, 198 (2012).
22. V. Vega, V. M. Prida, M. H. Velez, E. Manova, P. Aranda, E. R. Hitzky, and M. Vazquez, *Nanoscale Res. Lett.*, **2**, 355 (2007).
23. S. Chuangchote, J. Jitputti, T. Sagawa, and S. Yoshikawa, *ACS Appl. Mater. Interfaces*, **1**, 1140 (2009).
24. Z. M. Huang, Y. Z. Zhang, M. Kotaki, and S. Ramakrishna, *Compos. Sci. Technol.*, **63**, 2223 (2003).
25. Q. Li, D. J. G. Satur, H. Kim, and H. G. Kim, *Mater. Lett.*, **76**, 169 (2012).
26. W. Luo, X. Hu, Y. Sun, and Y. Huang, *J. Mater. Chem.*, **22**, 4910 (2012).
27. C. Wessel, R. Ostermann, R. Dersch, and B. M. Smarsly, *J. Phys. Chem.*, **C115**, 362 (2011).
28. H. Li, W. Zhang, and W. Pan, *J. Am. Ceram. Soc.*, **94**, 3184 (2011).
29. P. Zhu., A. S. Nair, P. Shengjie, Y. Shengyuan, and S. Ramakrishna, *ACS Appl. Mater. Interfaces*, **4**, 581 (2012).
30. W. K. Pang and I. M. Low, *J. Am. Ceram. Soc.*, **93**, 2871 (2010).
31. D. R. Hummer and P. J. Heaney, *Powder Diffraction*, **22**, 352 (2007).
32. G. K. Williamson and W. H. Hall, *Acta Metall.*, **1**, 22 (1953).
33. R. A. Spurr and H. Myers, *Anal. Chem.*, **29**, 760 (1957).
34. R. Nicula, M. Stir, C. Schick, and E. Burkel, *Thermo. Acta*, **403**, 129 (2003).

Chapter 11

Effect of Vanadium-Ion Implantation on the Crystallization Kinetics and Phase Transformation of Electrospun TiO_2 Nanofibers

The influence of vanadium-ion implantation on the thermal response of electrospun amorphous TiO_2 nanofibers was studied with reference to structural phase transformation behavior, using in situ synchrotron radiation diffraction (SRD) measurements from room temperature to 1000°C. Analysis of the SRD data provided activation energies for amorphous-to-crystalline TiO_2 (anatase and rutile), and anatase-to-rutile transformations, and also assessments of the influence of V-ion implantation on microstructure development during calcination using estimates of crystallite size and microstrain. Non-implanted nanofibers were initially amorphous, with crystalline anatase first appearing at 600°C, followed by rutile at 700°C. The corresponding activation energies were 69(17) kJ/mol for the amorphous-to-crystalline TiO_2 transformation and 129(5) kJ/mol for the anatase-to-rutile transformation. V-ion implantation lowered the temperature at which each crystalline phase first appeared, with both phases being initially observed at 500°C, and with the anatase-to-rutile transformation being accelerated relative to the

Nanostructured Titanium Dioxide in Photocatalysis
It-Meng Low, Hani Manssor Albetran, Victor Manuel de la Prida Pidal, and Fong Kwong Yam

ISBN 978-981-4877-07-7 (Hardcover), 978-1-003-14853-1 (eBook)
www.jennystanford.com

non-implanted sample. The effect of V-ion implantation is seen through the substantial reduction in activation energies, which are 25(3) kJ/mol for amorphous-to-crystalline TiO_2 and 16(3) kJ/mol for anatase-to-rutile transformations.

11.1 Introduction

The photocatalytic activity of titania (TiO_2) has become well known over the last three decades, and numerous investigations have been conducted to improve the efficiency of photocatalysis [1, 2]. TiO_2 is an excellent photocatalyst and a competitive candidate for many photocatalytic applications. These applications depend on the morphology, particle size, crystallinity, and phase composition of the photocatalyst [1]. The main attractions of using TiO_2 as a photocatalyst are its high photoactivity, photodurability, nontoxicity, high oxidizing power, high chemical stability, mechanical robustness, and relatively low cost [3–5].

TiO_2 has three polymorphs: anatase, rutile, and brookite [6–8]. The anatase polymorph generally has higher photoactivity than the other polymorphs, but mixtures of anatase and rutile have been generally reported to exhibit superior photoactivity than either anatase or rutile alone [9–11]. However, Chuangchote et al. claimed that the TiO_2 nanofiber photocatalyst reported lower activity when both rutile and anatase phases were present [12].

TiO_2 is an n-type semiconductor with a 3.0–3.2 eV band gap (E_g) [13–15]. It is activated by UV irradiation, about 4% of natural solar light, which is one of the major factors limiting the efficiency of TiO_2 photocatalysis [2, 16]. Considerable effort has been conducted over the last two decades to developing modified TiO_2 that becomes active under visible light irradiation [2, 17–20]. Both metal and nonmetal doping of TiO_2 are cost-effective methods for narrowing its band gap. Visible light photoactivity of metal- and nonmetal-doped TiO_2 can be achieved by creating a new energy level between the conduction band and the valence band [16].

Various methodologies have been proposed for doping titanium dioxide, including ion-assisted sputtering, hydrothermal, sol–

gel, and ion implantation [2, 21]. Metal-ion implantation using an accelerated metal-ion beam has been applied to modify the electronic structure of the TiO_2 semiconductor [19–23]. Depending on the type and amount of metal used, metal-ion implantation can improve the electronic properties of TiO_2 photocatalysts by extending the absorption range up to the visible light region [19–23].

The presence of foreign ions can affect the kinetics of the anatase-to-rutile transformation at elevated temperatures, which is considered in this study. In general, ions that increase the oxygen vacancy concentration will enhance the phase transition. By contrast, ions that increase the concentration of interstitial titanium will tend to inhibit the transformation [1, 7].

Nanostructured TiO_2 with various morphologies (e.g., nanopowders [9], nanoparticles [24], nanosquares [2], nanowires [25], thin films [20, 26], nanorods [25], nanosheets [15], nanotubes [19, 25, 27], and nanofibers [12, 28–30]) can provide a large specific surface area for increasing the photodegradation rate since photocatalytic reactions mainly take place on the surface of the TiO_2 photocatalyst [15, 26, 28]. Various techniques such as anodization, evaporation, self-assembly, and electrospinning have been used to synthesize nanostructured TiO_2 [29]. Among these techniques, TiO_2-based nanofibers fabricated by electrospinning are of much interest as the technique is a simple, straightforward, and cost-effective technique for using at the industrial level. This technique employs a sol–gel solution of the polymer and a solvent, and the amorphous TiO_2 precursor is ejected through a needle under a strong electric field [12]. Using the sol–gel method, highly photocatalytic electrospun TiO_2 nanofibers doped with V have been synthesized by Zhang et al. [31]. However, the use of ion implantation to dope electrospun TiO_2 nanofibers with V ions has yet to be attempted.

For this study, TiO_2 nanofibers were prepared from titanium isopropoxide sol–gel precursors, synthesized using electrospinning, and then doped with V ions using ion implantation. The effects of V-ion implantation on crystallization kinetics, phase transformations, and activation energies of electrospun TiO_2 nanofibers were investigated using in situ high-temperature SRD over the temperature range of

25–1000°C. The electrospun TiO_2 nanofibers were characterized by means of scanning electron microscopy (SEM), energy dispersive X-ray spectroscopy (EDS), high-resolution transmission electron microscopy (HRTEM), and X-ray photoelectron spectroscopy (XPS). This chapter reports the first attempt at preparing V-ion-implanted electrospun TiO_2 nanofibers and focuses on the TiO_2 phase transformation and crystallization kinetics.

11.2 Results and Discussion

11.2.1 Microstructures of Electrospun TiO_2 Nanofibers

A typical SEM image of the as-prepared TiO_2/PVP nanofibers before thermal treatment is shown in Fig. 11.1(a). The diameters of electrospun nanofiber diameters from the SEM images were measured by ImageJ® software (version 1.48e) developed by the National Institutes of Health (NIH), USA. The randomly oriented TiO_2/PVP composite nanofibers had smooth surfaces. The corresponding mean diameter and standard deviation for 40 nanofibers being 665 ± 330 nm.

After the in situ high-temperature SRD measurements at 900°C for non-implanted material and at 1000°C for the V-implanted material, the morphologies of the nanofibers were re-examined after cooling to room temperature, using the same SEM conditions. Figures 11.1(b) and (c) show the SEM images of non-implanted and V-implanted materials, respectively. The mean diameters and standard deviation for 40 nanofibers after thermal treatment were much smaller, being 351 ± 145 nm for non-implanted and 358 ± 155 nm for V-implanted nanofibers. It should be noted that the mean diameters of non-implanted and V-implanted nanofibers are about the same within the error or scatter of the standard deviation. The average diameters of thermally treated nanofibers were much smaller than observed for the as-prepared TiO_2/PVP nanofibers, representing diametric shrinkage of ~47% in non-implanted nanofibers and ~46% in the implanted nanofibers. The shrinkage is attributed to the evaporation of the solvents and loss of the PVP polymer, and titanium isopropoxide (TIP) during thermal treatment.

Figure 11.1 SEM micrographs of (a) electrospun TiO_2 nanofibers prior to heating; (b) after thermal treatment at 900°C and cooling to room temperature; and (c) after V-ion implantation and thermal treatment at 1000°C and cooling to room temperature.

The qualitative and quantitative EDS analyses were determined using Oxford Instruments INCA (version 5.05) software. The EDS was collected from a small point. Figure 11.2(a) gives an EDS spectrum for the as-prepared TiO_2/PVP material, showing strong signatures for Ti, O, and C, with the C content being due to the PVP binder. The quantitative EDS analysis is approximated for electrospun nanofibers because the EDS has a large interaction volume, which can be greater than the fiber diameter size. The quantitative EDS analysis for the as-prepared TiO_2/PVP indicated 42.48% C k, 38.38% O k, and 19.13% Ti k. Figure 11.2(b) shows an EDS spectrum for V-implanted nanofibers after cooling from 1000°C, which was essentially the same for non-implanted electrospun TiO_2 nanofibers after cooling from 900°C. The absence of the C peak in Fig. 11.2(b) indicates complete loss of organic material. The quantitative EDS analysis of V-ion-implanted nanofiber indicated 66.18% O k and 33.82% Ti k, thus confirming the presence of TiO_2 in the material. The V K_β signature (5.43 keV) was not detected for the ion-implanted material due to the low V concentration, noting also that the Ti K_β peak (4.93 keV) and V K_α peak (4.95 keV) have overlapping EDS peaks.

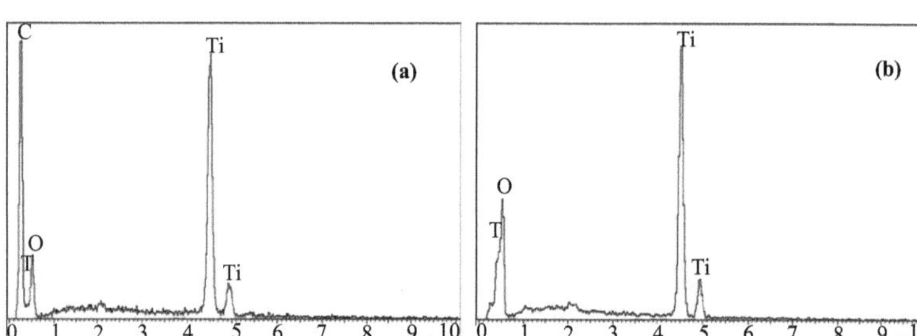

Figure 11.2 EDS spectra of (a) electrospun TiO_2 nanofibers prior to heating and (b) after V-ion implantation and cooling to room temperature from 1000°C.

11.2.2 HRTEM Imaging of Calcined TiO_2 Nanofibers

Figure 11.3(a) shows a typical TEM image of V-implanted material after thermal treatment at 1000°C. The diameters of nanofibers are in agreement with the values obtained from SEM analysis [Fig. 11.1(b)]. The presence of V ions is not obvious from the image because the atomic numbers for ^{22}Ti and ^{23}V are very similar.

Figures 11.3(b) and (c) show polycrystalline rutile composed of randomly oriented crystallites.

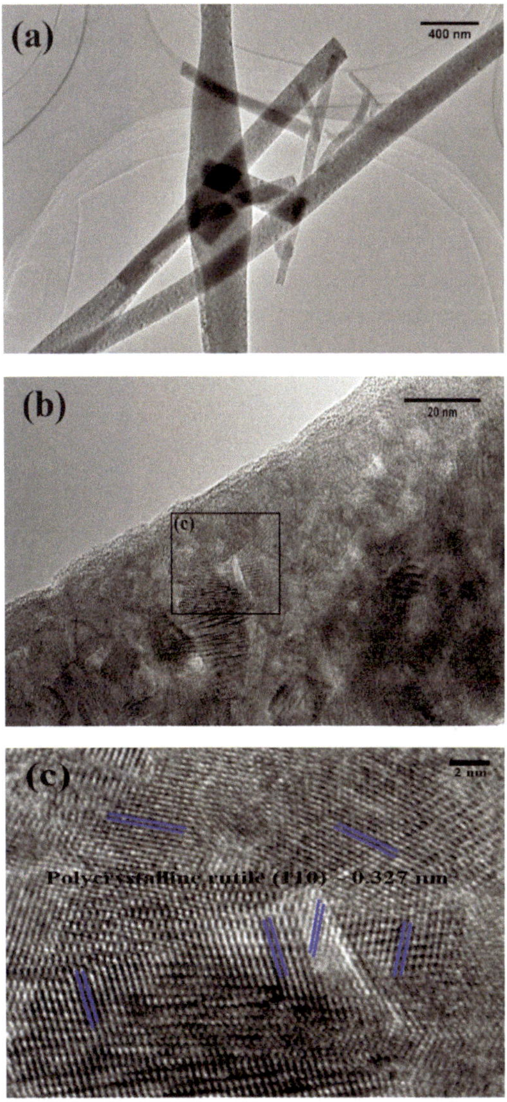

Figure 11.3 TEM images of V-ion-implanted electrospun TiO_2 nanofibers after the high-temperature SRD experiment, followed by cooling to room temperature from 1000°C: (a) low-resolution image, (b) and (c) lattice-resolved HRTEM images, which show the existence of polycrystalline rutile material, with the d-spacing between adjacent (110) with lattice planes being ~0.327 nm.

11.2.3 X-ray Photoelectron Spectroscopy

Figure 11.4 illustrates the XPS spectra of the non-implanted and V-implanted material at room temperature after the in situ SRD measurements. O, Ti, N, and C features are observed, according to the photoelectron peaks at binding energies of 532, 461, 399, and 284 eV, respectively. The peak located at 284 eV corresponds to residual carbon from the adventitious hydrocarbon coming from the XPS instrument and the N_{1s} peak is attributed to impurities in the XPS equipment. The appearance of Ti and O for the non-implanted material [Fig. 11.4(a)] is that expected for TiO_2. There are two spin orbital components for Ti, where $Ti_{2p1/2}$ is located at 461 eV and $Ti_{2p3/2}$ at 455 eV. The splitting between them is ~6 eV, indicating the presence of Ti^{4+} ions in the non-implanted TiO_2 [3]. The XPS plot for the implanted material in Fig. 11.4(b) clearly shows the O_{1s} feature, as in the corresponding peak for the non-implanted sample. However, the twin Ti_{2p} peaks are barely seen for the implanted material, which is attributed to a substantial reduction in Ti^{4+} concentration through V-ion implantation. The Ti_{2p} spectrum in Fig. 11.4(c), collected over a longer period, shows that there remain some Ti^{4+} ions in the implanted material but much less than the concentration seen in the non-implanted material. Figure 11.4(d) shows the XPS spectrum in the vicinity of the V_{2p} position for the V-implanted material. It is further evident from Figs. 11.4(c) and 11.4(d) that V-ion implantation has substantially diminished the Ti^{4+} concentration. In addition, it is seen from Fig. 11.4(d) that the V concentration is insufficient to unambiguously observe a V_{2p} peak.

XPS quantitative analysis, which is usually performed in the low-resolution mode, cannot provide evidence of the presence of V in the ion-implanted sample because XPS peaks that originate from elements with low concentration are not clearly visible. V was detected using high-resolution XPS, but this technique is applied mainly for qualitative analysis rather than quantitative analysis. The V_{2p} line spectrum of V appears at ~513 eV. If we assume that the inelastic mean free path (IMFP) of scattered electrons for V is ~1.25 nm, then a mean depth of analysis of approximately 3.75 nm is obtained.

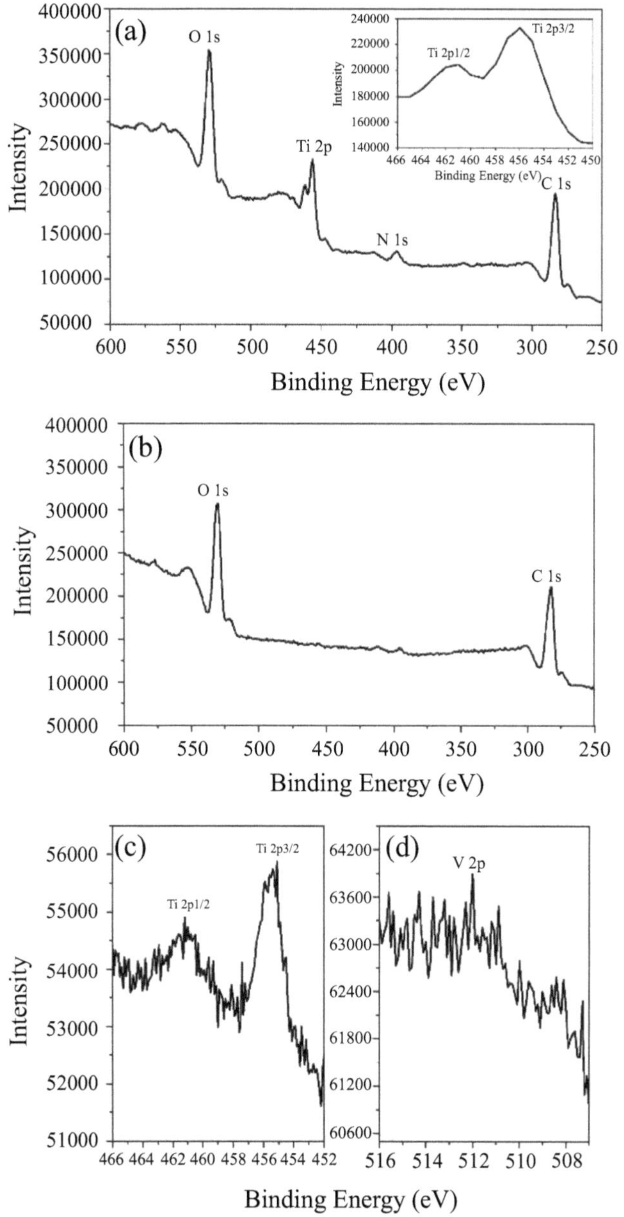

Figure 11.4 The XPS spectra of (a) non-implanted electrospun TiO_2 nanofibers after thermal treatment at 900°C, (b) V-ion-implanted material after thermal treatment at 1000°C, and high-resolution spectra for the implanted material showing the spectra for (c) Ti_{2p} and (d) V_{2p}.

The RBS results in Fig. 11.5 show the elemental concentration versus depth for the V-implanted TiO_2 fibers. A maximum V concentration of 4 at% has been implanted in TiO_2 fibers down to a depth of about 400 monolayers or 70 nm from SRIM prediction.

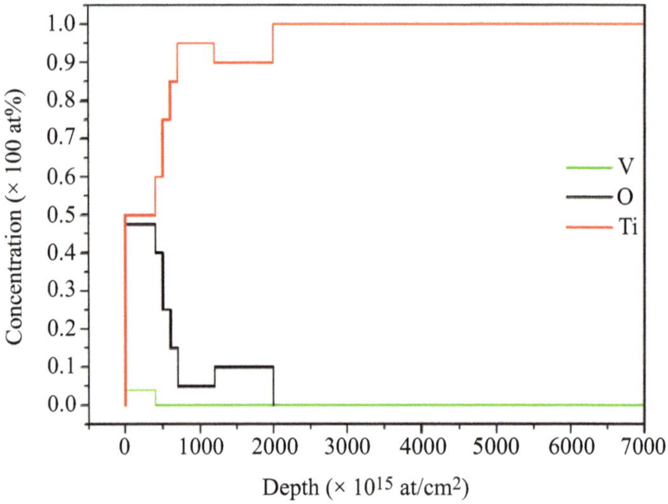

Figure 11.5 Composition depth profiles in V-implanted TiO_2 nanofibers.

11.2.4 Effect of Ion Implantation on Phase Transitions

Figure 11.6 illustrates stacked SRD plots for the non-implanted and V-implanted materials over the temperature range of 25–900°C. The Pt peaks in the SRD patterns come from the platinum holder. The SRD patterns for both the non-implanted and implanted samples exhibit prominent "amorphous" humps for the range between 25°C and 300°C, which disappear at 400°C. This behavior is attributed to amorphous PVP and other organic substances evaporating completely by 400°C, which was very similar to the interpretation of TGA spectra for TIP/PVP nanofibers from room temperature to 400°C by Park et al. [29, 30]. The SRD patterns for both non-implanted and V-ion-implanted materials indicate the presence of anatase and rutile crystalline phases at elevated temperatures, which are TiO_2 phases and contain Ti and O elements. This confirms the signatures for Ti and O elements in the EDS and XPS results.

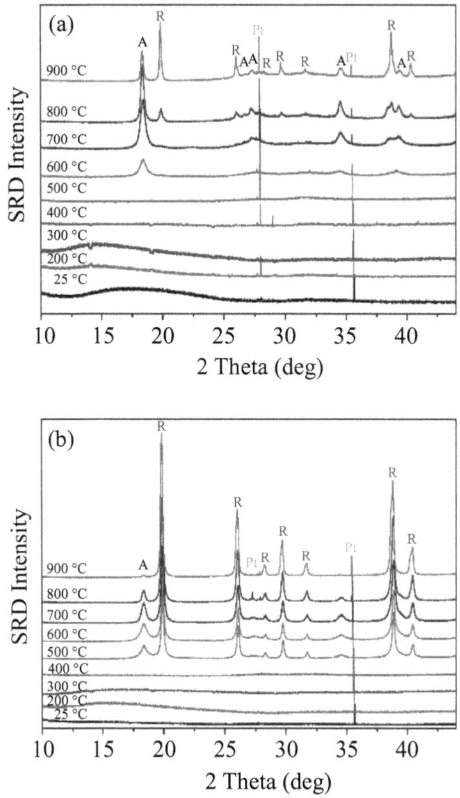

Figure 11.6 Stacked SRD plots for the temperature range 25–900°C of (a) non-implanted material and (b) after V-ion implantation [Legend: anatase (A), rutile (R), and platinum holder peaks (P_t)].

TiO_2 was initially amorphous but then partly crystallized to form anatase and rutile at elevated temperatures. In the non-implanted material [Fig. 11.6(a)], anatase and rutile were first observed at 600°C and 700°C, respectively; whereas, in V-implanted material [Fig. 11.6(b)], anatase and rutile were first observed at 500°C, which is 100°C lower for anatase and 200°C lower for rutile than observed in the non-implanted material. These results are consistent with previously reported observations of non-implanted and Cr-ion-implanted anodized TiO_2 nanotubes whereby Cr-ion implantation lowered the crystallization temperature of anatase from 600°C to 400°C and rutile from 600°C to 500°C [14]. The observations are consistent with V ions entering the TiO_2 crystal structure [33], through the implanted ions occupying the Ti sub-lattice substitutionally, rather than interstitial positions. The presence of interstitial ions inhibits the anatase-to-rutile transformation; however, substitutional ions may inhibit or accelerate the phase transformation depending on the substitutional ion types [7]. The relative ease by which anatase transforms to rutile in the V-implanted material appears to be due to the relatively small radius of the V ions (0.058 nm), which partly replace the Ti ions (0.061 nm), thus causing some relaxation of the Ti bonding environment through lessening of structural rigidity. This behavior of easy rearrangement of the atoms in the anatase and rutile lattices has also been reported with respect to their influence on oxygen vacancies when electrospun TiO_2 nanofibers were heated in an argon atmosphere [32].

Figure 11.7 shows typical SRD Rietveld residual plots for data measured in non-implanted and V-ion-implanted electrospun TiO_2 nanofibers at 800°C. In non-implanted material, R_{wp} = 13.0%, R_{exp} = 3.7%, R_B (anatase) = 4.6%, and R_B (rutile) = 7.3% and for the implanted material, R_{wp} = 13.6%, R_{exp} = 3.1%, R_B (anatase) = 5.9%, and R_B (rutile) = 5.8%. The goodness-of-fit (GOF) values are 12.2 for non-implanted and 19.2 for implanted nanofibers, which are acceptable refinements given the focus of the data analysis.

The absolute weight percentages of amorphous TiO_2 have been determined using the background intensities of SRD patterns measured at an angle of 17°. The background levels for the SRD patterns at 500°C for non-implanted TiO_2 and 400°C for V-ion-implanted TiO_2 were used as the initial background intensity B_o, corresponding to 100% amorphous TiO_2 and 0% crystalline TiO_2. The background intensities were corrected for minor variations in incident beam intensity during data collection. The SRD background level decreased with temperature due to the amorphous-to-crystalline TiO_2 transformation.

The value of final background intensity B_f and the wt_{am}% for both non-implanted and V-ion-implanted TiO_2 (Fig. 11.8) were estimated by fitting the plots of the intensities of the backgrounds versus temperature data with the function:

$$B(T) = B_f + ae^{-bT} \tag{11.1}$$

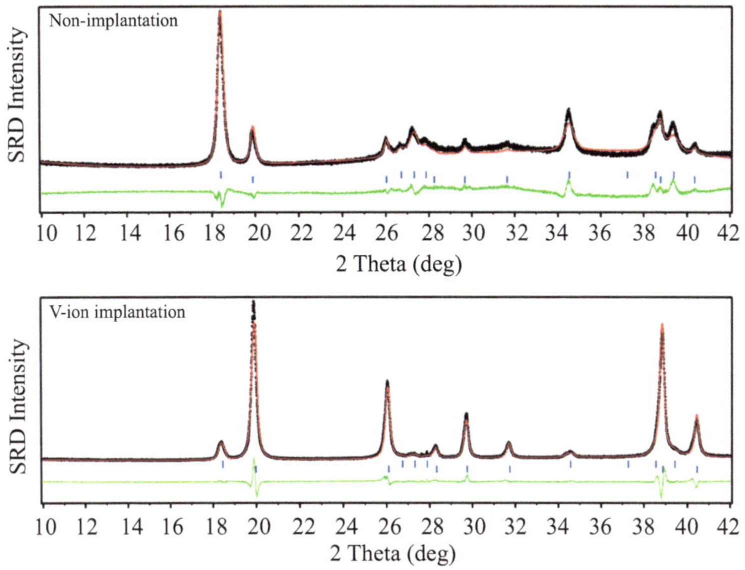

Figure 11.7 SRD Rietveld difference plots for data measured in non-implanted and V-ion-implanted material at 800°C. Measured patterns are indicated by black crosses, and calculated patterns by solid red lines. The green residual plot shows the difference between the calculated and the measured patterns. The peak positions for anatase and rutile are indicated by the top and bottom blue bars, respectively.

The final background intensity (B_f) estimates were 0 and 110 counts, and the amorphous transformation rates (b) were −0.0026°C^{-1} and −0.0042°C^{-1} for non-implanted and ion-implanted material, respectively.

Figure 11.9 shows the influence of V-ion implantation on the temperature dependence of the levels of amorphous TiO_2, crystalline anatase, and crystalline rutile. It is evident from the plots that implantation accelerates the transformations for amorphous material to anatase and for anatase to rutile also. In non-implanted nanofibers, only 6% of amorphous TiO_2 had transformed to anatase crystalline phase by 600°C, whereas ~48% of the amorphous material had transformed in the implanted material at this temperature. The temperature dependences of the anatase and rutile concentrations further underline the influence of implantation in accelerating the anatase-to-rutile transformation. These results are consistent with there being sequential transformation of

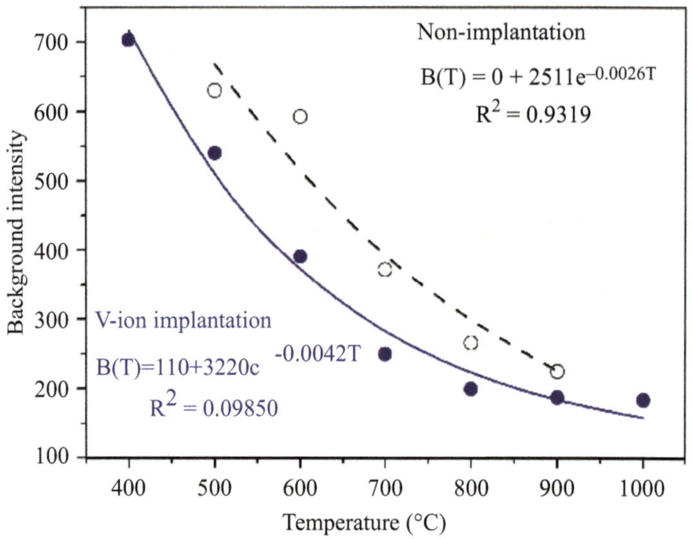

Figure 11.8 Variation with temperature of the SRD background intensity for the non-implanted and implanted material as fitted with exponential functions.

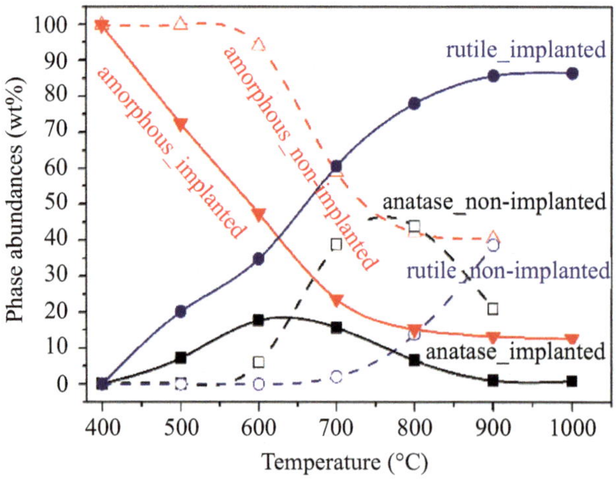

Figure 11.9 Phase abundances of amorphous TiO_2, anatase, and rutile in the temperature range of 400–1000°C for non-implanted and V-ion-implanted material.

amorphous to anatase and then anatase to rutile. In non-implanted material, anatase first appears at 600°C and its concentration then peaks (ca. 46% by weight) at 750°C; and the formation of rutile from anatase is first seen at 700°C and increased to ca. 39% at 900°C. Formation of anatase and then rutile occurs much more readily in the implanted material, with the anatase concentration peaking (ca. 18% by weight) at only 625°C. The rapid formation of the rutile phase after V-ion implantation is evident, and by 1000°C, the implanted material is principally rutile (ca. 87% by weight).

11.2.5 Crystallization Kinetics Modelling

Table 11.1 shows the average activation energies for the amorphous-to-crystalline TiO_2 (anatase plus rutile) and anatase-to-rutile transformations for non-implanted and implanted material, using Arrhenius and Matusita et al. equations (Fig. 11.10). For non-implanted material, the activation energy for the amorphous-to-crystalline TiO_2 transformation was determined to be 69(17) kJ/mol, and anatase-to-rutile transformation was 129(3) kJ/mol. The estimated activation energy for anatase-to-rutile transformation is substantially lower than reported for other pure TiO_2 samples, e.g., 264 kJ/mol for TiO_2 nanotubes, 213 kJ/mol for TiO_2 powder, and 213 kJ/mol for TiO_2 membrane [34–36]. For V-ion implantation, the apparent activation energies had reduced substantially, being only 25(3) kJ/mol for the amorphous-to-crystalline TiO_2 transformation and 16(3) for the anatase-to-rutile transformation. Thus, V implantation substantially lowered the activation energies for the amorphous-to-crystalline TiO_2 and anatase-to-rutile transformations. This is consistent with reports in the literature that the presence of impurity ions and the ion concentration can affect the activation energy for TiO_2 transformations [35, 36].

Table 11.1 Phase transformation activation energies (kJ/mol) for non-implanted and V-ion-implanted materials from non-isothermal SRD data analyses using the Arrhenius and Matusita et al. equations

	Amorphous-to-Crystalline TiO_2			Anatase-to-Rutile		
	Arrhenius Eq	Matusita et al.	Average	Arrhenius Eq	Matusita et al.	Average
Non-implanted	65(23)	73(24)	69(17)	122(9)	136(2)	129(5)
V-ion Implanted	19(4)	30(5)	25(3)	7(2)	24(5)	16(3)

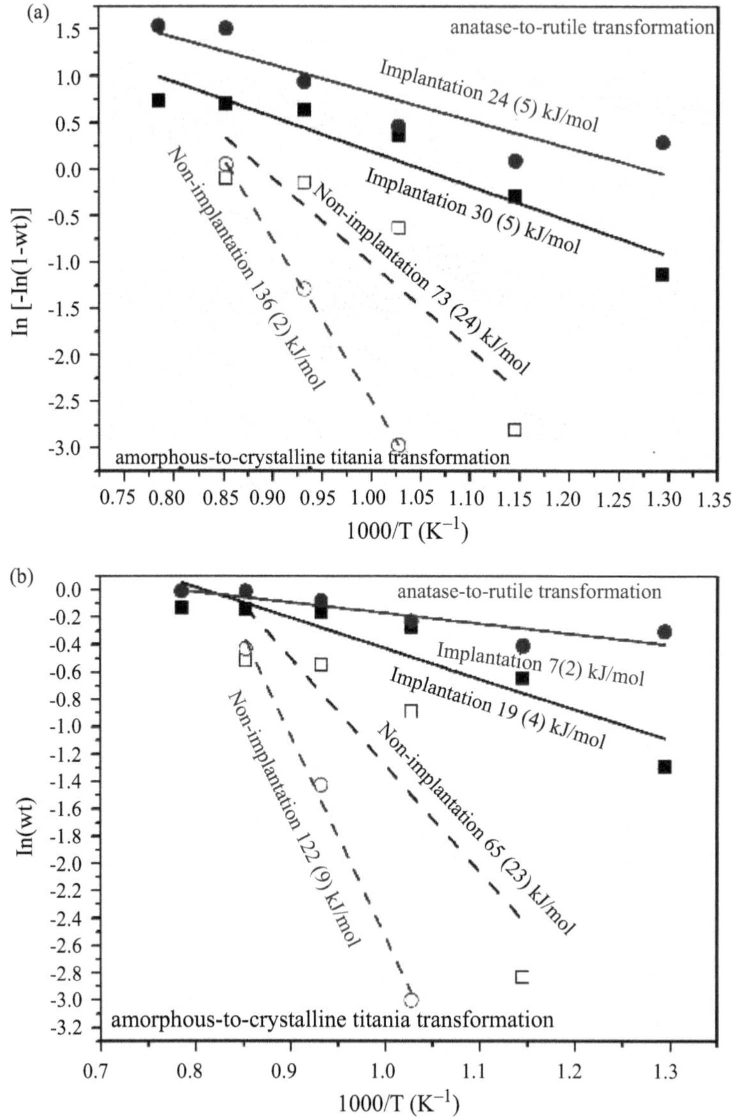

Figure 11.10 The estimated values of non-isothermal activation energies for amorphous-to-crystalline TiO_2 and anatase-to-rutile transformations were calculated from the corresponding slopes of (a) variation of ln [−ln (1 − wt)] against reciprocal temperature $1/T$ using Matusita et al. and (b) variation of ln (wt) against reciprocal temperature $1/T$ using the Arrhenius equation.

11.2.6 Microstructure Development

The results from the Williamson–Hall plot analyses are provided for both crystallite size and nonlinear strain. The plots of crystallite size versus temperature are given in Fig. 11.11, and the nonlinear strains using SRD integral breadth measurements are provided in Table 11.2. Both crystallite size and strain contribute substantially to line broadening. The crystallite size results show that the anatase and rutile crystalline phases for the non-implanted material are in the nanoscale range of 9–34 nm for the temperatures considered. The plots for anatase show that implantation does not have a discernible influence on the temperature dependence of crystalline anatase growth following nucleation from the amorphous phase. The temperature dependence of the rutile crystallite size and, therefore, the grain growth are very similar to that for anatase from 800°C, with there being no indication of rutile crystallite growth below 800°C. Comparing the SRD size results with the TEM images in Fig. 11.3 for the implanted material heated to 1000°C, the detail within the fiber microstructure in Fig. 11.3(a) points to there being rutile grains in the vicinity of 30–50 nm, which is consistent with the SRD size estimate of 58 nm in Table 11.2. The estimated rate of grain growth with temperature points to grain growth is being driven by crystallization of material onto existing grains from the surrounding amorphous TiO_2 material as the temperature is raised.

Table 11.2 Nonlinear strain estimates from Williamson–Hall plots, using SRD integral breadth measurements

Temperature (°C)	Nonlinear Strain (%)			
	Non-Implanted		Implanted	
	Anatase	Rutile	Anatase	Rutile
500			<0.35	0.07
600	0.51		0.39	0.15
700	0.27		0.24	0.11
800	0.23	0.09	0.10	0.07
900	0.32	0.09	0.29	0.17
1000				0.21
Mean	0.33	0.09	0.26	0.13

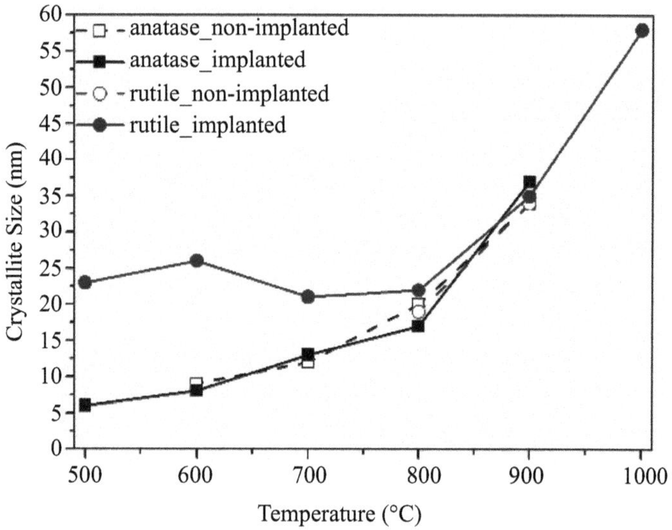

Figure 11.11 Williamson–Hall plot estimates of the effect of ion implantation on the average crystallite size for the anatase and rutile crystalline phases over the temperature range of 500–1000°C for non-implanted and V-ion-implanted material.

The strain estimates show (i) substantially higher nonlinear strain in anatase (ca. 0.3%) than in rutile (ca. 0.1%) for both non-implanted and V-implanted materials, but (ii) no increase in strain for both phases in the implanted material at elevated temperatures. The greater strains observed for anatase are consistent with the findings of Nicula et al., who reported that anatase exhibits larger strain values than rutile in TiO_2 powders from 400°C to 850°C [37].

11.3 Conclusion

The effect of V-ion implantation on the crystallization of electrospun amorphous TiO_2 nanofibers was investigated using in situ high-temperature SRD. The SRD results were used to study the crystallization kinetics in terms of the influence of implantation on the activation energies for the amorphous-to-crystalline TiO_2 and anatase-to-rutile transformations.

The following principal conclusions are drawn from the study:

- The anatase crystalline phase was first observed at 600°C and rutile at 700°C for non-implanted TiO_2, whereas anatase and rutile were both initially observed at 500°C for V-implanted TiO_2.

- The temperature dependence of both the amorphous-to-crystalline TiO_2 and anatase-to-rutile transformations was much more pronounced in V-implanted material, with the composition of the implanted material predominantly comprising crystalline rutile by 1000°C.
- The average activation energies for the amorphous-to-crystalline TiO_2 transformation decreased from 69(17) kJ/mol to 25(3) kJ/mol following implantation, and from 129(5) kJ/mol to 16(3) kJ/mol, for the anatase-to-rutile transformation.
- XPS analysis showed that the concentration of Ti^{4+} ions in TiO_2 is substantially reduced by implantation, which appears to account for the relative ease for the transformations in the implanted material.
- The crystallite sizes for both anatase and rutile for non-implanted and anatase for V-ion-implanted electrospun TiO_2 nanofibers increased steadily with the temperature, indicating that grain growth is driven by crystallization of material onto existing grains from the surrounding amorphous TiO_2 as the temperature is elevated, rather than unit cell expansion.
- The strain estimates show substantially higher nonlinear strain in anatase than in rutile for both non-implanted and implanted material. There is no discernible increase in strain in the implanted material at elevated temperatures for either phase.

It was concluded that V-ion implantation had lowered the crystallization temperatures due to substitution of relatively small V ions for Ti in the TiO_2 crystal structure. Implantation of V ions is responsible for the dramatic reductions in activation energies.

Acknowledgments

The authors acknowledge financial support from the Australian Synchrotron (Powder Diffraction Beamline) (AS122/PDFI/5075) and the Australian Institute of Nuclear Science and Engineering (ALNGRA11135). H. Albetran is grateful to the College of Education, University of Dammam, for the financial support in the form of a

PhD scholarship. Spanish MINECO grants under projects MAT2010-20798-C05-04 and MAT2013-48054-C2-2-R are also gratefully acknowledged. Scientific support from the XPS laboratory of the University of Oviedo SCTs is recognized. The authors would like to thank Dr. Y. Dong for allowing the use of his electrospun machine, E/Prof. B. H. O'Connor for guidance on the Rietveld work, Ms. E. Miller for assistance with the SEM imaging, and Dr. X. Wang for assistance with TEM, Dr. J. Kimpton at the Australian Synchrotron for advising on instrumentation at Powder Diffraction Beamline, and Dr. M. Ionescu of ANSTO for helpful assistance with ion implantation and SRIM simulation of RBS data.

References

1. D. A. Hanaor and C. C. Sorrell, *J. Mater. Sci.*, **46**, 855 (2011).
2. C. Yu, L. Wei, X. Li, J. Chen, Q. Fan, and J. Yu, *Mater. Sci. Eng. B*, **178**, 344 (2013).
3. P. Zhang, C. Shao, X. Li, M. Zhang, X. Zhang, Y. Sun, and Y. Liu, *J. Hazard Mater.*, **237–238**, 331 (2012).
4. Q. Li, D. J. G. Satur, H. Kim, and H. G. Kim, *Mater. Lett.*, **76**, 169 (2012).
5. J. H. Huang and M. S. Wong, *Thin Solid Films*, **520**, 1379 (2011).
6. P. Manurung, Y. Putri, W. Simanjuntak, and I. M. Low, *Ceramics International*, **39**, 255 (2013).
7. R. D. Shannon and J. A. Pask, *J. Am. Ceram. Soc.*, **48**, 391 (1965).
8. D. W. Kim, N. Enomoto, Z. Nakagawa, and K. Kawamura, *J. Am. Ceram. Soc.*, **79**, 1095 (1996).
9. A. Natoli, A. Cabeza, A. G. Torre, M. A. Aranda, and I. Santacruz, *J. Am. Ceram. Soc.*, **95**, 502 (2012).
10. S. Kment, P. Kluson, V. Stranak, P. Virostko, J. Krysa, M. Cada, J. Pracharova, M. Kohout, M. Morozova, P. Adamek, and Z. Hubicka, *Electrochimica Acta*, **54**, 3352 (2009).
11. H. Li, W. Zhang, and W. Pan, *J. Am. Ceram. Soc.*, **94**, 3184 (2011).
12. S. Chuangchote, J. Jitputti, T. Sagawa, and S. Yoshikawa, *ACS Appl. Mater. Interfaces*, **1**, 1140 (2009).
13. S. Shang, X. Jiao, and D. Chen, *ACS Appl. Mater. Interfaces*, **4**, 860 (2012).
14. I. M. Low, H. Albetran, V. M. Prida, V. Vega, P. Manurung, and M. Ionescu, *J. Mater. Res.*, **28**, 304 (2013).

15. J. Liu, E. J. Nichols, J. Howe, and S. T. Misture, *J. Mater. Res.*, **28**, 424 (2013).

16. V. J. Babu, A. S. Nair, Z. Peining, and S. Ramakrishna, *Mater. Lett.*, **65**, 3064 (2011).

17. R. Asahi, T. Morikawa, T. Ohwaki, K. Aoki, and Y. Taga, *Science*, **293**, 269 (2001).

18. S. U. Khan, M. Al-Shahry, and W. B. Ingler, *Science*, **297**, 2243 (2002).

19. A. Ghicov, J. M. Macak, H. Tsuchiya, J. Kunze, V. Haeublein, L. Frey, and P. Schmuki, *Nano Lett.*, **6**, 1080 (2006).

20. M. Takeuchi, H. Yamashita, M. Matsuoka, M. Anpo, T. Hirao, N. Itoh, and N. Iwamoto, *Catalysis Lett.*, **67**, 135 (2000).

21. H. Yamashita, M. Harada, J. Misaka, M. Takeuchi, B. Neppolian, and M. Anpo, *Catalysis Today*, **84**, 191 (2003).

22. H. Yamashita, M. Harada, J. Misaka, M. Takeuchi, K. Ikeue, and M. Anpo, *J. Photochem. Photobiol. A: Chem.*, **148**, 257 (2002).

23. G. Impellizzeri, V. Scuderi, L. Romano, P. M. Sberna, E. Arcadipane, R. Sanz, M. Scuderi, G. Nicotra, M. Bayle, R. Carles, F. Simone, and V. Privitera, *J. Appl. Phys.*, **116**, 173507 (2014).

24. C. Wessel, R. Ostermann, R. Dersch, and B. M. Smarsly, *J. Phys. Chem. C*, **115**, 362 (2011).

25. I. M. Low, B. Curtain, M. Philipps, Q. Z. Liu, and M. Ionescu, *J. Aust. Ceram. Soc.*, **48**, 198 (2012).

26. J. S. Lee, T. J. Ha, M. H. Hong, and H. H. Park, *Thin Solid Films*, **529**, 98 (2013).

27. R. Liu, L. S. Qiang, W. D. Yang, and H. Y. Liu, *Mater. Res. Bull.*, **48**, 1458 (2013).

28. W. Luo, X. Hu, Y. Sun, and Y. Huang, *J. Mater. Chem.*, **22**, 4910 (2012).

29. J. Y. Park and S. S. Kim, *Met. Mater. Int.*, **15**, 95 (2009).

30. J. Y. Park, J. J. Yun, C. H. Hwang, and I. H. Lee, *Mater. Lett.*, **64**, 2692 (2010).

31. Z. Zhang, C. Shao, L. Zhang, X. Li, and Y. Liu, *J. Colloid. Interface Sci.*, **351**, 57 (2010).

32. H. Albetran, H. Haroosh, Y. Dong, V. M. Prida, B. H. O'Connor, and I. M. Low, *Appl. Phys. A*, **116**, 161 (2014).

33. G. C. Bond, A. J. Sarkany, and G. D. Parfitt, *J. Catalysis*, **57**, 476 (1979).

34. O. K. Varghese, D. Gong, M. Paulose, C. A. Grimes, and E. C. Dickey, *J. Mater. Res.*, **18** 156 (2003).

35. K. N. P. Kumar, J. Engell, J. Kumar, K. Keizer, T. Okubo, and M. Sadakata, *J. Mater. Sci. Lett.*, **14**, 1784 (1995).
36. K. N. P. Kumar, K. Keizer, and A. J. Burggraaf, *J. Mater. Chem.*, **3**, 917 (1993).
37. R. Nicula, M. Stir, C. Schick, and E. Burkel, *Therm. Acta*, **403**, 129 (2003).

Chapter 12

A Comparative Study on Crystallization Behavior, Phase Stability, and Binding Energy in Pure and Cr-Doped TiO_2 Nanotubes

Visible-light photocatalysis in doped TiO_2 is a cost-effective and sustainable technology. However, to make this an attractive viable technology, the design of TiO_2 photocatalyst is required to be capable of harnessing the energy of visible light. One possible solution is the doping of TiO_2 to reduce its band gap. In this chapter, the effect of Cr doping by ion implantation on the in situ crystallization and phase stability of anodic TiO_2 nanotubes at elevated temperature is described. Cr doping has dramatically reduced the fraction of anatase-to-rutile transformation and lowered the crystallization temperature of anatase from 600 to 400°C and rutile from 600 to 500°C. Ion-beam analysis by Rutherford backscattering spectrometry has confirmed the existence of Cr ions composition gradation in doped TiO_2 nanotubes. The real doping of Ti lattices with Cr ions was evidenced by the analyses of surface compositions and chemical states of the nanotubes using X-ray photoelectron spectroscopy.

Nanostructured Titanium Dioxide in Photocatalysis
It-Meng Low, Hani Manssor Albetran, Victor Manuel de la Prida Pidal, and Fong Kwong Yam

ISBN 978-981-4877-07-7 (Hardcover), 978-1-003-14853-1 (eBook)
www.jennystanford.com

12.1 Introduction

Titanium dioxide (TiO_2) is a wide-band-gap semiconductor with an energy of 3.0–3.2 eV. It is widely used in applications such as hydrogen production, gas sensors, photocatalytic activities, dye-sensitized solar cells, and photoelectrochemical cells because of its relative high efficiency and high stability [1]. However, due to its wide-band-gap energy, TiO_2 is active only under near-ultraviolet irradiation. Therefore, numerous studies have been carried out over the last 20 years to develop modified TiO_2 catalysts so that they are active under visible-light irradiation (>400 nm). One of the most studied methods is doping the TiO_2 materials with metal ions (Fe, Ni, Co, Va, and Cr) [2–5] or nonmetallic elements (N, S, and C) [6–15]. For instance, Cr or N doping of TiO_2 can result in visible-light response by virtue of a narrowing in the band gap due to the mixing of p states of nitrogen with O 2p states. Among these doping methods, doping with transitional metals is one of the most efficient methods. Recent theoretical calculations showed that 3d transition metals such as Cr and Mn have good potential as a substitutional dopant for TiO_2 because of their ability to induce significant spin polarization and reduce the energy gap of rutile by forming intermediate bands with adequate curvatures and density of states [16–18]. Unlike defect states in the form of nearly straight horizontal lines within the energy gap, these curvy intermediate bands can act as stepping-stones to relay valence electrons to the conduction band [16, 17]. As a result, the optical absorption edge of TiO_2 is shifted to the visible or even IR spectral range of solar irradiance [18].

X-ray photoelectron spectroscopy (XPS) results showed that the doped nanotube samples contained N, C, B, and F elements, and the doped TiO_2 showed shift in the band gap transition down to 2.98 eV. The photocatalytic activity of doped TiO_2 was 1.61 times better than undoped TiO_2 [15]. Metals compounded on semiconductor materials increase charge-collection efficiency due to a much slower electron–hole recombination, giving rise to longer electron lifetime, which will result in an increasing interfacial electron-transfer process [1, 12].

Hitherto, the doping of TiO_2 has been widely synthesized using the sol–gel method [19–23]. However, ion implantation has now emerged as an alternative but an effective doping method to improve the separation of the photogenerated electron–hole pairs or to extend the wavelength range of the TiO_2 photoresponses into the visible region [24–29]. XPS measurements revealed that the implanted nitrogen species were mainly interstitial ones. The nitrogen concentration was increased with increasing the flux of ions, which could be controlled by adjusting the gas flow rate of the ion source, resulting in improved visible-light photocatalytic activities [25]. Higher visible-light photocatalytic efficiency was achieved with higher implanted nitrogen concentration [25, 26].

In this study, we present X-ray diffraction (XRD) results of the role of Cr doping on the in situ crystallization behavior and phase stability of TiO_2 nanotubes in the temperature range of 25–1000°C. The effects of Cr doping on the resultant microstructure, composition depth profile, and binding energy are discussed in terms of scanning electron microscopy (SEM), ion-beam analysis by Rutherford backscattering spectrometry, and XPS.

12.2 Results and Discussion

12.2.1 Crystallization Behavior

Figure 12.1 shows the XRD plots of undoped TiO_2 before and after thermal annealing in air at different temperatures ranging between 20 and 1000°C. The anodic TiO_2 nanotube arrays before thermal treatment were amorphous but eventually crystallized at 400°C to form anatase, which started to decrease in abundance but persisted up to 1000°C. A small amount of rutile started to form at 600°C, and its abundance increased rapidly after 800°C, which corresponds to the rapid decrease in anatase. This suggests that the anatase formed was metastable and began to transform into rutile at about 600°C.

Figure 12.2 shows the phase abundances of anatase and rutile as a function of temperature. A typical Rietveld profile fit for the XRD data at 800°C is shown in Fig. 12.3. The good refinement quality of diffraction data can be discerned by the acceptable fitting parameters of R_p =17.4, R_{wp} = 26.0, R_{exp} = 3.5, R_B (anatase) = 5, and R_B (rutile) = 10.

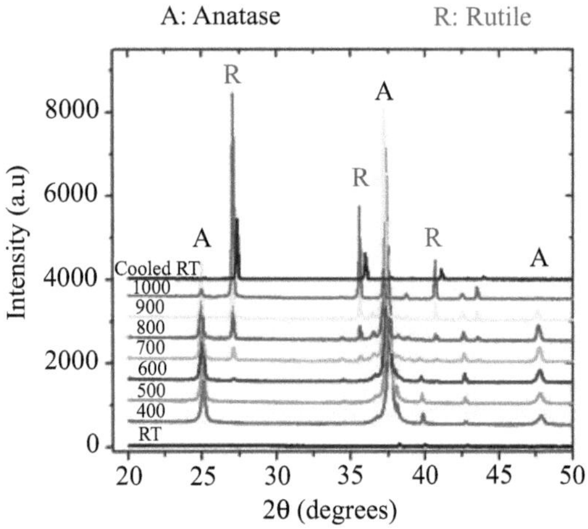

Figure 12.1 High-temperature XRD plots for pure TiO_2 showing the in situ crystallization of anatase and rutile from the amorphous TiO_2 in the temperature range of 20–1000°C.

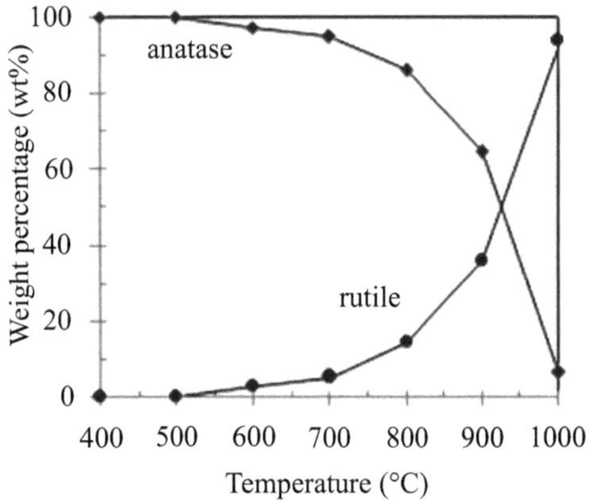

Figure 12.2 Phase abundances of anatase and rutile for pure TiO_2 in the temperature range of 20–1000°C.

There was also a distinct narrowing and sharpening in the (101) peak for anatase and (110) peak for rutile. The corresponding increase in crystallite size calculated from the Scherrer equation is shown in Fig. 12.4. The crystallite size of anatase was about 30 nm at 400°C and increased gradually with temperature to ~65 nm at 1000°C. The crystallite size of rutile formed at 600°C was about 35 nm, which grew rapidly and levelled off at 1000°C with a size of 70 nm.

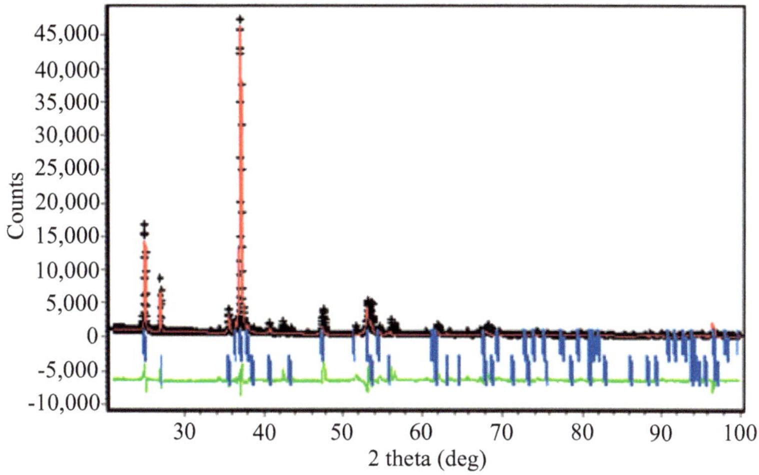

Figure 12.3 XRD profile plot for pure TiO_2 nanotubes at 800°C. Measured and calculated patterns are indicated by crosses (black) and solid line (red), respectively. Intensity differences between the two patterns are shown along the bottom of the plot (green). Vertical bars (blue) represent the allowable peak position for each of the phases (from top to bottom: anatase and rutile).

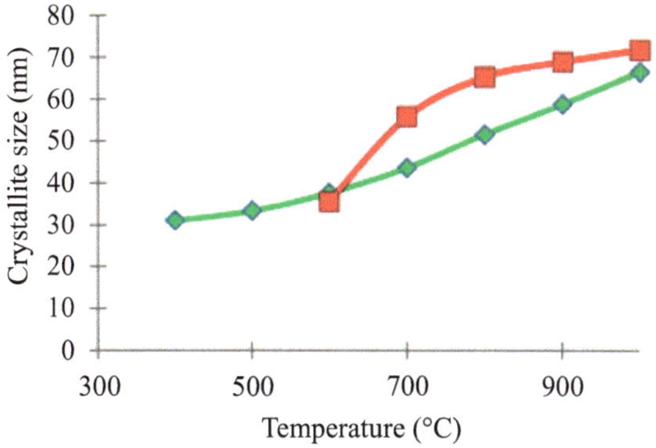

Figure 12.4 Variations in the crystallite size of anatase (♦) and rutile (■) as a function of temperature for pure TiO_2.

The fraction of anatase-to-rutile transformation in the samples annealed between 600 and 1000°C can be calculated from the followed equation [30]:

$$x = \frac{1}{1+0.8\left(\frac{I_A}{I_R}\right)}$$

where I_A is the X-ray integrated intensities of the (101) reflection of anatase at about 25.5°, while I_R is that of the (110) reflection of rutile at about 27.7°; x is the weight fraction of rutile in the nanotubes, which is calculated to be 4.4% for the nanotubes annealed at 600°C, 16.4% at 700°C, 30.3% at 800°C, 59.6% at 900°C, and 96.6% at 1000°C (see Table 12.1).

Table 12.1 Fractions of anatase-to-rutile transformation in pure and Cr-doped TiO_2 at various temperatures

Temperature (°C)	x (%) TiO_2	x (%) Cr-doped TiO_2
500	—	0.6
600	4.4	3.6
700	16.4	4.6
800	30.3	5.6
900	59.6	—
1000	96.6	—

Figure 12.5 shows the synchrotron radiation diffraction plots of Cr-doped TiO_2 nanotubes before and after thermal annealing in argon at 20–900°C. The TiO_2 nanotube arrays were initially amorphous but eventually crystallized to anatase at 300°C. The peak intensities of anatase increased rapidly when the annealing temperature rises from 300 to 400°C and continued to increase until 900°C. The formation of rutile commenced at 500°C, but its growth was sluggish where the peak intensities increased very slightly from 500 to 800°C. However, rutile disappeared at 900°C with the concomitant appearance of TiO_{2-x} where $x = 0.2$ (see Fig. 12.6). This suggests that rutile was unstable in argon atmosphere at 900°C and decomposed to form a nonstoichiometric titanium oxide and oxygen vacancies (O_v) as follows:

$$5TiO_2 \xrightarrow{900^\circ C/\text{argon}} Ti_5O_9 + O_v$$

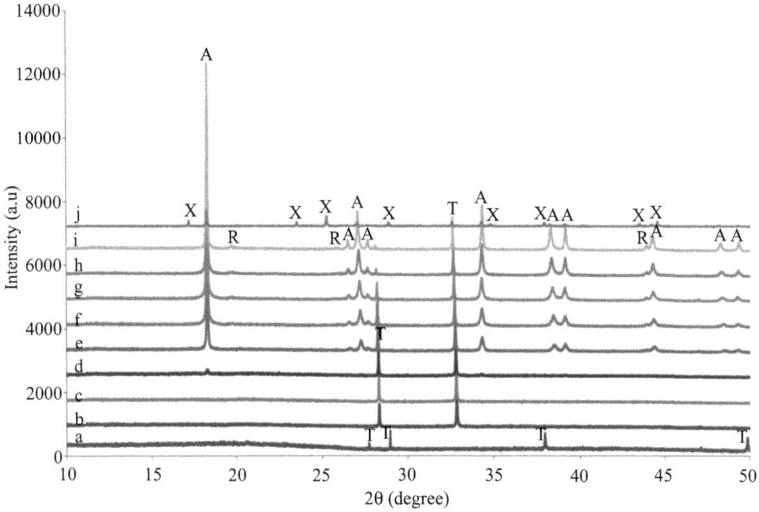

Figure 12.5 In situ synchrotron radiation diffraction plots of Cr-doped TiO_2 nanotubes showing the existence of anatase (A), rutile (R), Ti_5O_9 (X), and titanium (T) in the temperature range of 20–900°C (Legend: a = 20°C, b = 100°C, c = 200°C, d = 300°C, e = 400°C, f = 500°C, g = 600°C, h = 700°C, I = 800°C, j = 900°C).

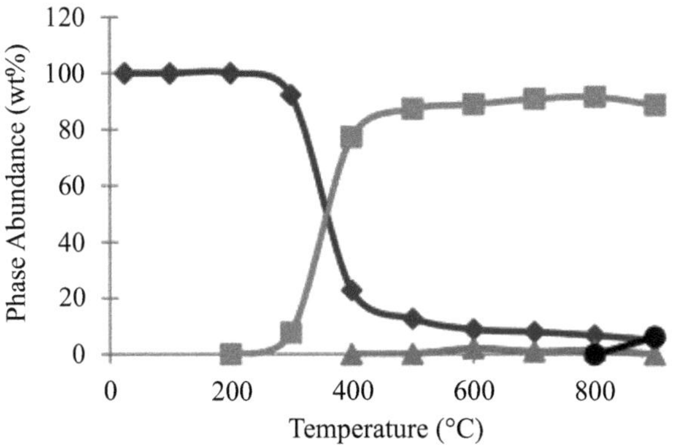

Figure 12.6 Variations in phase abundances as a function of temperature for Cr-doped TiO_2 nanotubes [Legend: titanium (♦); anatase (■); rutile (▲); TiO_{2-x} (●)].

In contrast, no decomposition of rutile was observed in undoped TiO_2 nanotubes (see Fig. 12.1). When compared to Cr-doped TiO_2, both anatase and rutile in pure TiO_2 crystallized at a higher temperature, i.e., 400 and 600°C, respectively. The fraction of anatase-to-rutile transformation in Cr-doped TiO_2 at 600–1000°C was computed to be 0.6% at 500°C, 3.6% at 600°C, 4.6% at 700°C, and 5.6% at 800°C. As shown in Table 12.1, these values are significant when compared to pure TiO_2.

Figure 12.7 shows the influence of annealing temperature on crystallite size for anatase and rutile in Cr-doped TiO_2. The crystallite size of anatase was about 55 nm at 400°C and gradually increased with temperature to 100 nm at 900°C. In contrast, the crystallite size of rutile formed was much smaller than that of anatase and peaked at 700°C with only 33 nm in size.

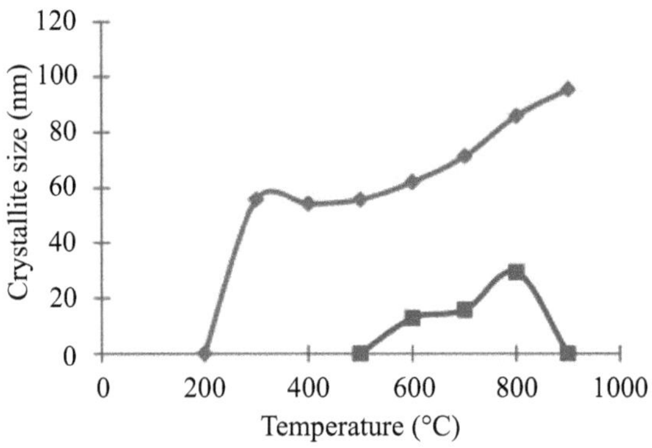

Figure 12.7 Variations in crystallite size (L) of anatase (♦) and rutile (■) as a function of temperature for Cr-doped TiO_2 nanotubes.

It appears that the presence of Cr doping has accelerated the formation of anatase and rutile at a lower temperature, probably through the facilitation of defects such as vacancies. It is widely accepted that the onset temperature of the anatase-to-rutile phase transformation and the rate at which it proceeds can be affected significantly by dopants, firing atmosphere, microstructure, sample morphology, and the presence of impurities in the material [1]. The process of the anatase-to-rutile transformation is known to occur via (a) the nucleation of rutile at point defects, oxygen vacancies, secondary phase inclusions, particle surfaces, and/or at (112) twin interfaces in anatase and (b) the subsequent consumption of the anatase phase by the growing rutile phase. The formation of rutile from the metastable phase anatase is reconstructive and so takes place through atomic rearrangement involving the breaking of two of the six Ti–O bonds in the TiO_6 octahedra [31]. As anatase transforms into rutile, significant grain growth takes place, resulting in lower surface area and thus poorer photocatalytic performance.

Transition metals (e.g., Cr) of variable valence are reported to enter the titania lattice and create oxygen vacancies through reduction effects as follows [1, 32]:

$$2M^{n+} + O^{2-} \rightarrow 2M^{(n-1)+} + \frac{1}{2}O_2 + O_v$$

where M denotes a transition metal atom and O_v denotes an oxygen vacancy. Once substituted for Ti^{4+}, the reduction of such species gives rise to new oxygen vacancies and so enhances rutile formation from metastable anatase through easing of structural rearrangement. These oxygen vacancies are believed to be responsible for crystallization of anatase at a lower temperature in Cr-doped TiO_2 as well as destabilization and subsequent decomposition of rutile to form nonstoichiometric TiO_{2-x}.

12.2.2 Microstructures and Formation Mechanisms of Nanostructured TiO_2

After anodizing in the electrolyte of NH_4F/ethylene glycol at 25°C for 20 h, a homogeneous layer of TiO_2 nanotubes formed on the Ti foil. Figure 12.8(a) shows the SEM image of the microstructure of amorphous TiO_2 nanotubes in the as-anodized sample. Upon annealing at 400°C for 2 h, vertically oriented and highly ordered arrays of anatase nanotubes with diameter of ~80 nm formed [Fig. 12.8(b)], which concurs well with previous studies on anodized TiO_2 [33, 34]. The EDS analysis of nanotubes in Spectrum 1 of Fig. 12.8(c) indicated 31.04% Ti and 68.96% O, thus confirming the presence of TiO_2 in the sample. Upon annealing at 900°C for 1 h, some anatase transformed to rutile, which appeared as elongated grains randomly distributed among the nanotubes shown in Fig. 12.8(d).

The microstructure of Cr-doped TiO_2 is shown in Fig. 12.9(a) where the morphology of nanotubes has been retained during the process of ion implantation. The existence of implanted Cr ions has been verified by the EDS analysis of a bright region in Spectrum 1 as shown in Fig. 12.9(b).

Figure 12.8 Scanning electron micrographs showing the microstructure of pure TiO_2 nanotubes at (a) room temperature, (b) 900°C, (c) 400°C for 2 h, and (d) the corresponding EDS analysis of composition in Spectrum 1.

Three simultaneously occurring processes can be ascribed to the formation of the nanotube arrays in fluoride-containing electrolyte [35]. The first involves the field-assisted oxidation of Ti metal to form TiO_2, which is followed by field-assisted dissolution of Ti metal ions in the electrolyte, and finally the chemical dissolution of Ti and TiO_2 in the presence of hydrogen and fluoride ions. Plausible pathways of chemical reaction occurring during anodization are as follows:

$$\mathrm{Ti\ (s)} \rightarrow \mathrm{Ti^{4+}\ (aq)} + 4\mathrm{e}^-$$

$$\mathrm{Ti^{4+}\ (aq)} + 4\mathrm{F^-\ (aq)} \rightarrow \mathrm{TiF_4\ (aq)}$$

$$\mathrm{TiF_4\ (aq)} + 2\mathrm{H_2O\ (aq)} \rightarrow \mathrm{TiO_2\ (s)} + 4\mathrm{HF\ (aq)}$$

However, it should be noted that the above-mentioned pathways are somewhat different to the model proposed by Macak et al. [36] for the oxide formation of anodic TiO_2 nanotubes. According to their model, Ti oxidation takes place at the metal/oxide interface through the following process:

$$Ti \rightarrow Ti^{4+} + 4e^-$$

Some of the Ti^{4+} ions react with O^{2-} ions that reach the metal/oxide interface by ionic migration across the TiO_2 barrier layer to produce titanium dioxide:

$$Ti^{4+} + 2O^{2-} \rightarrow TiO_2$$

Whereas some of them are directly complexated by F^- ions:

$$Ti^{4+} + 6F^- \rightarrow [TiF_6]^{2-}$$

Field-assisted chemical dissolution of TiO_2 takes place at the oxide/electrolyte interface as a result of the formation of water-soluble fluoride complexes:

$$TiO_2 + 6F^- \rightarrow [TiF_6]^{2-}$$

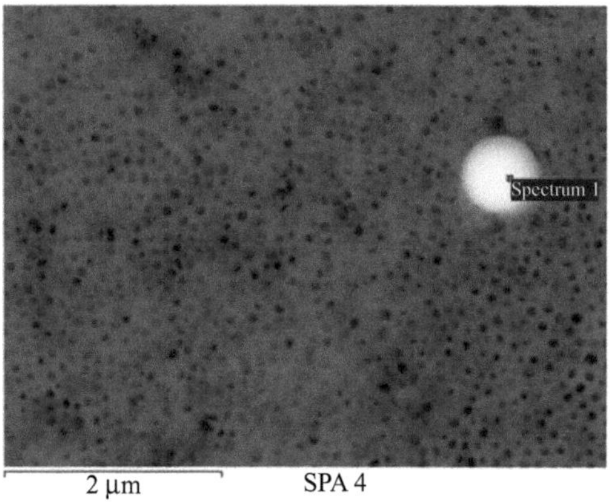

(a)

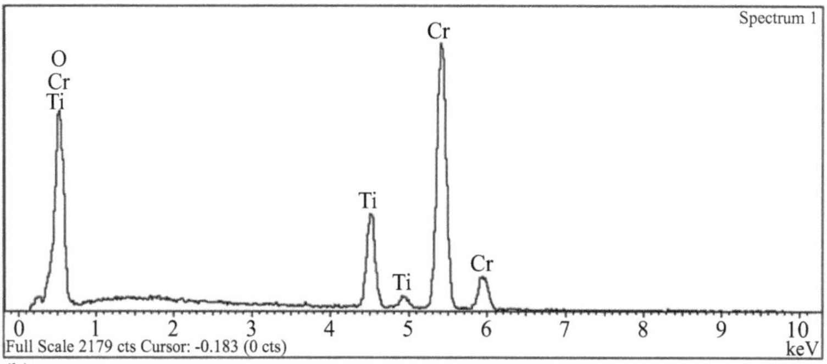

(b)

Figure 12.9 Scanning electron micrographs showing the microstructure of (a) Cr-doped TiO_2 and (b) EDS analysis of composition in Spectrum 1.

12.2.3 Composition Depth Profiles and Binding Energies

The RBS results in Fig. 12.10 show the concentration of elements (at%) versus depth (at/cm^2) for both the undoped and Cr-doped samples. The depth profiles of Ti and O for undoped TiO_2 are shown in Fig. 12.10(a) where the composition at the near-surface layer is

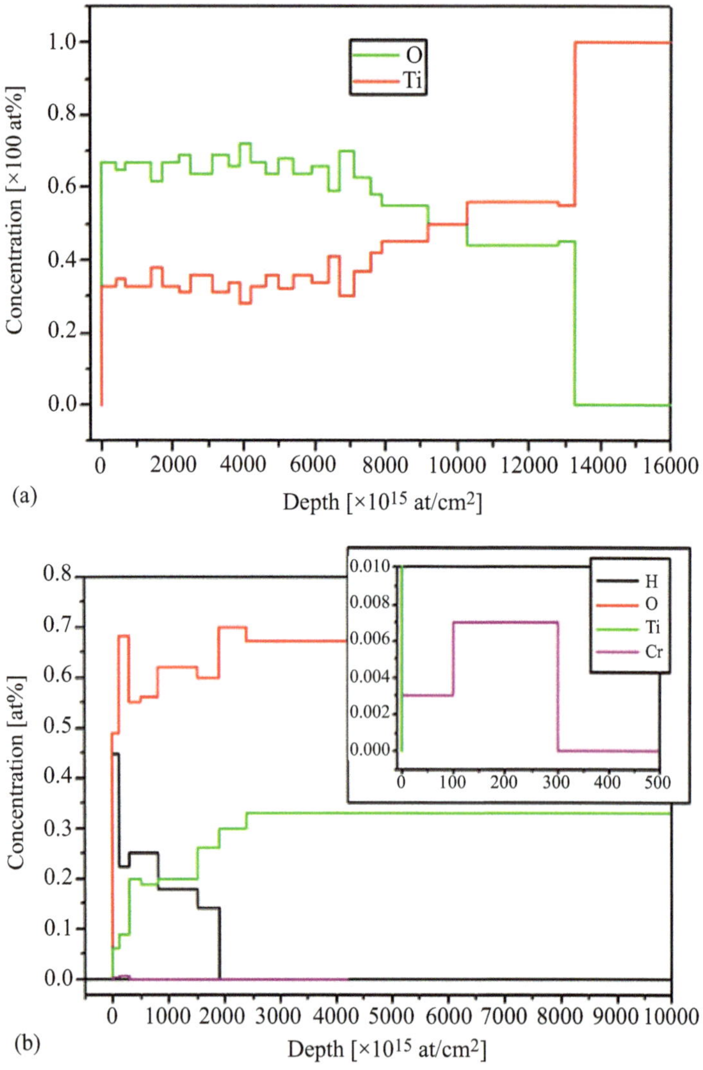

Figure 12.10 Composition depth profiles in (a) pure TiO_2 and (b) Cr-doped TiO_2 nanotubes. Inset shows the close-up view of Cr concentration.

fairly constant, but a composition gradation in both O and Ti exists at the interface between the substrate and the oxide layer. In the Cr-doped sample [Fig. 12.10(b)], it is shown that a maximum Cr concentration of 0.7 at% has been implanted in TiO_2 down to a depth of about 300 mono layers. A composition gradation also appears to exist at the interface between the substrate and the oxide layer.

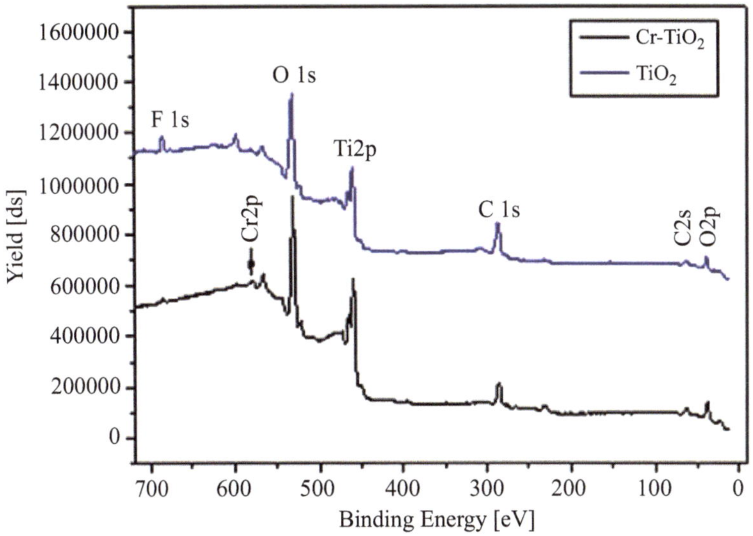

Figure 12.11 XPS plots showing the chemical states and binding energies for pure and Cr-doped TiO_2 nanotubes.

The XPS plots in Fig. 12.11 show the binding energies of elements present in both undoped and Cr-doped TiO_2 nanotubes. The presence of Cr in the doped sample is evident by the display of Cr 2p peak. When compared to Cr-doped TiO_2, significantly more fluorine is present in pure TiO_2. The cause of this diminished fluorine in Cr-doped TiO_2 is uncertain, although it is quite possible that the process of ion implantation may have caused some fluorine to escape. Also noticeable is the slight peak shifts of Ti_{2p} and O_{1s} to the lower binding energy for Cr-doped TiO_2. Such peak shifts have also been observed for Cu-doped TiO_2 [37]. This suggests that the Ti_{2p} binding energy in Cr-doped TiO_2 nanotubes increased when compared to that of pure TiO_2. The peak shift also illustrates the incorporation of Cr ions into the TiO_2 lattices. These results suggest that Cr^{3+} ions are doped in the TiO_2 lattice forming a substitutional solid solution. When TiO_2 is doped with Cr^{3+} ions, the binding energy change in each atom occurs due to the electron transfer from Ti^{4+} to Cu^{3+}. In essence, the Fermi level of Cr_2O_3 is lower than that of TiO_2, thus resulting in the outer electron cloud density of Ti to decrease and that of Cr to increase.

12.3 Conclusion

High-temperature synchrotron radiation diffraction has been used to study the effect of Cr doping on the in situ crystallization of anatase and rutile in the as-anodized TiO_2 nanotubes. The TiO_2 was initially amorphous but crystallized to anatase and rutile at elevated temperatures. Cr doping lowered the crystallization temperature of anatase from 600 to 400°C and rutile from 600 to 500°C. In addition, it reduced dramatically the fraction of anatase-to-rutile transformation. The rutile formed in TiO_2 was very stable after annealing at 1000°C but became unstable in Cr-doped TiO_2, which decomposed to form TiO_{2-x} at 900°C. Ion-beam analysis by Rutherford backscattering spectrometry has confirmed the existence of Cr ions composition gradation in doped TiO_2. The real doping of Ti lattices with Cr ions was evidenced by the analyses of surface compositions and chemical states of the nanotubes using XPS.

Acknowledgments

This work was supported by funding from the Australian Synchrotron (PD-3611), the Australian Institute of Nuclear Science and Engineering (AINGRA-11134), and Spanish MAT2010-20798-C05-04. We thank E/Prof. B. H. O'Connor for guidance on the Rietveld work, Ms. E. Miller of Curtin Centre for Materials Research for assistance with the SEM work. The scientific support from the University of Oviedo SCT's, particularly to the Nanoporous Membranes Unit and Photo-electronic Spectroscopy Laboratories, is also acknowledged.

References

1. D. Hanaor and C. Sorrell. Review of the anatase to rutile phase transformation. *J. Mater. Sci.*, **46**, 855 (2011).

2. M. Anpo, Y. Ichihashi, M. Takeuchi, and H. Yamashita. Design of unique titanium oxide photocatalysts by an advanced metal ion-implantation method and photocatalytic reactions under visible light irradiation. *Res. Chem. Intermed.*, **24**, 143 (1998).
3. A. Zaleska, J. W. Sobczak, E. Grabowska, and J. Hupka. Preparation and photocatalytic activity of boron-modified TiO_2 under UV and visible light. *Appl. Catal. B*, **78**, 92 (2008).
4. G. B. Saupe, T. U. Zhao, J. Bang, N. R. Desu, G. A. Carballo, R. Ordonem, and T. Bubphamala. Evaluation of a new porous titanium-niobium mixed oxide for photocatalytic water decontamination. *Microchem. J.*, **81**, 156 (2005).
5. D. B. Hamal and K. J. Klabunde. Synthesis, characterization, and visible light activity of new nanoparticle photocatalysts based on silver, carbon, and sulfur-doped TiO_2. *J. Colloid Interface Sci.*, **311**, 514 (2007).
6. J. Virkutyte, B. Baruwati, and R. S. Varma. Visible light induced photobleaching of methylene blue over melamine-doped TiO_2 nanocatalyst. *Nanoscale*, **2**, 1109 (2010).
7. T. Umebayashi, T. Yamaki, and S. Tanaka. Visible light-induced degradation of methylene blue on S-doped TiO_2. *Chem. Lett.*, **32**, 330 (2003).
8. S. Sakthivel, M. Janczarek, and H. Kisck. Visible light activity and photo-electrochemical properties of nitrogen-doped TiO_2. *J. Phys. Chem. B*, **108**, 19384 (2004).
9. Y. Choi, T. Umebayashi, and M. Yoshikawa. Fabrication and characterization of C-doped anatase TiO_2 photocatalysts. *J. Mater. Sci.*, **39**, 1837 (2004).
10. H. Irie, Y. Watanabe, and K. Hashimoto. Carbon-doped anatase TiO_2 powders as a visible-light sensitive photocatalyst. *Chem. Lett.*, **32**, 772 (2003).
11. S. Sakthivel and H. Kisch. Daylight photocatalysts by carbon-modified titanium dioxide. *Angew. Chem. Int. Edn.*, **42**, 4908 (2003).
12. R. Asahi, T. Morikawa, and T. Ohwaki. Visible-light photocatalysis in nitrogen-doped titanium oxides. *Science*, **293**, 269 (2001).
13. W. Ren, Z. Ai, F. Jia, L. Zhang, X. Fan, and Z. Zou. Low temperature preparation and visible light photocatalytic activity of mesoporous carbon-doped crystalline TiO_2. *Appl. Catal. B*, **69**, 138 (2007).
14. T. Ohno, M. Akiyoshi, T. Umebayashi, K. Asai, T. Mitsui, and M. Matsumura. Preparation of S-doped TiO_2 photocatalysts and their

photocatalytic activities under visible light. *Appl. Catal. A*, **265**, 115 (2004).

15. B. C. Bai, J. S. Im, J. G. Kim, and Y. S. Lee. Photo-catalytic degradation on B, C, N, and F element co-doped TiO_2 under visible-light irradiation. *J. Ind. Eng. Chem.*, **21**, 29 (2010).
16. G. Shao. Red shift in manganese- and iron-doped TiO_2: A DFT+U analysis. *J. Phys. Chem. C*, **113**, 6800 (2009).
17. G. Shao. Electronic structures of manganese-doped rutile TiO_2 from first principles. *J. Phys. Chem. C*, **112**, 18677 (2008).
18. X. H. Xia, L. Lu, A. S. Walton, M. Ward, X. P. Han, R. Brydson, J. K. Luo, and G. Shao. Origin of significant visible-light absorption properties of Mn-doped TiO_2 thin films. *Acta Mater.*, **60**, 1974 (2012).
19. Y. Kuroda, T. Mori, K. Yagi, N. Makihata, Y. Kawahara, M. Nagao, and S. Kittaka. Preparation of visible-light-responsive $TiO_{2-x}N_x$ photocatalyst by a sol-gel method: Analysis of the active center on TiO_2 that reacts with NH_3. *Langmuir*, **21**, 8026 (2005).
20. A. Hattori, M. Yamamoto, H. Tada, and S. Ito. A promoting effect of NH_4F addition on the photocatalytic activity of sol-gel TiO_2 films. *Chem. Lett.*, **27**, 707 (1998).
21. Y. Shen, T. Xiong, T. Li, and K. Yang. Tungsten and nitrogen co-doped TiO_2 nano-powders with strong visible light response. *Appl. Cataly. B: Environ.*, **83**, 177 (2008).
22. T. C. Jagadale, S. P. Takale, R. S. Sonawane, H. M. Joshi, S. I. Patil, B. B. Kale, and S. B. Ogale. N-doped TiO_2 nanoparticle based visible light photocatalyst by modified peroxide sol-gel method. *J. Phys. Chem. C*, **112**, 14595 (2008).
23. Y. Tseng, C. Kuo, C. Huang, Y. Li, P. Chou, C. Cheng, and M. Wong. Visible-light-responsive nano-TiO_2 with mixed crystal lattice and its photocatalytic activity. *Nanotechnology*, **17**, 2490 (2006).
24. T. Yamaki, T. Sumita, and S. Yamamoto. Formation of $TiO_{2-x}F_x$ compounds in fluorine-implanted TiO_2. *J. Mater. Sci. Lett.*, **21**, 33 (2002).
25. H. Shen, L. Mi, P. Xu, W. Shen, and P. N. Wang. Visible-light photocatalysis of nitrogen-doped TiO_2 nanoparticulate films prepared by low-energy ion implantation. *Appl. Surf. Sci.*, **253**, 7024 (2007).
26. J. Park, J. Y. Lee, and J. H. Cho. Ultraviolet-visible absorption spectra of N-doped TiO_2 film deposited on sapphire. *J. Appl. Phys.*, **100**, 113534 (2006).

27. H. Yamashita, Y. Ichihashi, and M. Takeuchi. Characterization of metal ion-implanted titanium oxide photocatalysts operating under visible light irradiation. *J. Synchrotron Radia.*, **6**, 451 (1999).

28. T. Yamaki, T. Umebayashi, and T. Sumita. Fluorine-doping in titanium dioxide by ion implantation technique. *Nucl. Instr. Meth. Phys. Res. B*, **206**, 254 (2003).

29. H. Yamashita, M. Harada, J. Misaka, M. Takeushi, and M. Anpo. Degradation of propanol diluted in water under visible light irradiation using metal ion-implanted titanium dioxide photocatalysts. *J. Photochem. Photobiol. A*, **148**, 257 (2002).

30. R.A. Spurr and H. Myers. Quantitative analysis of anatase-rutile mixtures with an x-ray diffractometer. *Anal. Chem.*, **29**, 760 (1957).

31. S. Riyas, G. Krishnan, and P. N. Mohandas. Anatase-rutile transformation in doped titania under argon and hydrogen atmospheres. *Adv. Appl. Ceram.*, **106**, 255 (2007).

32. K. J. D. Mackenzie. Calcination of titania V. Kinetics and mechanism of the anatase-rutile transformation in the presence of additives. *Trans. J. Bri. Ceram. Soc.*, **74**, 77 (1975).

33. V. M. Prida, M. Hernandez-Velez, K. R. Pirota, A. Menendez, and M. Vazquez. Synthesis and magnetic properties of Ni nano-cylinders in self-aligned and randomly disordered grown titania nanotubes. *Nanotechnology*, **16**, 2696 (2005).

34. O. K. Varghese, D. Gong, M. Paulose, C. A. Grimes, and E. C. Dickeya. Crystallization and high-temperature structural stability of TiO_2 nanotube arrays. *J. Mater. Res.*, **18**, 156 (2003).

35. K. Mor, O. K. Varghese, M. Paulose, K. Shankar, and C. A. Grimes. A review on highly ordered, vertically oriented TiO2 nanotube arrays: Fabrication, material properties, and solar energy applications. *Solar Energy Mat. Solar Cells*, **90**, 2011 (2006).

36. J. M. Macak, H. Tsuchiya, A. Ghicov, K. Yasuda, R. Hahn, S. Bauer, and P. Schmuki. TiO_2 nanotubes: Self-organized electrochemical formation, properties and applications. *Curr. Opin. Solid State Mater. Sci.*, **11**, 3 (2007).

37. M. You, T. G. Kim, and Y. M. Sung. Synthesis of Cu-doped TiO_2 nanorods with various aspect ratios and dopant concentrations. *Crys. Growth Des.*, **10**, 983 (2010).

Chapter 13

Effect of Indium-Ion Implantation on Crystallization Kinetics and Phase Transformation of Anodized Titania Nanotubes

Titania nanotube arrays were synthesized electrochemically by anodization of titanium foils, and the synthesized titania nanotubes were then implanted with indium ions. The effect of In-ion implantation on crystallization and phase transformation of titania was investigated using in situ high-temperature X-ray diffraction (XRD) and synchrotron radiation diffraction (SRD) from room temperature to 1000°C. Diffraction results show that crystalline anatase first appeared at 400°C in both the non-implanted and the In-implanted materials. The temperature at which crystalline rutile appeared was 600°C for non-implanted materials and 700°C for In-implanted materials, and the indium implantation inhibited the anatase-to-rutile transformation. Although In^{3+} is expected to increase oxygen vacancy concentration and then the rate of titania transformation, the observations are consistent with implanted In ions occupying the Ti sublattice substitutionally and then inhibiting the transformation. The relatively difficult anatase-to-rutile transformation in the In-implanted material appears to result from the relatively large In^{3+} radius (0.080 nm). The In^{3+} partly replaces

Nanostructured Titanium Dioxide in Photocatalysis
It-Meng Low, Hani Manssor Albetran, Victor Manuel de la Prida Pidal, and Fong Kwong Yam

ISBN 978-981-4877-07-7 (Hardcover), 978-1-003-14853-1 (eBook)
www.jennystanford.com

the Ti^{4+} (0.061 nm), which provides a greater structural rigidity and prevents relaxation in the Ti bonding environment.

13.1 Introduction

Titania is known to be a very useful nontoxic, environmentally friendly photocatalyst [1–4]. It is a stable, photodurable, inexpensive, and natural aid in the efficient realization of many applications, such as bio-application, dye-sensitized solar cells, photoelectrochemical cells, gas sensors, and hydrogen production [5–9]. Titania has three crystal structures: anatase, rutile, and brookite [10–12]. In titania phase transformation, amorphous titania crystallizes into the most common titania phases, namely, anatase, rutile, or a mixture of them, by way of thermal treatment [13, 14]. The amorphous-to-anatase and subsequent anatase-to-rutile phase transformation behaviors depend on the impurities, morphology, preparation method, dopants, heating rate, calcining temperature, calcining time, and atmospheres [15–19].

One-dimensional (1D) titania nanostructures, for example nanowires, nanorods, nanobelts, nanofibers, and nanotubes, provide unique electronic characteristics, such as a high surface-area-to-volume ratio, high electron mobility or quantum-confinement effects, and high mechanical strength [1, 3, 7, 8]. These characteristics increase the decomposition rate of organic pollutants significantly because photocatalytic reactions occur rapidly and intensely on the catalyst surface [20, 21]. Various techniques have been used to synthesize 1D nanostructured titania, including sol–gel [22], microwaves [23], electrochemical anodization [6, 24–27], hydrothermal methods [4, 28–30], and electrospinning [31]. Thus, there is a great deal of interest in synthesizing 1D titania nanostructures using inexpensive and simple methods.

Among fabrication methods, anodization is currently one of the most commonly used for fabrication of 1D titania nanostructures [1, 6, 31], and the general nanostructure obtained through this process is nanotube [17, 22]. Nanotubes are 1D nanometer-scale tubes such as nanowires, nanorods, and nanofiber structures. In the case of nanotubes, there is a hollow center as compared to the other 1D structures. The anodized titania nanotube arrays are synthesized by electrochemical anodization of titanium-based alloy in an electrolyte with Pt grid used as an anode for a short period at room temperature [6, 24–27].

In this chapter, titania nanotube arrays were synthesized electrochemically by anodization of titanium foils, and the synthesized titania nanotubes were then implanted with indium ions. The effect of In-ion implantation on the crystallization kinetics and phase transformations of anodized titania nanotubes was investigated using in situ high-temperature XRD and SRD from room temperature to 1000°C and were then characterized by field emission scanning electron microscopy (FESEM) and associated energy dispersive spectroscopy (EDS).

13.2 Results and Discussion

13.2.1 Microstructural Imaging

A homogeneous white layer is formed on the Ti foil after the anodizing process in the ethylene glycol/NH_4F electrolyte at 25°C for 20 h (optical image not shown). Figure 13.1(a) shows a typical secondary electron FESEM image of the as-anodized titania nanotubes, which were amorphous. The as-anodized nanotube arrays were more uniform and highly ordered than the titanium foil. The corresponding mean inner diameter and standard deviation for 100 nanotubes was 26 ± 4 nm.

After the in situ high-temperature XRD at 1000°C for the as-anodized titania nanotube arrays and for SRD at 900°C for In-ion-implanted titania nanotube arrays, the morphologies were re-examined at room temperature using the same FESEM conditions. Figures 13.1(b) and (c) illustrate the FESEM images of non-implanted and In-ion-implanted materials, respectively. For the non-implanted sample, the nanotube array structures were destroyed and covered with rutile elongated grains. However, for In-ion-implanted titania nanotubes, some grains were randomly distributed among the titania nanotubes, which indicates that implantation had inhibited or slowed the grain growth of titania nanotube arrays. These grains are crystalline rutile because metastable anatase transformed into stable crystalline rutile at elevated temperatures (see Section 13.2.2).

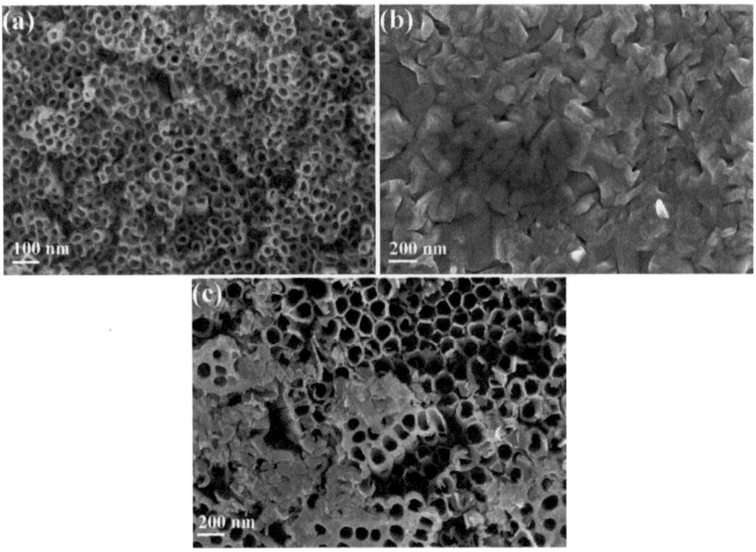

Figure 13.1 FESEM micrographs of (a) as-anodized titania nanotube arrays prior to heating; (b) after thermal treatment for the XRD experiment at 1000°C and cooling to room temperature; and (c) after In-ion implantation and thermal treatment for SRD experiments at 900°C and cooling to room temperature.

Figure 13.2(a) shows the EDS spectrum for the as-anodized titania nanotube arrays in Spectrum 1, showing strong signatures for Ti, O, and Pt, with the Pt content begin due to the platinum coating. Assuming that the Pt peak from the platinum coating is negligible, the quantitative EDS analysis indicated 66.18% O K_α and 33.82%Ti K_α, thus confirming the presence of titania. It should be noted that quantitative EDS is performed for flat samples, which is approximated for the nanotubes. The EDS for the nanotubes was collected from a small square to minimize the non-flat sample issue, using EDS software (AZtec Software, Oxford). The qualitative EDS spectrum for non-implanted and In-ion-implanted nanotubes after cooling was essentially the same for the as-anodized nanotubes, whereas the quantitative EDS analysis indicated approximately 33.82% Ti K_α and 66.18% O K_α, thus confirming and providing evidence of titania with a TiO_2 chemical formula structure. The existence of In ions in implanted titania nanotubes was verified by the quantitative EDS analysis in Spectrum 2, as shown in Fig. 13.2(b) indicating 0.8(4)% In. This occurs because the surface composition is detected to a greater depth by SEM.

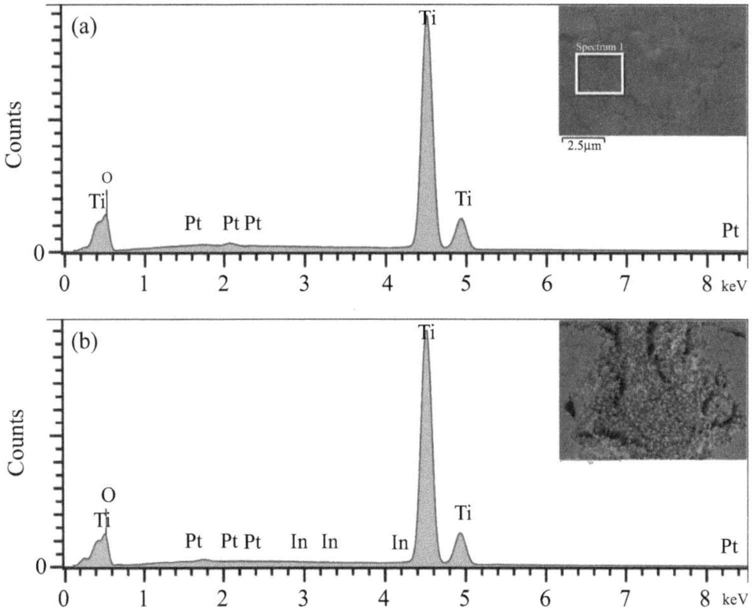

Figure 13.2 EDS spectra of (a) anodized TiO_2 nanotubes prior to heating and (b) after thermal treatment and In-ion implantation and cooling to room temperature.

13.2.2 Crystallization Behavior

Figures 13.3(a) and (b) show the stacked XRD and SRD plots during thermal annealing in air from room temperature to 1000°C for non-implanted titania nanotubes and In-ion-implanted titania nanotubes, respectively. Diffraction results showed that both titania nanotube arrays were initially amorphous, but crystalline anatase formed at 400°C for both samples. The anatase-to-rutile transformation started at 600°C by a small crystalline rutile peak observed for non-implanted material, compared with 700°C for In-ion-implanted material. The reason for the 100°C delay in the initial crystallization of rutile for the In-ion-implanted titania nanotubes was that the In-ion implant provided greater structural rigidity and

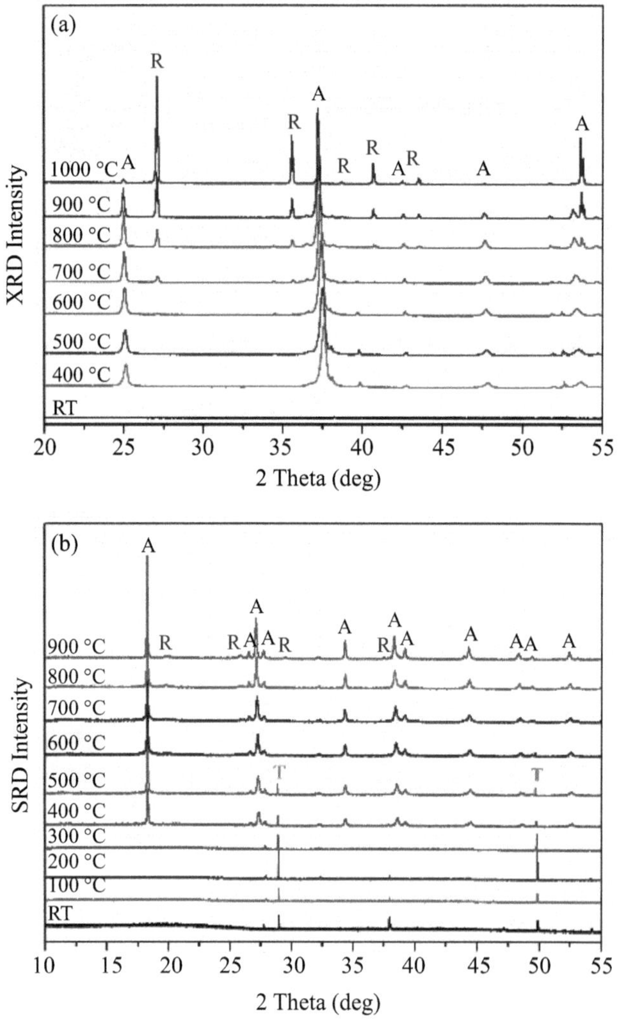

Figure 13.3 Stacked XRD and SRD plots for the temperature range of room temperature to 1000°C of (a) non-implanted material and (b) after In-ion implantation, respectively. [Legend: anatase (A), rutile (R), titanium peaks (T), and room temperature (RT)].

prevented relaxation in the Ti bonding environment. This behavior of severe/difficult rearrangement of the atoms in the anatase and rutile lattices could be due to the presence of In ions interstitially or substitutionally. The presence of interstitial ions can inhibit the anatase-to-rutile transformation, whereas substitutional ions can also inhibit phase transformation if the substitutional ion is larger and does not differ significantly from Ti^{4+} [32–35]. The In^{3+} radius (0.080 nm) is relatively larger than the Ti radius (0.061 nm). (For a discussion of whether In ions enter the titania interstitially or substitutionally, see Section 13.2.3.)

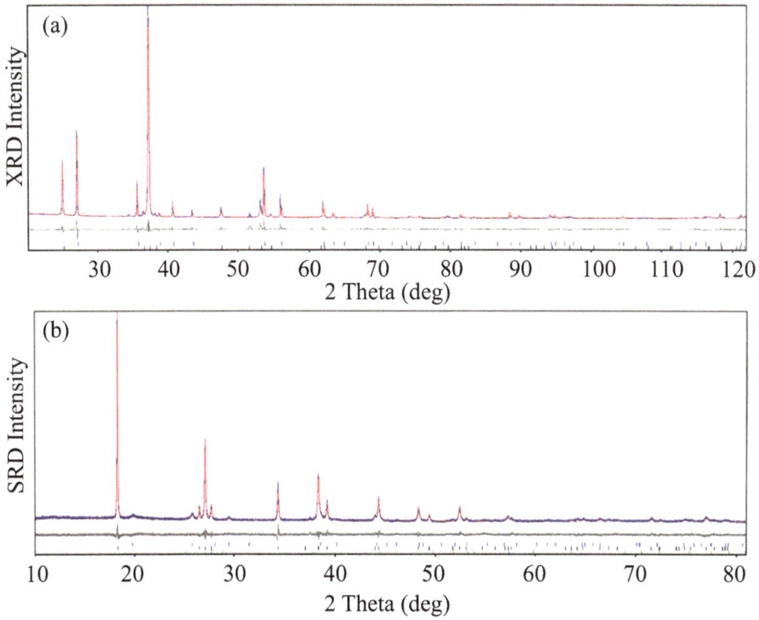

Figure 13.4 XRD and SRD Rietveld difference plots for data measured in (a) non-implanted and (b) In-ion-implanted titania nanotubes at 900°C, respectively. Measured patterns are indicated by black crosses and calculated patterns by solid red lines. The gray residual plot shows the difference between calculated and measured patterns. Peak positions for titanium, anatase, and rutile are indicated by green, black, and blue bars, respectively.

The relative phase levels of crystalline titanium, anatase, and rutile in the anodized nanotubes were calculated using the Rietveld method. Figures 13.4(a) and (b) show typical XRD and SRD Rietveld residual plots at 900°C for data measured in non-implanted and In-ion-implanted titania nanotube arrays, respectively. In non-implanted material, R_{wp} = 13.23, R_{exp} = 3.75, R_B (anatase) = 4.791, and R_B (rutile) = 6.416%, and for the In-ion-implanted material, R_{wp} = 12.14, R_{exp} = 9.94, R_B (titanium α) = 2.817, R_B (anatase) = 2.908, and R_B (rutile) = 2.951%. The goodness-of-fit (GOF) values are 3.53 for non-implanted nanotube arrays and 1.22 for In-ion-implanted nanotube arrays. The ideal value for the GOF is 1.0, but 4 or higher can be accepted for abundance data analysis [36].

Table 13.1 shows the variation in relative phase concentrations of crystalline titanium, amorphous anatase, and rutile in the temperature range of 400–1000°C for non-implanted and In-ion-implanted material. At 600°C, the relative phase concentrations were 1.40(12)% Ti, 96.55(64)% anatase, and 2.05(63)% for non-implanted nanotubes, whereas they were 1.56(47)% Ti, and 98.44(47)% anatase without any crystalline rutile formed for In-ion-implanted titania nanotubes. The metastable anatase crystallizes from amorphous titania, which formed from the anodized Ti foils at relatively low temperature, and the anatase subsequently transforms into stable crystalline rutile at a higher temperature. For the non-implanted sample, it is evident that the relative phase concentration of the crystalline rutile increased dramatically as the temperature increased from 600 to 1000°C. However, the titania transformations were observed to increase more slowly in In implantation than in non-implantation samples. At 900°C, the relative phase concentration was 0.2194(68)% Ti, 84.16(32)%, and just 15.66(31)% rutile for the In-ion-implanted nanotubes, whereas all of the crystalline Ti and more than 53% of the crystalline anatase transformed into crystalline rutile for the non-implanted sample. The relatively larger radius of In^{3+} than Ti^{4+} would inhibit titania's transformation, since it involves greater structural rigidity and prevents relaxation in the Ti-bonding environment.

13.2.3 Influence of In-Ion Implantation on Lattice Parameters

Figure 13.5 shows the temperature variations with the anatase and rutile lattice parameters and cell volumes for non-implanted and In-ion-implanted nanotubes. The unit-cell parameters and cell volumes for both anatase and rutile in both non-implanted and In-ion-implanted titania nanotube arrays increased linearly with temperature, but the lattice parameters and cell volumes showed an increase in the In-ion-implanted sample relative to the non-implanted sample. For example, at 900°C, the anatase lattice

Table 13.1 The effect of In-ion implantation on the relative weight percentages of anatase and rutile as a function of temperature, using the Rietveld method

Temperature (°C)	Non-implantation			In-ion Implantation		
	Ti (%)	Anatase (%)	Rutile (%)	Ti (%)	Anatase (%)	Rutile (%)
400	3.25(12)	96.75(12)	0	4.48(13)	95.52(13)	0
500	2.85(93)	97.15(93)	0	2.61(67)	97.63(58)	0
600	1.40(12)	96.55(64)	2.05(63)	1.56(47)	98.44(47)	0
700	0.87(11)	86.60(88)	12.53(88)	0.85(68)	85.27(58)	13.88(11)
800	0.65(16)	76.06(43)	23.29(41)	0.4202(20)	84.14(34)	15.64(56)
900	0	46.81(44)	53.19(44)	0.2194(68)	84.16(32)	15.66(31)
1000	0	5.06(62)	94.94(62)	NA	NA	NA

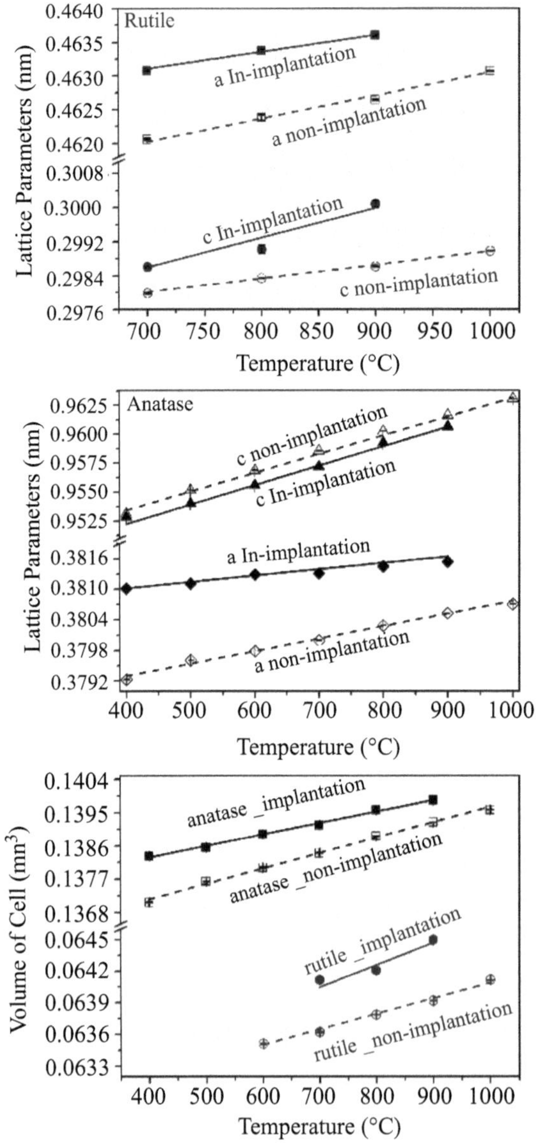

Figure 13.5 Variation with temperature of the lattice parameters and cell volumes for the anatase and rutile phases in non-implanted and In-ion-implanted anodized titania nanotube arrays.

parameters ***a*** and ***c*** for the non-implanted titania nanotube arrays were 0.380523(18) and 0.961642(26) nm, whereas they were

0.381530(75) and 0.960651(81) nm for the In-ion-implanted samples, respectively. Moreover, the rutile lattice parameters ***a*** and ***c*** at the same temperature were 0.4626490(98) and 0.2986160(75) nm for the non-implanted sample, whereas they were 0.46360(13) and 0.300084(79) nm for In-ion-implanted samples, respectively. Thus, the cell volumes of anatase and rutile were 0.139243(13) and 0.0639170(31) nm^3 for non-implanted samples and 0.139838(27) 0.064495(41) nm^3 for In-ion implantation at 900°C, respectively. As can be seen, the volume of cells for the anatase and rutile was higher for the In-ion-implanted samples than the non-implanted samples. The current lattice parameter increase in the In-ion implantation suggests that added ionic In substitutes for Ti^{4+} rather than occupying interstitial sites, which causes the ions to be incorporated as dopants in the titania lattice [34, 35]. The lattice deformation of both titania phases caused by the incorporation of foreign Indium dopants affects the lattice parameter for In-ion-implanted samples with temperature. The relatively difficult anatase-to-rutile transformation in the In-implanted titania nanotubes appears from an In-ion radius occupying the Ti sublattice substitutionally.

13.3 Conclusion

The effect of In-ion implantation on the dynamic crystallization behavior of anodized titania nanotubes was investigated using in situ high-temperature XRD and SRD in the range from room temperature to 1000°C. Diffraction results show that the diffraction peaks of the Ti foils were only observed before calcination until 300°C, which indicated the amorphous titania nanotubes. The amorphous-to-anatase transformation started at 400°C for non-implanted and In-implanted materials, including the characteristic crystalline anatase peaks. The crystalline rutile temperature increased from 600°C to 700°C for In implantation, with the anatase-to-rutile transformation being inhibited relative to the non-implanted material. It was concluded that In ions occupy the Ti sublattice substitutionally and then inhibit the anatase-to-rutile transformation. The relatively difficult anatase-to-rutile transformation in the In-implanted material appears to result from the relatively larger radius of In^{3+}

than Ti^{4+}, which provides greater structural rigidity and prevents relaxation in the Ti-bonding environment.

Acknowledgments

This research was supported by funding from the Australian Institute of Nuclear Science and Engineering (AINGRA-11134) and the Australian Synchrotron (PD-3611). The authors would like to thank E/Prof. B. H. O'Connor for guidance on the Rietveld work and Prof. V. M. Prida from the Department of Physics, University of Oviedo, for XRD collection.

References

1. H. Li, L. Cao, W. Liu, G. Su, and B. Dong. Synthesis and investigation of TiO_2 nanotube arrays prepared by anodization and their photocatalytic activity. *Ceram. Int.*, **38**, 5791 (2012).
2. D. V. Bavykin, A. N. Kulak, V. V. Shvalagin, N. S. Andryushna, and O. L. Stroyuk. Photocatalytic properties of rutile nanoparticles obtained via low temperature route from titanate nanotubes. *J. Photochem. Photobiol. A: Chem.*, **218**, 231 (2011).
3. O. K. Varghese, D. Gong, M. Paulose, C. A. Grimes, and E. C. Dickey. Crystallization and high-temperature structural stability of titanium oxide nanotube arrays. *J. Mater. Res.*, **18**, 156 (2003).
4. M. Inagakia, N. Kondoa, R. Nonakaa, E. Itob, M. Toyodac, K. Sogabec, and T. Tsumura. Structure and photoactivity of titania derived from nanotubes and nanofibers. *J. Hazard. Mater.*, **161**, 1514 (2009).
5. M. Senna, N. Myers, A. Aimable, V. Laporte, C. Pulgarin, O. Baghriche, and P. Bowen. Modification of titania nanoparticles for photocatalytic antibacterial activity via a colloidal route with glycine and subsequent annealing. *J. Mater. Res.*, **28**, 354 (2013).
6. J. M. Macak, H. Tsuchiya, A. Ghicov, K. Yasuda, R. Hahn, S. Bauer, and P. Schmuki. TiO_2 nanotubes: Self-organized electrochemical formation, properties and applications. *Curr. Opin. Solid State Mater. Sci.*, **11**, 3 (2007).
7. H. Xiong, M. D. Slater, M. Balasubramanian, C. S. Johnson, and T. Rajh. Amorphous TiO_2 nanotube anode for rechargeable sodium ion batteries. *J. Phys. Chem. Lett.*, **2**, 2560 (2011).

8. D. Yanga, H. Parka, S. Choa, H. Kima, and W. Choi. TiO_2-nanotube-based dye-sensitized solar cells fabricated by an efficient anodic oxidation for high surface area. *J. Phys. Chem. Solids*, **69**, 1272 (2008).

9. D. Hanaor and C. Sorrell. Review of the anatase to rutile phase transformation. *J. Mater. Sci.*, **46**, 855 (2011).

10. S. Chuangchote, J. Jitputti, T. Sagawa, and S. Yoshikawa. Photocatalytic activity for hydrogen evolution of electrospun TiO_2 nanofibers. *ACS Appl. Mater. Interfaces*, **1**, 1140 (2009).

11. D. Kim, N. Enomoto, Z. Nakagawa, and K. Kawamura. Molecular dynamic simulation in titanium dioxide polymorphs: Rutile, brookite, and anatase. *J. Am. Ceram. Soc.*, **79**, 1095 (1996).

12. G. Liu, L. Wang, H. G. Yang, H. M. Cheng, and G. Q. M. Lu. Titania-based photocatalysts-crystal growth, doping and heterostructuring. *J. Mater. Chem.*, **20**, 831 (2010).

13. H. Albetran, H. Haroosh, Y. Dong, V. M Prida, B. H. O'Connor, and I. M. Low. Phase transformations and crystallization kinetics in electrospun TiO_2 nanofibers in air and argon atmospheres. *Appl. Phys. A*, **116**, 161 (2014).

14. H. Albetran, H. Haroosh, Y. Dong, B. H. O'Connor, and I. M. Low. Effect of atmosphere on crystallisation kinetics and phase relations in electrospun TiO_2 nanofibres. *Ceram. Environ. Appl. II: Ceram. Trans.*, **246**, 125 (2013).

15. A. Ghicov, H. Tsuchiya, J. M. Macak, and P. Schmuki. Annealing effects on the photoresponse of TiO_2 nanotubes. *Phys. Status Solidi A*, **203**, 28 (2006).

16. Z. Liu, X. Yan, W. Chu, and D. Li. Effects of impurities containing phosphorus on the surface properties and catalytic activity of TiO_2 nanotube arrays. *Appl. Surf. Sci.*, **257**, 1295 (2010).

17. R. P. Antony, T. Mathews, P. K. Ajikumar, D. N. Krishna, S. Dash, and A. K. Tyagi. Electrochemically synthesized visible light absorbing vertically aligned N-doped TiO_2 nanotube array films. *Mater. Res. Bull.*, **47**, 4491 (2012).

18. Y. Iida and S. Ozaki. Grain growth and phase transformation of titanium oxide during calcination. *J. Am. Ceram. Soc.*, **44**, 120 (1961).

19. R. D Shannon and J. A Pask. Kinetics of the anatase-rutile transformation. *J. Am. Ceram. Soc.*, **48**, 391 (1965).

20. H. Li, W. Zhang, and W. Pan. Enhanced photocatalytic activity of electrospun TiO_2 nanofibers with optimal anatase/rutile ratio. *J. Am. Ceram. Soc.*, **94**, 3184 (2011).

21. J. Lee, T. Ha, M. Hong, and H. Park. The effect of porosity on the CO sensing properties of TiO_2 xerogel thin films. *Thin Solid Films*, **529**, 98 (2013).

22. Z. Zainal and C. Y. Lee. Properties and photoelectrocatalytic behaviour of sol-gel derived TiO_2 thin films. *J. Sol-Gel Sci. Technol.*, **37**, 19 (2006).

23. D. Monti, A. Ponrouch, M. Estruga, M. R. Palacin, J. A. Ayllon, and A. Roig. Microwaves as a synthetic route for preparing electrochemically active TiO_2 nanoparticles. *J. Mater. Res.*, **28**, 340 (2013).

24. C. Arunchandran, S. Ramya, R. P. George, and U. K. Mudali. Corrosion inhibitor storage and release property of TiO_2 nanotube powder synthesized by rapid breakdown anodization method. *Mater. Res. Bull.*, **48**, 635 (2012).

25. N. Baram, D. Starosvetsky, J. Starosvetsky, M. Epshtein, R. Armon, and Y. Ein-Eli. Enhanced photo-efficiency of immobilized TiO_2 catalyst via intense anodic bias. *Electrochem. Commun.*, **9**, 1684 (2007).

26. K. P. Beh, F. K. Yam, S. S. Tneh, and Z. Hassan. Fabrication of titanium dioxide nanofibers via anodic oxidation. *Appl. Surf. Sci.*, **257**, 4706 (2011).

27. J. Liao, S. Lin, N. Pan, D. Li, S. Li, and J. Li. Free-standing open-ended TiO_2 nanotube membranes and their promising through-hole applications. *Chem. Eng. J.*, **211–212**, 343 (2012).

28. R. Sanchez-Tovar, K. Lee, J. Garcia-Anton, and P. Schmuki. Formation of anodic TiO_2 nanotube or nanosponge morphology determined by the electrolyte hydrodynamic conditions. *Electrochem. Commun.*, **26**, 1 (2012).

29. D. V. Bavykin, A. N. Kulak, and F. C. Walsh. Control over the hierarchical structure of titanate nanotube agglomerates. *Langmuir*, **27**, 5644 (2011).

30. D. V. Bavykin, A. A. Lapkin, P. K. Plucinski, J. M. Friedrich, and F. C. Walsh. Reversible storage of molecular hydrogen by sorption into multilayered TiO_2 nanotubes. *J. Phys. Chem. B*, **109**, 19422 (2005).

31. A. W. Tan, B. Pingguan-Murphy, R. Ahmad, and S. A. Akbar. Review of titania nanotubes: Fabrication and cellular response. *Ceram. Int.*, **38**, 4421 (2012).

32. I. M. Low, H. Albetran, V. M. Prida, V. Vega, P. Manurung, and M. Ionescu. A comparative study on crystallization behavior, phase stability, and binding energy in pure and Cr-doped TiO_2 nanotubes. *J. Mater. Res.*, **28**, 304 (2013).

33. H. Albetran, B. H. O'Connor, and I. M. Low. Effect of vanadium ion implantation on the crystallization kinetics and phase transformation of electrospun TiO_2 nanofibers. *Appl. Phys. A*, **120**, 623 (2015).

34. R. D. Shannon and J. A. Pask. Kinetics of the anatase-rutile transformation. *J. Am. Ceram. Soc.*, **48**, 391 (1965).

35. K. Okada, N. Yamamoto, Y. Kameshima, A. Yasumori, and K. J. D. MacKenzie. Effect of silica additive on the anatase-to-rutile phase transition. *J. Am. Ceram. Soc.*, **84**, 1591 (2001).

36. E. H. Kisi. Rietveld analysis of powder diffraction patterns. *Mater. Forum*, **18**, 135 (1994).

Chapter 14

Ni Nanowires Grown in Anodic TiO_2 Nanotube Arrays as Diluted Magnetic Semiconductor Nanocomposites

In this study, we report on novel synthesis of ferromagnetic/semiconductor oxide nanomaterial consisting of ferromagnetic Ni nanowires electrodeposited and grown inside the semiconductor oxide matrix of self-aligned TiO_2 nanotube arrays. Microstructure and morphological and magnetic properties of the nanocomposite have been investigated. The outer diameter of the thus-obtained titanium dioxide nanotube ranges between 90 and 150 nm, wall thickness is about 25–40 nm, and depth is 300 nm. The electrodeposited Ni nanowires reach above 100 nm in diameter and 240 nm in length, giving rise to coercive fields of 98 and 200 Oe, within the hysteresis loops perpendicular or parallel to the axis of nanowires, respectively, which could be ascribed to the formation of magnetic vortex domain states. This novelty diluted magnetic semiconducting nanocomposite can be a promising candidate for many applications in a broad range of scientific and technological areas, such as ultrahigh density magnetic storage media or spin-based electronic devices.

Nanostructured Titanium Dioxide in Photocatalysis
It-Meng Low, Hani Manssor Albetran, Victor Manuel de la Prida Pidal, and Fong Kwong Yam

ISBN 978-981-4877-07-7 (Hardcover), 978-1-003-14853-1 (eBook)
www.jennystanford.com

14.1 Introduction

Nanodimensional structures, including nanotubes, nanowires, and nanoporous architecture materials based on semiconducting metal oxides, are nowadays of fundamental interest to the development of functional nanomaterials, nanodevices, and nanosystems [1]. Recently, the synthesis of nanostructured functional oxides based on transition metals, with controlled structure and morphology, has attracted huge interest due to their potential applications in broad research fields such as nanoelectronics, spintronic, fuel cells, and nano-biotechnological or magneto-optoelectronic devices [2, 3]. These new materials have shown a broad range of novel and enhanced mechanical, optical, magnetic, and electronic features with respect to those exhibited by their bulk counterparts [4–6]. Actually, great efforts have been made in order to obtain self-assembled nanostructures based on TiO_2 nanoporous membranes prepared by sol–gel coating [7], nanoimprint [8], or electrochemical processes [9], due to the higher corrosion and oxidation resistance, low density, high yield strength at low and high temperatures, and excellent biocompatibility exhibited by this oxide semiconductor nanomaterial. The search is focused on low-cost and efficient fabrication techniques of nanostructured transition metal oxides with high-quality nanoporous structures over large surface areas and an accurate pore size control together with long-range ordering to enhance the efficiencies of devices based on nanoporous titania (TiO_2) templates [10–13]. The main advantages of using pure titanium and its alloys are, among others, their high corrosion and good oxidation resistances, low density, high yield strength in a wide temperature range, and excellent biocompatibility, which makes this metal an outstanding candidate for its application in wide scientific and technological areas. Otherwise, some of these properties when adequately combined with the large band gap of semiconductors, high photocatalytic activity, and excellent biocompatibility exhibited by TiO_2 make it a very promising material for applications in bone implants or hydrogen sensors, semiconductor memory alloy devices, diluted magnetic semiconductors, micro-optoelectronic applications, transparent oxides, or gas/humidity or conductivity sensors [14–19].

TiO_2 nanotubes can also be a highly promising encapsulating material for a magnetic core as a high degree of biocompatibility functionalities [14]. TiO_2 is a well-known highly active photocatalyst due to its ability to produce electron–hole pairs upon light irradiation, which allows to create highly reactive radical species. The photocatalytic ability of TiO_2 can be enhanced by using nanosized materials, such as nanoparticles or nanowires, because of their large specific surface area [15, 19, 20].

It has been recently demonstrated that TiO_2 nanotubes can be used as a magnetically guided photocatalyst for the decomposition of organic matter, and also the photocatalytic mechanism can be enhanced by doping the TiO_2 nanomaterials with magnetic nanoparticles, such as embedding Co or Fe oxide nanoparticles into TiO_2 nanotubes, which enables the photon-absorbing zone of TiO_2 to be extended from UV toward visible radiation light range, as well as to reduce its band-gap energy from 3.2 eV down to 2.67 eV, therefore increasing the photocatalytic activities of the TiO_2 samples doped with magnetic nanoparticles of the order of several times higher [15, 19–21].

Among the various synthetic routes used to prepare TiO_2 nanotubes [7], anodization approaches have gained significant attention as they lead to highly self-aligned nanotubular arrangements [22, 23].

On the other hand, patterned media made of ferromagnetic nanowires can act as information storage devices, where each single-domain nanowire represents one bit, depending on its magnetization state. High remanence states for providing the readback process and high-enough coercivity to avoid spontaneous demagnetization of the nanowire array due to the magnetostatic interactions among them are both required to be satisfied at the same time. Therefore, patterned media can be achieved, in which data are stored in a large surface area of periodic arrays of single-domain magnetic nanowires, which could enable data-recording densities exceeding 1 $Tbit/cm^2$ [10].

The cost-effective obtention of nanoarchitectured semiconducting metal oxides with high-quality nanoporous structures over large surface areas and with precise control of pore size and periodic ordering degree distribution still remains an open task. There is no doubt that the control of all these requirements must be fulfilled at the same time in order to optimize the efficiency of novel devices based on TiO_2 nanotube arrays [21]. All these make it an outstanding candidate for its application in wide scientific and technological areas. It is expected that this so-formed diluted magnetic semiconductor nanocomposite made of titanium oxide nanotubes/ferromagnetic Ni nanowires will lead to new applications and device designs in the research fields of high-density magnetic data storage and spintronic devices.

14.2 Results and Discussion

Self-aligned titania nanotubes were grown by anodizing high-purity titanium foils (99.99+%, metal basis) supplied by Alfa Aesar (Johnson Matthey, GmbH, Germany), which were cut into pieces of 2.25 mm^2 surface area and 25 mm thickness. The electrochemical anodizations to obtain the Ti oxide nanotubes were carried out potentiostatically under ambient conditions (20°C) and using different electrolytes based on hydrofluoric acid (0.15–0.5 wt%) mixed with H_2SO_4 (1 M) acidic media in a two-electrode electrochemical cell, as explained in Chapter 3. A direct current (DC) voltage source was kept to a constant value of 20 V during the whole treatment time ranging from 20 min to 18 h.

The titanium-anodized samples in an H_2SO_4 aqueous solution mixed with HF acid yield a regular self-aligned and randomly disordered nanotube array grown over large surface areas of the Ti sample substrate, over more than a hundred microns, as can be seen from the SEM top-view images shown in Fig. 14.1 and their magnification in the inset for the sample anodized at 20 V for 18 h in a mixture of H_2SO_4 (1 M) plus HF (0.15 wt%), at room temperature. The tubular structure exhibited in the inset of the SEM image shown in Fig. 14.1 has an average inner pore diameter of 100 ± 10 nm and a nearly uniform wall thickness above 40 ± 10 nm [21].

Therefore, the resulting average outer diameter of the TiO_2 nanotubes ranges between 140 and 150 nm in size.

Energy-dispersive X-ray spectroscopy (EDX) in the SEM was used to confirm the atomic compositional analysis of the Ti oxide nanotube template, which exhibit an atomic concentration of approximately 60% titanium and 36% oxygen for the nanotubular layer, which confirms that the nanotube array composition consists of TiO_2.

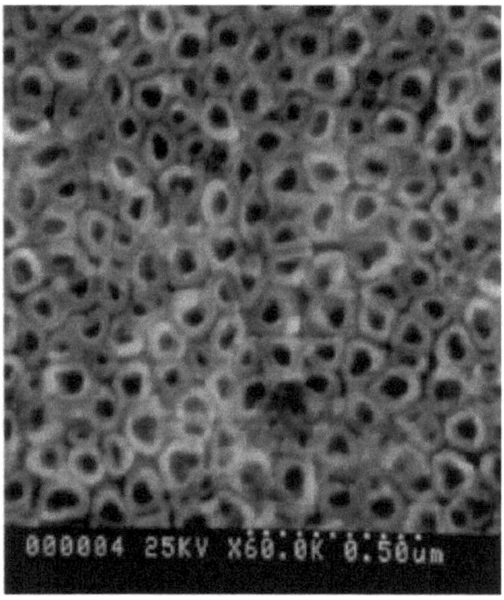

Figure 14.1 SEM top-view image of the self-aligned and randomly disordered TiO_2 nanotube array grown by a single anodization process in a mixture of HF (0.15 wt%.) + H_2SO_4 (1 M) acidic media for 18 h at a constant voltage of 20 V and room temperature.

TiO_2 nanotube arrays were further filled with ferromagnetic Ni nanowires that were grown inside the innermost self-aligned TiO_2 nanotubes by electrochemical deposition from aqueous electrolytes by means of modulated pulsed electrodeposition method [10], for a 15 min Ni electrodeposition process, where a TiO_2 nanotube membrane was employed as template, as shown in Fig. 14.2.

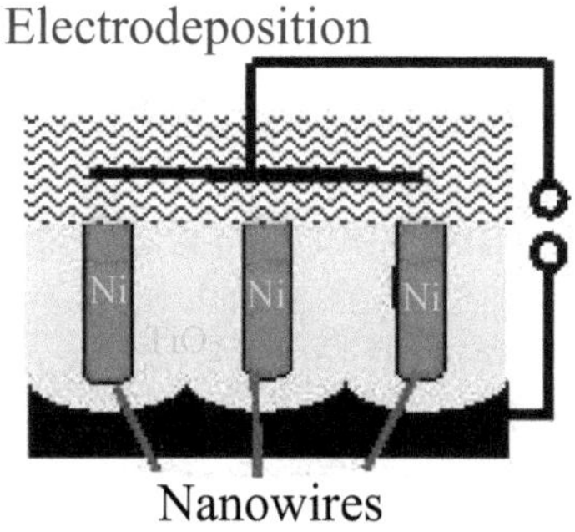

Figure 14.2 Schematic drawing of the template-assisted electrochemical deposition cell employed for the growth of Ni nanowires inside the TiO_2 nanotube array.

The depth-profiling analysis was carried out using a radiofrequency glow discharge optical emission spectroscopy (rf-GD-OES) technique by using a commercial JY 5000 RF instrument working at 13.56 MHz in an argon atmosphere at 600 Pa and an applied power of 40 W. Data were collected for a sample area of 4 mm diameter. The wavelengths of spectral lines employed for detecting the sputtered specimens of Ti, Ni, S, O, and Ar were 365, 341, 181, 130, and 404 nm, respectively. The sputtering rate and penetration depths were obtained from calibration by direct profilometer measurements (Perth-ometer S5P, Mahr Perthen) [24].

The rf-GDOES qualitative depth profile of a self-aligned TiO_2 nanotube array filled with electrodeposited Ni nanowires is shown in Fig. 14.3, for the Ti sample anodized in HF mixed with H_2SO_4 acidic electrolyte. The direct analysis performed by means of this technique reveals its ability to control the quality and testing the composition of the nanoporous structure along the depth profile of the samples. The distributions of titanium and nickel contents, along with those of oxygen and sulfur, which arises from SO_4^{-2} anions incorporated into the titanium oxide during the anodization process, and Ar traces that are provided for the trapped gas during plasma ionization, can be clearly revealed.

As can be seen in Fig. 14.3, the signal intensities of nickel, oxygen, and carbon remain relatively steady along the first hundred nanometers in depth. On reaching the Ti substrate, the signal intensity for Ti increases sharply up to its maximum value. At the same time the corresponding signals from nickel, oxygen, and sulfur also decrease, respectively, thus allowing a well-defined oxide/substrate interface. Assuming a constant rate of Ar sputtering of 4 μm/min across the thickness of the Ti film, the estimated length for the grown TiO_2 nanotubes turns out to be about 300 nm. This would correspond to a Ti oxide layer growth rate about 15 nm/V, formed far below the dielectric breakdown voltage, typically above 100 V for these materials [21, 24].

Meanwhile, the Ni nanowires electrodeposited into the TiO_2 nanotube array have an estimated uniform thickness around 230–240 nm in length [21, 25, 26].

Magnetic characterization measurements of the Ni magnetic nanowires embedded in the TiO_2 nanotube arrays by means of the hysteresis loops in the configurations parallel (out of plane, OP) and perpendicular to the Ni nanowire axis (in plane, IP) were carried out at room temperature using a vibrating sample magnetometer (VSM), measured at room temperature and in the applied magnetic field range up to 10^4 Oe.

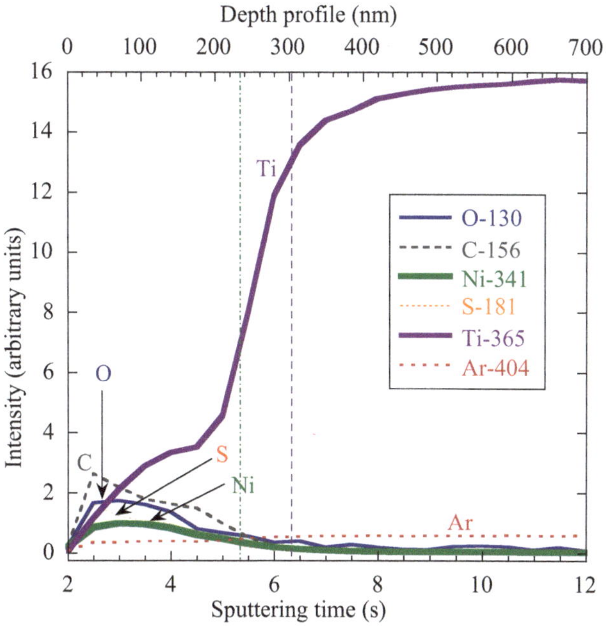

Figure 14.3 Qualitative and quantitative rf-GDOES depth profile analysis of an array of Ni nanowires embedded in a self-aligned nanotube titania template by pulsed electrodeposition method. The wavelengths (in nm) of each spectral emission line of the relevant elements investigated are listed in the legend. Adapted from Ref. [26].

Figure 14.4 shows the room-temperature VSM hysteresis loops measured for the magnetic field parallel applied to the Ni nanowire axis (OP) shown in Fig. 14.4(a), and when the magnetic field is perpendicularly applied to the Ni nanowire axis (IP), in Fig. 14.4(b). In both cases, the hysteresis loops were obtained for an array of Ni nanowires electrodeposited for 20 min into a TiO_2 nanotubular membrane. Additionally, Fig. 14.4(c) shows the paramagnetic contribution of the TiO_2 nanotube film substrate and X4(d) represents the hysteresis loop of an Ni calibration bulk sample for comparison. The OP hysteresis loop corresponds to a hard magnetization axis of the Ni nanowire array showing a coercive field $H_C \approx 199$ Oe; meanwhile, the IP hysteresis loop is typical of an easy magnetization axis, with a coercivity of about 98 Oe. As the pure TiO_2 does not show any ferromagnetic behavior [25], it appears evident that the peculiar ferromagnetic long-range order exhibited is due to the Ni electrodeposition doped TiO_2 nanotubes. The magnetic behavior exhibited by the peculiar shape of the hysteresis loops of the Ni nanowire arrays embedded into the self-aligned and randomly disordered amorphous TiO_2 nanotubes is quite similar to the Ni nanowire arrays with large-diameter nanowires and quite high aspect ratio (length/diameter) about 2.3–2.4, as reported by others [26]. The coercivity and remanence values would correspond to a vortex state, where the magnetization follows a helical structure beginning to tilt in the circumferential direction, starting at the ends of the nanowire, giving rise to a circumferential magnetization at the wire-shaped perimeter, while the magnetization at the center of the wire remains oriented along its axis. The vortex-type remanent magnetization distribution in these flat nanowires can be due to the strong magnetostatic interactions between the larger-diameter nanowires, helped by the minor role played by the weak magnetocrystalline anisotropy of the polycrystalline Ni at room temperature [26].

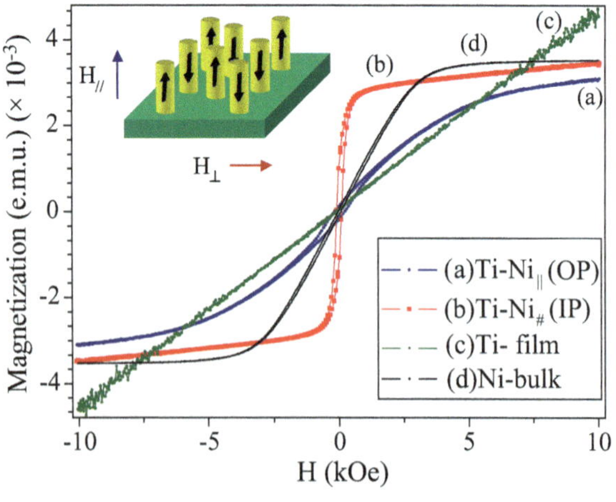

Figure 14.4 Room-temperature VSM hysteresis loops of the Ni nanowire arrays electrodeposited inside the TiO_2 nanotube template: (a) applied magnetic field parallel to the Ni nanowire axis (out of plane, OP), (b) applied field perpendicular to the nanowire axis (in plane, IP), (c) anodic TiO_2 template, and (d) Ni bulk sample. Adapted from Ref. [26].

When the inter-nanowire distance is shortened as compared to the nanowire diameter, the magnetostatic interactions between the coupled nanowires can be so strong as to make the IP direction like the easy magnetization axis. Magnetic nanowires exhibiting a vortex domain state show typical switching field values around 100 Oe for the easy magnetization axis, as a result of differences in the shapes, heights, microstructures, roughness or inhomogeneities of nanowires, which act as pinning centers of the magnetization as it rotates in plane, such as in the case of these Ni nanowires embedded into the TiO_2 nanotube arrays. Additionally, the OP coercivity of these Ni nanowires is low because the vortex magnetic configuration can cause partial flux closure, which reduces it since it turns out to be relatively easy to move a vortex domain wall through the nanowires [26].

14.3 Conclusion

Novel ferromagnetic/semiconductor oxide nanocomposites consisting of ferromagnetic cylindrical Ni nanowires that are about 100 nm in diameter and 240 nm in length have been synthesized. These magnetic nanowires were embedded in a semiconducting oxide template, formed by self-aligned and randomly ordered titania nanotubes grown through a simple anodization process employing an electrolytic bath of H_2SO_4 solution and small amounts of HF acid. The resulting TiO_2 nanotubes have an outer diameter of about 150 nm and length of 300 nm. It has been revealed that the self-organized TiO_2 nanotubular structure appears after several hours of competition between porous oxide formation at the inner interface combined with oxide dissolution at the outer interface. The magnetic behavior exhibited by this so-obtained ferromagnetic nanocomposite corresponds to a vortex domain state of the large-diameter Ni nanowires, showing a great interaction between them due to the shorter inter-nanowire distance compared to the nanowire diameter, and also enhanced by the paramagnetic contribution of the titania nanotube matrix, leading to the easy magnetization axis lying along the perpendicular direction to the nanowire length. This so-formed ferromagnetic/semiconductor oxide nanocomposite could be a promising candidate for magnetic memory and spin-based electronic devices.

References

1. T. M. Whitney, J. S. Jiang, P. C. Searson, and C. L. Chien. Fabrication and magnetic properties of arrays of metallic nanowires. *Science*, **261** (1993) 1316.
2. D. Routkevitch, A. A. Tager, J. Haruyama, D. AlMawlawi, M. Moskovits, and J. M. Xu. Nonlithographic nano-wire arrays: Fabrication, physics, and device applications. *IEEE Trans. Electron Devices*, **43** (1996) 1646.
3. P. V. Kamat. TiO_2 nanostructures: Recent physical chemistry advances. *J. Phys. Chem. C*, **116** (2012) 11849.
4. H. Masuda and K. Fukuda. Ordered metal nanohole arrays made by a two-step replication of honeycomb structures of anodic alumina. *Science*, **268** (1995) 1466.
5. A.-P. Li, F. Müller, A. Birner, K. Nielsch, and U. Gösele. Fabrication and microstructuring of hexagonally ordered two-dimensional nanopore arrays in anodic alumina. *Adv. Mater.*, **11** (1999) 483.
6. C. A. Ross. Patterned magnetic recording media. *Annu. Rev. Mater. Res.*, **31** (2001) 203.
7. C. L. Luu, Q. T. Nguyen, and S. T. Ho. Synthesis and characterization of Fe-doped TiO_2 photocatalyst by the sol–gel method. *Adv. Nat. Sci. Nanosci. Nanotechnol.*, **1** (2010) 015008.
8. H. Masuda, K. Nishio, and N. Baba. Fabrication of porous TiO_2 films using two-step replication of microstructure of anodic alumina. *Jpn. J. Appl. Phys.*, **31** (1992) L1775.
9. D. Wong, C. A. Grimes, O. K. Varghese, W. Hu, R. S. Singh, Z. Chen, and E. C. Dickey. Titanium oxide nanotube arrays prepared by anodic oxidation. *J. Mater. Res.*, **16** (2001) 3331.
10. V. M. Prida, K. R. Pirota, D. Navas, A. Asenjo, M. Hernández-Vélez, and M. Vázquez. Self-organized magnetic nanowire arrays based on alumina and titania templates. *J. Nanosci. Nanotechnol.*, **7** (2007), 272.
11. V. Vega, J. M. Montero-Moreno, J. García, V. M. Prida, W. Rahimi, M. Waleczek, C. Bae, R. Zierold, and K. Nielsch. Long-range hexagonal arrangement of TiO_2 nanotubes by soft lithography-guided anodization. *Electrochim. Acta*, **203** (2016) 51.
12. H. Sopha, T. Samoril, E. Palesch, L. Hromadko, R. Zazpe, D. Skoda, M. Urbanek, S. Ng, J. Prikryl, and J. M. Macak. Ideally hexagonally ordered TiO_2 nanotube arrays. *ChemistryOpen*, **6** (2017) 480.
13. T. Yanagishita, H. Inada, T. Kondo, N. T. Nguyen, P. Schmuki, and H. Masuda. Efficient preparation process for TiO_2 through-hole

membranes with ordered hole arrangements. *J. Electrochem. Soc.*, **165** (2018) E763.

14. S. Maher, A. Mazinani, M. R. Barati, and D. Losic. Engineered titanium implants for localized drug delivery: Recent advances and perspectives of titania nanotubes arrays. *Expert Opin. Drug Deliv.*, **15** (2018) 1021.
15. L. Yuxiang, Z. Mei, G. Min, and W. Xidong. Preparation and properties of a nano TiO_2/Fe_3O_4 composite superparamagnetic photocatalyst. *Rare Metals*, **28** (2009) 423.
16. L.-L. Li, Y.-J. Chen, H.-P. Wu, N. S. Wang, and E. Wei-Guang Diau. Detachment and transfer of ordered TiO_2 nanotube arrays for front-illuminated dye-sensitized solar cells. *Energy Environ. Sci.*, **4** (2011) 3420.
17. B. Tan, Y. Zhang, and M. Long. Large-scale preparation of nanoporous TiO_2 film on titanium substrate with improved photoelectrochemical performance. *Nanoscale Res. Lett.*, **9** (2014) 190.
18. S. Ali, H. Granbohm, J. Lahtinen, and S.-P. Hannula. Titania nanotubes prepared by rapid breakdown anodization for photocatalytic decolorization of organic dyes under UV and natural solar light. *Nanoscale Res. Lett.*, **13** (2018) 179.
19. J. Liang, C. Hao, K. Yu, and Y. Li. Excellent photocatalytic performance of cobalt-doped titanium dioxide nanotubes under ultraviolet light. *Nanomater. Nanotechnol.*, **6** (2016) 1.
20. N. K. Shrestha, J. M. Macak, F. Schmidt-Stein, R. Hahn, C. T. Mierke, B. Fabry, and P. Schmuki. Magnetically guided titania nanotubes for site-selective photocatalysis and drug release. *Angew. Chem. Int. Ed.*, **48** (2009) 969.
21. V. Vega, V. M. Prida, M. Hernández-Vélez, E. Manova, P. Aranda, E. Ruiz-Hitzky, and M. Vázquez. Influence of anodic conditions on self-ordered growth of highly aligned titanium oxide nanopores. *Nanoscale Res. Lett.*, **2** (2007) 355.
22. V. M. Prida, E. Manova, V. Vega, M. Hernandez-Velez, P. Aranda, K. R. Pirota, M. Vázquez, and E. Ruiz-Hitzky. Temperature influence on the anodic growth of self-aligned titanium dioxide nanotube arrays. *J. Magn. Magn. Mater.*, **316** (2007) 110.
23. V. Vega, M. A. Cerdeira, V. M. Prida, D. Alberts, N. Bordel, R. Pereiro, F. Mera, S. García, M. Hernández-Vélez, and M. Vázquez. Electrolyte influence on the anodic synthesis of TiO_2 nanotube arrays. *J. Non-Cryst. Solids*, **354** (2008) 5233.

24. M. Bustelo, B. Fernández, J. Pisonero, R. Pereiro, N. Bordel, V. Vega, V. M. Prida, and A. Sanz-Medel. Pulsed radiofrequency glow discharge time-of-flight mass spectrometry for nanostructured materials characterization. *Anal. Chem.*, **83** (2011) 329.

25. V. M. Prida, M. Hernández-Vélez, M. Cervera, K. Pirota, R. Sanz, D. Navas, A. Asenjo, P. Aranda, E. Ruiz-Hitzky, F. Batallán, M. Vázquez, B. Hernando, A. Menéndez, N. Bordel, and R. Pereiro. Magnetic behaviour of arrays of Ni nanowires by electrodeposition into self-aligned titania nanotubes. *J. Magn. Magn. Mater.*, **294** (2005) e69.

26. V. M. Prida, M. Hernández-Vélez, K. R. Pirota, A. Menéndez, and M. Vázquez. Synthesis and magnetic properties of Ni nanocylinders in self-aligned and randomly disordered grown titania nanotubes. *Nanotechnology*, **16** (2005) 2696.

Chapter 15

Applications of TiO_2 Nanostructures

TiO_2 nanostructures have a wide range of environmental and energy applications [1]. Nanostructured TiO_2 is a promising original production material for many applications because of its high specific area and electronic semiconductor properties, including photocatalytic, photochromic, photovoltaic, electroluminescence, electrochromic devices and sensors [1–3]. Examples of current and potential applications of these nanostructures are described as follows.

15.1 Photocatalytic Applications

Various industries produce dyes, which have become a major source of environmental contamination. Most industries use dyes to color their products, including textile, food, tannery, paper, printing, pulp, and carpet industries. The discharge of dye wastewater into the environment is harmful. It adds deleterious color to the receiving stream and inhibits sunlight penetration into the stream, as well as it is toxic to aquatic lives and food chain organisms [4]. Some of the dyes are mutagenic and carcinogenic; any debasement that does occur may produce molecules that are alien to the environment, such as amines, which also are toxic [4].

Nanostructured Titanium Dioxide in Photocatalysis
It-Meng Low, Hani Manssor Albetran, Victor Manuel de la Prida Pidal, and Fong Kwong Yam

ISBN 978-981-4877-07-7 (Hardcover), 978-1-003-14853-1 (eBook)
www.jennystanford.com

The technique of removing pollutants from wastewater through light is called photolysis. Photocatalysis reaction employs a catalyst that absorbs light and accelerates chemical reactions, or in other words, the conversion of solar energy for oxidation and reduction processes [5]. When a semiconducting photocatalyst is illuminated by a light source greater than the band-gap energy of the semiconductor, an electron is excited from the valence band to the conduction band, and a hole is left in the valence band. The photogenerated electron–hole pairs yield oxidation and reduction processes. The photocatalytic process often occurs in an aqueous environment, on the photoanode; the photogenerated holes react with H_2O and OH^- to form hydroxyl radical $OH^\bullet$, which is the major component in the oxidation process and follows a series of repetitive reactions to finally achieve complete oxidation [5]. Whereas the photogenerated electrons react with O_2 and produce superoxide radical anion $O_2^{\bullet -}$. These two processes must be balanced for the photocatalysis reactions to occur [6].

The photocatalytic process is a qualified technique for water purification due to its benefits. First, in the presence of UV or near-UV illumination, many organic pollutants are decomposed into less harmful substances. Second, several nontoxic materials can be used as semiconductor photocatalysts, especially metal oxides, for instance ZnO, Al_2O_3, WO_3, Fe_2O_3, In_2O_3, and V_2O_5 and most especially TiO_2 [7, 8]. TiO_2 semiconductor photocatalyst has been widely used compared to CdS, SrO_2, ZrO_2, and ZnO, among others [9]. Third, photocatalysis is relevant to long-term situations as photocatalyst material is not consumed during the process. Last but not the least, the photocatalytic process has low operating cost due to the use of sunlight as the source of irradiation in addition to serving as an environmental cleaning process [10].

TiO_2 was discovered as a photocatalyst in 1972 by two Japanese scientists Fujishima and Honda, who demonstrated the potential of titania semiconductor materials to split water into oxygen and hydrogen in a photochemical cell [11]. Since then, TiO_2 has received much attention as a photocatalyst and has become very attractive in the application of self-cleaning, air cleaning, and antibacterial effects in the construction industry. The merits of TiO_2, including high photostability, low cost, and nontoxicity, have allowed its wider usage in photocatalytic applications.

The photocatalytic activity of TiO_2 depends mainly on its physicochemical properties, such as particle size, morphology, porosity, surface area, and other physical characteristics [12]. The photocatalytic applications of TiO_2 include the degradation of organic pollutants in aqueous and gaseous phases, hydrogen gas generation from photocatalytic water splitting, and removal of heavy metals from contaminated water, among others. TiO_2 semiconductor interacts with light of sufficient energy to produce reactive oxidizing species, which can lead to the photocatalytic transformation of an air or water pollutant. At least two events must occur simultaneously during the photocatalytic process/reaction for the successful production of reactive oxidizing species: first, the oxidation of dissociative adsorbed H_2O by photogenerated holes; and second, the reduction of an electron acceptor by photoexcited electrons. These reactions lead to the production of a hydroxyl and superoxide radical anion of the two events [13].

The anatase and rutile phases of TiO_2 are commonly applied as photocatalysts, with the anatase type showing a higher photocatalytic activity than rutile [14]. However, mixtures of anatase and rutile have been reported to exhibit superior photoactivity than either anatase or rutile alone [15–17]. Degussa P25 TiO_2 photocatalyst reveals multiphasic material consisting of an amorphous state, together with the approximate crystalline phases 80% anatase and 20% rutile. The photocatalytic activity of Degussa P25 is reported as being greater than the activities of either of the pure titania crystalline phases (anatase and rutile). An interpretation of this reflection has also been given in terms of the development in the magnitude of the space–charge potential, which is created by contact between anatase and rutile crystalline phases and by the presence of localized electronic states from the amorphous phase [18].

A heterogeneous photocatalytic system consists of semiconductor photocatalysts like TiO_2, which are in close contact with gaseous or liquid reaction. Doping, heterostructuring, and crystal growth of semiconductor photocatalysts should be studied substantially. The functional properties of TiO_2 photocatalyst are important for its photoreactive properties, including electronic structure, surface properties, and charge transport.

Doh et al. [19] developed a photocatalyst of anatase particles/nanofibers. TiO_2 nanofibers (average 236 nm thick) were fabricated using the electrospinning method. Using the sol–gel method, the photocatalytic TiO_2 particles were coated on the TiO_2 nanofibers to increase the effective surface area and photocatalytic activity. They claimed that the degradation rate of composite TiO_2 was much higher than that of TiO_2 nanoparticles and TiO_2 nanofibers. Therefore, they suggested that the composite TiO_2 of nanoparticles and nanofibers would be suitable for the degradation of organic pollutants [19].

A TiO_2 photocatalyst is a semiconducting substance that can be chemically activated under UV light radiation, which results in an oxidation–reduction (redox) reaction, i.e., it has a photocatalysis action or photoactivity. Thus, various modifications were carried out to construct the photocatalytic application to be active under visible light. Metal- and nonmetal-doped TiO_2 photocatalysts were developed to enable photocatalysis under visible light. Doping can exert a substantial influence on modifying the electronic structure and the construction of heteroatomic surface structures that allow high-efficiency TiO_2 photocatalysis under solar light irradiation. In 2001, the nitrogen-doped TiO_2 photocatalyst was developed to enable photocatalysis under visible light [20]. The nitrogen-doped TiO_2 narrowed the band gap to visible light by mixing O_{2p} states with the N_{2p} states of the substituted nitrogen atoms in the newly formed valence band (VB). Since then, it has attracted much attention with other metal and nonmetal doping. Metal-ion dopants such as Cr, V, Fe, Mn, Ni, Co, etc. in the TiO_2 lattices can influence the performance of photocatalyst through the effects on electron/hole recombination dynamics and interfacial charge transfer [21].

15.1.1 Antifogging and Self-Cleaning

Fogging on eyeglasses, car windows, and bathroom mirrors is caused by condensation of water, which forms small droplets on the glass surface. The wetting of a solid surface with water where air is the surrounding medium is dependent on the contact angle. In general, 0° contact angle means completely wet, whereas 180° means completely non-wet. Cleaning by soap is based on low contact angle by reducing the water surface tension. The contact angle of water on glass and other inorganic materials is in the range of 20–30° but is not less than 10° for any surfaces under any conditions. The contact angle can be decreased without detergents but by putting a thin layer on the material surface. A hydrophobic surface is the main method to prevent this fogging by repelling water molecules and then removing water drops by blowing or shaking. Materials such as ceramic tiles, glass, or plastics can develop super-hydrophilic materials by surface coating with the photocatalyst TiO_2.

The antifogging, antibacterial, and self-cleaning functions of TiO_2-coated materials are obtained without using any chemicals, but with only UV sunlight and rainwater. The water contact angle of the illuminated surface with UV light of anatase TiO_2 is less than 1°. The super-hydrophilic effect of TiO_2 occurs when the surface is exposed to light and the water contact angle approaches zero after a certain time of moderate illumination [22].

The application of TiO_2 photocatalysis on surface materials has enabled the degradation of a range of organic and inorganic compounds that are aggressive toward material properties and the environment. TiO_2 photocatalysis has been applied in different cementation materials, and the applications varied from self-cleaning to air cleaning especially for indoor. TiO_2 coating on ceramic tiles is used for its antibacterial properties to increase the lifecycle of cement-based materials and to substantially reduce the concentration of air pollutants in hospital rooms, care facilities, kitchens, baths, and schools [23]. A particularly interesting aspect is clear synergy between TiO_2 and cement composites, which makes this material an ideal substrate for environmental photocatalysis. Photocatalytic building materials used outdoors, e.g., PVC fabric, aluminum wall, glass, and exterior tiles, are coated with TiO_2, where they could be easily washed by water, raining, rainwater, and rainfall. This is because water soaks between the highly hydrophilic TiO_2 surface and the materials.

Rutile has been reported to have a higher surface enthalpy and higher surface free energy than anatase due to different crystal structures and associated exposed planes [24]. Therefore, it is expected that anatase wetting by water is less than that of rutile since higher surface free energies commonly contribute to hydrophilicity [25]. Rutile (100) and (110) faces exhibit a higher hydrophilicity than the (001) face, which are favorable for increasing the number of OH groups. The (100) and (110) faces have twofold oxygens, which are higher in position and energetically more reactive than their surrounding atoms and are called "bridging site oxygen." On the other hand, the rutile (001) has threefold oxygens, which are lower in position and energetically more inactive [26].

15.1.2 Photocatalysts for Water Treatment and Air Purification

In the 1970s, Frank and Bard discovered that illuminated TiO_2 and other metal oxides can be served as a photocatalyst for photocatalytic decomposition of pollutants in water [27–29]. A wide range of organic and inorganic as well as toxic compounds, including cyanide, sulfite, phenol, and aniline, were decomposed (oxidized) by the illuminated polycrystalline TiO_2 [27, 28, 30]. These reports demonstrated that water and air purification can be achieved by photocatalysis reaction, which contributes to the decomposition of pollutants for a cleaner and greener environment. Regarding air purification, another critical environmental issue is the substantial climate change primarily caused by the increased concentration of greenhouse gases. In particular, the excess emission of CO_2 is attributed to the large amount of human activities because of rapid industrial revolution and excessive combustion of fossil fuels. In the 1990s, Anpo et al. designed an efficient TiO_2 photocatalyst via metal-ion implantation within TiO_2 to address the decomposition of pollutants in water, which has been mentioned above, as well as for the photoreduction of CO_2 [31–33]. The illuminated TiO_2 again serves as a photocatalyst to transform CO_2 into other nontoxic or reusable compounds such as H_2, H_2O, and other hydrocarbons (methane, ethane, methanol, acetic acid, formic acid) [34–36]. Consecutively, some of the derived compounds can be utilized in another green solar fuel, which in turn reduces the use of fossil fuels [37, 38].

The three-dimensional (3D) spaces such as water, air, and soil are much more difficult than that of the two-dimensional (2D) surfaces of photocatalytic building materials. In general, the total amount of reactants in 3D spaces is higher than on the 2D surface, indicating that much more light energy is necessary for the purification of 3D space [39]. Moreover, the photocatalytic reactions are surface reactions, and thus the reactants must be captured by the photocatalyst surface. Fortunately, the construction of practical purification systems for wastewater from agriculture and polluted soil has been successful with volatile organic compounds (VOCs), which were based on TiO_2 photocatalysts and use only solar light. Nanosized TiO_2 photocatalyst powders were dispersed on substrates with extremely large surface areas, spreading them on the ground widely to collect sunlight [39].

15.1.3 TiO_2 Photobioreactor

In the food and environmental industries, anatase photocatalysts as a photobioreactor have attracted great attention for sterilizing selected food-borne pathogenic bacteria such as *Listeria monocytogenes*, *Vibrio parahaemolyticus*, and *Salmonella choleraesuis* subsp [40]. Various anatase concentrations and UV illumination time have been used to study the photocatalytic reaction. On all bacterial suspensions, the bactericidal effect of UV/anatase was much higher than without anatase. As the anatase concentration increased to 1.0 mg/ml, bactericidal effect increased, but it was rapidly abbreviated at higher than 1.25 mg/ml anatase concentration to all selected bacteria. The time of anatase/UV illumination affected the viability of bacteria with different death rates drastically. After 3 h of UV illumination, *S. choleraesuis* subsp and *V. parahaemolyticus* were completely killed, whereas about 87% death ratio was for the killed *Listeria monocytogenes* [40].

15.2 Photovoltaic Applications

Rutile TiO_2 was selected as a photoanode due to its sufficiently positive valence band edge to convert H_2O to O_2 through oxidation [39]. It was also a coincidence that during the same time, crude oil prices boosted tremendously due to the ever-depleting oil reservoirs. Therefore, photoelectrochemistry and photocatalysts became closely related and immediately captured the attention of many, as TiO_2 photocatalyst in photoelectrochemical cells appeared as a promising approach to develop a clean energy source [39, 41].

Photovoltaics is the conversion of sunlight to electrical power. In the past decade, the cost of solar cells or photovoltaics reduced greatly because the conversion efficiency sharply increased with the introduction and application of sophisticated technologies. However, the production cost of a solar cell is still higher than that of fossil fuels. The high cost of silicon, which is the main solar energy material, is the main reason behind the present limitation in the use of solar energy. Thus, there is an increasingly urgent need for developing environmentally friendly energy with relatively cheap price. Low-cost TiO_2 semiconductor is a promising original production material, which is often related to the cost of solar energy technologies. Furthermore, recent developments in the synthesis of nanocrystalline TiO_2 with various morphologies/forms have opened several new opportunities for the construction of efficient photovoltaic cells and a wide range of applications.

Solar energy is available in abundance and should be used more extensively. Semiconductors have important application in renewable energy and environment. Nanocrystalline TiO_2 and its conductivity are being established and challenged for use as a new generation of photovoltaic cells. Sensitized nanostructured solar cells have been assembled into ruthenium dye (N-719) using photoanode TiO_2 nanotube arrays. A solar cell in backside illumination mode has been formed using anodic TiO_2 nanotube arrays, yielding a 7.4% power conversion efficiency (PCE) [42].

15.2.1 Lithium Batteries

Lithium-ion batteries (LIBs) are widely used in portable electric vehicles and electronic devices and toys. For high-performance second LIB generation, recent progress in advanced nanoscience and nanotechnology has paved the way for the development of diverse nanostructured electrodes. It has led to improved energy density, higher cycling rate, and capability, because it has large surface area, short mass and charge-diffusion distance, numerous active sites, and efficient accommodation of volume changes [43]. Nanostructured TiO_2 is widely used in rechargeable lithium batteries as electrodes because of their superior safety, low cost, chemical stability, and nontoxicity. Elongated TiO_2 with various morphologies (e.g., nanotubes [44–46], nanowires [47–49], nanofibers [50], and nanorods [51]) has been utilized for lithium storage. These TiO_2 nanostructures attract attention as a possible negative electrode material for lithium cells, because they have open mesoporous structure, efficient transport of lithium ions, and effective ion-exchange/properties. In terms of safety, TiO_2 is a promising anode material for lithium-ion storage, as it alleviates overcharge by moving toward a higher potential, thereby avoiding lithium plating and parasitic electrochemical reaction [52].

15.2.2 Photoelectrochemical Cells

Hydrogen is likely to be an alternative energy and an ideal fuel in the future, which is produced from clean and green energy sources. It is an environmentally friendly gas and recognized as an efficient source of fuel. Renewable hydrogen production is not common yet due to its high production cost. Photovoltaic water electrolysis can become more competitive by splitting off water using semiconductor photocatalyst. The cost of hydrogen continues to decrease with technological advancement, but a considerable use of some semiconducting materials can cause serious lifecycle impacts on the environment. Fortunately, a promising semiconductor for clean, low-cost, and environmentally friendly production of hydrogen is offered by photocatalytic water splitting [53, 54].

In early 1972, the first photocatalytic decomposition of water into oxygen and hydrogen used photoexcited TiO_2 as the anode and Pt as the electrode in the photoelectrochemical cell (PEC) technology [55]. The PEC technology is an environmentally safe renewable resource since it is based on solar energy and can be used in both small and large scales and in a relatively uncomplicated manner. There are three options in PEC technology: cathode made of metal and photoanode made of n-type semiconductor; photocathode made of p-type semiconductor and photoanode made of n-type semiconductor; and anode made of metal and photocathode made of p-type semiconductor [56].

In the first PECs by Fujishima [55], the surface of n-type TiO_2 as a photoanode was irradiated with light consisting of energy larger than its band gap and photocurrent flowed from the platinum counter photocathode to the titania through the external circuit. The direction of the current revealed that the oxidation reaction (oxygen evolution) and the reduction reaction (hydrogen evolution) occur at the TiO_2 photoanode and the Pt photocathode, respectively. Water can be decomposed into hydrogen and oxygen using UV light, without the application of an external voltage, according to the following scheme [22, 39].

TiO_2 excitation by light:

$$TiO_2 + h\nu \longrightarrow e^- + h^+ \quad (15.1)$$

At the TiO_2 electrode:

$$2H_2O + 4h^+ \longrightarrow O_2 + 4H^+ \quad (15.2)$$

At the Pt electrode:

$$2H^+ + 2e^- \longrightarrow H_2 \quad (15.3)$$

The overall reaction is

$$2H_2O + 4h\nu \longrightarrow O_2 + 2H_2 \quad (15.4)$$

The water-splitting reaction over TiO_2 semiconductor-photocatalyst materials has attracted much attention, although TiO_2 has several advantages and remains one of the most promising because of its photostability, chemical inertness, environmentally friendly solar hydrogen production to support the future hydrogen economy, and low cost [11]. To maximize photocatalytic decomposition of water using TiO_2 as the photoanode, the TiO_2 band gap should utilize visible-light energy and high contact area with the electrolyte to increase the splitting of the e^-/h^+ pairs. Thus, metal- and nonmetal-doped nanostructured TiO_2 has great potential for photocatalytic water-splitting technology with high aspect ratios for maximizing the photocleavage of water under visible-light irradiation. The total photocurrent of carbon-doped TiO_2 nanotube arrays was 20 times higher than the pure P-25 TiO_2 under visible-light illumination [57].

15.2.3 Dye-Sensitized Solar Cells

Dye-sensitized solar cells (DSSCs) are a relatively new class of low-cost solar cells for renewable electricity generation [58, 59]. Semiconductor-sensitized solar cells have attracted growing interest, starting from quite low (4–5%) conversion efficiencies [60]. Nanotechnology and nanoscience are believed to revolutionize the industry shortly and will have a significant economic impact allowing a considerable reduction in the cost of photovoltaic energy, by both increasing efficiencies and reducing the cost of devices.

In 1991, nanofilm titania was reported as an efficient DSSC by O'regan and Graetzel [61]. Unlike the traditional photovoltaic technology in silicon-based solar cells, the establishment of a p-n junction is not required in the third-generation solar cells. Moreover, it serves to enhance the poor electrical performance of second-generation solar cells such as CdTe and copper indium gallium selenide (CIGS), yet maintaining low material and fabrication costs [62, 63]. A DSSC structure consists of four major components: a dye sensitizer, a TiO_2 anode, and a highly transparent counter electrode (Pt or C coated on glass), chemically connected by a liquid electrolyte as the redox mediator (commonly I^-/I_3^- electrolyte) [63, 64]. TiO_2 nanoparticles are used in DSSCs, which are based on solar energy by employing sensitizers adsorbed on the surface of n-type semiconductor metal oxide [65]. Moreover, TiO_2-films-based DSSCs that use ruthenium sensitizers offer power conversion efficiency of only 9.57–11.18% and primarily absorb light in the region of 350–700 nm [66].

Although most of the reported works on DSSCs are based on TiO_2 thin films, various nanostructured TiO_2 films have been used for DSSC fabrication. Nanocrystalline anatase TiO_2 is usually used for DSSCs, and various modifications have been carried out on other TiO_2 phases to construct an energy-efficient photovoltaic device [67].

TiO_2 nanotubes have been examined for use as an electrode for DSSCs. The potential advantage of TiO_2 nanotubes compared to TiO_2 nanopowder is realized by exploiting the phenomenon of improved adsorption of the positively charged dyes from aqueous solution onto the surface of negatively charged TiO_2 nanotubes [68]. The other advantage of the nanotubes is the elongated structures of these semiconductors, which can improve the electron transport and the charge collection efficiency [2].

15.3 Sensing Applications

Contamination of air by several pollutants is a consequence of human activities. An extensive research is necessary for the design and fabrication of more efficient gas-sensing systems that can detect pollutants contained in the air. Many metal oxide semiconductors such as SnO_2, TiO_2, and ZnO have been used because these are the most common ceramics used for gas sensor application. To be able to control various air pollutants, sensors must be very sensitive, selective, and reliable. Below 250°C, SiO_2-based sensors are widely used as gas-sensitive resistors because of their good sensitivity and selectivity. However, above 250°C, TiO_2-based sensors have been reported to become the preferred material for high-temperature gas sensors because of their good sensitivity and stability [69].

15.4 Coatings

TiO_2 is the most important white pigment used in the coating industry because of its brightness and opacity. It effectively scatters visible light and thereby imparts whiteness when incorporated in a coating. Table 15.1 shows the refractive indices for anatase, rutile, and other pigments. Anatase and rutile have the two highest refractive indices with 2.55 and 2.73, respectively. Rutile pigments scatter light more efficiently and are more stable and durable than anatase pigments.

Table 15.1 Refractive indices for some pigments (www.chemours.com)

Pigments	Refractive indices
Diatomaceous earth	1.45
Silica	1.45–1.49
Calcium carbonate	1.63
Barytes	1.64
Clay	1.65
Magnesium silicate	1.65
Lithopone	1.84
Zinc oxide	2.02
Antimony oxide	2.09–2.29
Zinc sulfide	2.37
Anatase	2.55
Rutile	2.73

15.5 Drug Delivery and Bioapplications

The nanostructured TiO_2 has recently been studied as a possible element in amperometric biosensors due to its high surface area and affinity toward positively charged ions in aqueous solutions [2]. TiO_2 nanotubes can be developed as capsules for drug delivery and controlled release based on a combination of several nanotube properties. The high surface area and large pore volume of TiO_2 nanotubes provide an increased load capacity required for drug storage. It can also be potentially beneficial in drug delivery to the targeted tissue. Moreover, TiO_2 nanowires can be used as bio-scaffolds for cell cultures, providing enough rigidity and a large macropod structure suitable for cell growth and nutrition [70].

Co-adsorption of horseradish peroxidase (HRP) with thionine on anodic TiO_2 nanotubes for biosensor design has been investigated. The TiO_2 nanotube provides excellent matrices for the co-adsorption of HRP and thionine, where the adsorbed HRP effectively retains its bioactivity on these TiO_2 nanotube arrays [71]. Moreover, the titanium dioxide thin film has shown an ability to decompose organic compounds, including *Escherichia coli* cells in water treatment [72, 73].

Integrating the applications discussed in this chapter, TiO_2 has proven its vast purposes as a photocatalyst. It is worth pointing that TiO_2 is also functional in electrochemical energy and storage such as ion-intercalation batteries and supercapacitors, which makes it a promising material in solar-based and energy-related applications, to resolve current global challenges in energy and environmental areas.

References

1. Tang, Z. R., Li, F., Zhang, Y. H., Fu, X. Z., and Xu, Y. J. (2011). Composites of titanate nanotube and carbon nanotube as photocatalyst with high mineralization ratio for gas-phase degradation of volatile aromatic pollutant, *J. Phys. Chem. C*, **115**, pp. 7880–7886.
2. Bavykin, D. V. and Walsh, F. C. (2010). *Titanate and Titania Nanotubes Synthesis: Properties and Applications*. Cambridge: RSC Publishing.
3. Park, J. Y., Yun, J. J., Hwang, C. H., and Lee, I. H. (2010). Influence of silver doping on the phase transformation and crystallite growth of electrospun TiO_2 nanofibers, *Mater. Lett.*, **64**, pp. 2692–2695.
4. Chu, W. (2001). Dye removal from textile dye wastewater using recycled alum sludge, *Water Res.*, **35**, pp. 3147–3152.

5. Casbeer, E., Sharma, V. K., and Li, X. Z. (2012). Synthesis and photocatalytic activity of ferrites under visible light: A review, *Sep. Purif. Technol.*, **87**, pp. 1–14.
6. Fujishima, A., Zhang, X., and Tryk, D. A. (2008). TiO_2 photocatalysis and related surface phenomena, *Surf. Sci. Rep.*, **63**, pp. 515–582.
7. Kment, S., Riboni, F., Pausova, S., Wang, L., Wang, L., Han, H., Hubicka, Z., Krysa, J., Schmuki, P., and Zboril, R. (2017). Photoanodes based on TiO_2 and α-Fe_2O_3 for solar water splitting-superior role of 1D nanoarchitectures and of combined heterostructures, *Chem. Soc. Rev.*, **46**, pp. 3716–3769.
8. Ishchenko, O. M., Rogé, V., Lamblin, G., and Lenoble, D. (2016). TiO_2-and ZnO-based materials for photocatalysis: Material properties, device architecture and emerging concepts, in *Semiconductor Photocatalysis-Materials, Mechanisms and Applications*. InTech.
9. Serpone, N., Maruthamuthu, P., Pichat, P., Pelizzetti, E., and Hidaka, H. (1995). Exploiting the interparticle electron transfer process in the photocatalysed oxidation of phenol, 2-chlorophenol and pentachlorophenol: Chemical evidence for electron and hole transfer between coupled semiconductors, *J. Photochem. Photobiol. A: Chem.*, **85**, pp. 247–255.
10. Jantawasu, P., Sreethawong, T., and Chavadej, S. (2009). Photocatalytic activity of nanocrystalline mesoporous-assembled TiO_2 photocatalyst for degradation of methyl orange monoazo dye in aqueous wastewater, *Chem. Eng. J.*, **155**, pp. 223–233.
11. Fujishima, A. and Honda, K. (1972). Photolysis-decomposition of water at the surface of an irradiated semiconductor, *Nature*, **238**, pp. 37–38.
12. Guo, D. D., Lin, S. S., Zhong, F. X., Zhu, Y. N., and Meng, Z. P. (2012). High-efficiency degradation sugar wastewater by Co-TiO_2 nanotube arrays, *Integr. Ferroelectr.*, **137**, pp. 10–17.
13. Mills, A. and Hunte, L. S. (1997). An overview of semiconductor photocatalysis, *J. Photochem. Photobiol. A: Chem.*, **108**, pp. 1–35.
14. Linsebigler, A. L., Lu, G., and Yates Jr, J. T. (1995). Photocatalysis on TiO_2 surfaces: Principles, mechanisms, and selected results, *Chem. Rev.*, **95**, pp. 735–758.
15. Kment, S., Kluson, P., Stranak, V., Virostko, P., Krysa, J., Cada, M., Pracharova, J., Kohout, M., Morozova, M., and Adamek, P. (2009). Photo-induced electrochemical functionality of the TiO_2 nanoscale films, *Electrochim. Acta*, **54**, pp. 3352–3359.

16. Natoli, A., Cabeza, A., Torre, A. G., Aranda, M. A., and Santacruz, I. (2012). Colloidal processing of macroporous TiO_2 materials for photocatalytic water treatment, *J. Am. Ceram. Soc.*, **95**, pp. 502–508.

17. Li, H., Zhang, W., and Pan, W. (2011). Enhanced photocatalytic activity of electrospun TiO_2 nanofibers with optimal anatase/rutile ratio, *J. Am. Ceram. Soc.*, **94**, pp. 3184–3187.

18. Bickley, R. I., Gonzalez-Carreno, T., Lees, J. S., Palmisano, L., and Tilley, R. J. (1991). A structural investigation of titanium dioxide photocatalysts, *J. Solid State Chem.*, **92**, pp. 178–190.

19. Doh, S. J. (2008). Development of photocatalytic TiO_2 nanofibers by electrospinning and its application to degradation of dye pollutants, *J. Hazard. Mater.*, **154**, pp. 118–127.

20. Asahi, R., Morikawa, T., Ohwaki, T., Aoki, K., and Taga, Y. (2001). Visible-light photocatalysis in nitrogen-doped titanium oxides, *Science*, **293**, pp. 269–271.

21. Choi, W., Termin, A., and Hoffmann, M. R. (1994). The role of metal ion dopants in quantum-sized TiO_2: Correlation between photoreactivity and charge carrier recombination dynamics, *J. Phys. Chem.*, **98**, pp. 13669–13679.

22. Fujishima, A., Rao, T. N., and Tryk, D. A. (2000). Titanium dioxide photocatalysis, *J. Photochem. Photobiol. C: Photochem. Rev.*, **1**, pp. 1–21.

23. Benedix, R., Dehn, F., Quaas, J., and Orgass, M. (2000). Application of titanium dioxide photocatalysis to create self-cleaning building materials, *Lacer*, **5**, pp. 157–168.

24. Zhang, H. and Banfield, J. F. (2005). Size dependence of the kinetic rate constant for phase transformation in TiO_2 nanoparticles, *Chem. Mater.*, **17**, pp. 3421–3425.

25. Rupp, F., Scheideler, L., Olshanska, N., De Wild, M., Wieland, M., and Geis-Gerstorfer, J. (2006). Enhancing surface free energy and hydrophilicity through chemical modification of microstructured titanium implant surfaces, *J. Biomed. Mater. Res. Part A*, **76**, pp. 323–334.

26. Wang, R., Sakai, N., Fujishima, A., Watanabe, T., and Hashimoto, K. (1999). Studies of surface wettability conversion on TiO_2 single-crystal surfaces, *J. Phys. Chem. B*, **103**, pp. 2188–2194.

27. Frank, S. N. and Bard, A. J. (1977). Heterogeneous photocatalytic oxidation of cyanide ion in aqueous solutions at titanium dioxide powder, *J. Am. Chem. Soc.*, **99**, pp. 303–304.

28. Frank, S. N. and Bard, A. J. (1977). Semiconductor electrodes. 12. Photoassisted oxidations and photoelectrosynthesis at polycrystalline titanium dioxide electrodes, *J. Amer. Chem. Soc.*, **99**, pp. 4667–4675.

29. Frank, S. N. and Bard, A. J. (1977). Heterogeneous photocatalytic oxidation of cyanide and sulfite in aqueous solutions at semiconductor powders, *J. Phys. Chem.*, **81**, pp. 1484–1488.

30. Vinodgopal, K., Hotchandani, S., and Kamat, P. V. (1993). Electrochemically assisted photocatalysis: Titania particulate film electrodes for photocatalytic degradation of 4-chlorophenol, *J. Phys. Chem.*, **97**, pp. 9040–9044.

31. Anpo, M. and Chiba, K. (1992). Photocatalytic reduction of CO_2 on anchored titanium oxide catalysts, *J. Mol. Catal.*, **74**, pp. 207–212.

32. Anpo, M., Yamashita, H., Ichihashi, Y., and Ehara, S. (1995). Photocatalytic reduction of CO_2 with H_2O on various titanium oxide catalysts, *J. Electroanal. Chem.*, **396**, pp. 21–26.

33. Anpo, M., Dohshi, S., Kitano, M., Hu, Y., Takeuchi, M., and Matsuoka, M. (2005). The preparation and characterization of highly efficient titanium oxide-based photofunctional materials, *Annu. Rev. Mater. Res.*, **35**, pp. 1–27.

34. Mahmodi, G., Sharifnia, S., Madani, M., and Vatanpour, V. (2013). Photoreduction of carbon dioxide in the presence of H_2, H_2O and CH_4 over TiO_2 and ZnO photocatalysts, *Solar Energy*, **97**, pp. 186–194.

35. Razzaq, A., Sinhamahapatra, A., Kang, T. H., Grimes, C. A., Yu, J. S., and In, S. I. (2017). Efficient solar light photoreduction of CO_2 to hydrocarbon fuels via magnesiothermally reduced TiO_2 photocatalyst, *Appl. Catalysis B: Environ.*, **215**, pp. 28–35.

36. Delavari, S. and Amin, N. A. S. (2016). Photocatalytic conversion of CO_2 and CH_4 over immobilized titania nanoparticles coated on mesh: Optimization and kinetic study, *Appl. En.*, **162**, pp. 1171–1185.

37. Habisreutinger, S. N., Schmidt-Mende, L., and Stolarczyk, J. K. (2013). Photocatalytic reduction of CO_2 on TiO_2 and other semiconductors, *Angew. Chem. Int. Ed.*, **52**, pp. 7372–7408.

38. Low, J., Cheng, B., and Yu, J. (2017). Surface modification and enhanced photocatalytic CO_2 reduction performance of TiO_2: A review, *Appl. Surf. Sci.*, **392**, pp. 658–686.

39. Hashimoto, K., Irie, H., and Fujishima, A. (2005). TiO_2 photocatalysis: A historical overview and future prospects, *Jap. J. Appl. Phys.*, **44**, pp. 8269.

40. Kim, B., Kim, D., Cho, D., and Cho, S. (2003). Bactericidal effect of TiO_2 photocatalyst on selected food-borne pathogenic bacteria, *Chemosphere*, **52**, pp. 277–281.

41. Fujishima, A. and Honda, K. (1972). Electrochemical photolysis of water at a semiconductor electrode, *Nature*, **238**, pp. 37.

42. Wang, J. and Lin, Z. (2009). Dye-sensitized TiO_2 nanotube solar cells with markedly enhanced performance via rational surface engineering, *Chem. Mater.*, **22**, pp. 579–584.

43. Nam, K. T., Kim, D. W., Yoo, P. J., Chiang, C. Y., Meethong, N., Hammond, P. T., Chiang, Y. M., and Belcher, A. M. (2006). Virus-enabled synthesis and assembly of nanowires for lithium ion battery electrodes, *Science*, **312**, pp. 885–888.

44. Zhou, Y. K., Cao, L., Zhang, F. B., He, B. L., and Li, H. L. (2003). Lithium insertion into TiO_2 nanotube prepared by the hydrothermal process, *J. Electrochem. Soc.*, **150**, pp. 1246–1249.

45. Fang, H. T., Liu, M., Wang, D. W., Sun, T., Guan, D. S., Li, F., Zhou, J., Sham, T. K., and Cheng, H. M. (2009). Comparison of the rate capability of nanostructured amorphous and anatase TiO_2 for lithium insertion using anodic TiO_2 nanotube arrays, *Nanotechnology*, **20**, pp. 225701.

46. Li, J., Tang, Z., and Zhang, Z. (2005). H-titanate nanotube: A novel lithium intercalation host with large capacity and high rate capability, *Electrochem. Commun.*, **7**, pp. 62–67.

47. Armstrong, A. R., Armstrong, G., Canales, J., Garcia, R., and Bruce, P. G. (2005). Lithium-ion intercalation into TiO_2-B nanowires, *Adv. Mater.*, **17**, pp. 862–865.

48. Li, J., Tang, Z., and Zhang, Z. (2005). Layered hydrogen titanate nanowires with novel lithium intercalation properties, *Chem. Mater.*, **17**, pp. 5848–5855.

49. Wang, Q., Wen, Z., and Li, J. (2006). Solvent-controlled synthesis and electrochemical lithium storage of one-dimensional TiO_2 nanostructures, *Inorg. Chem.*, **45**, pp. 6944–6949.

50. Nam, S. H., Shim, H. S., Kim, Y. S., Dar, M. A., Kim, J. G., and Kim, W. B. (2010). Ag or Au nanoparticle-embedded one-dimensional composite TiO_2 nanofibers prepared via electrospinning for use in lithium-ion batteries, *ACS Appl. Mater. Interfaces*, **2**, pp. 2046–2052.

51. Gao, X., Zhu, H., Pan, G., Ye, S., Lan, Y., Wu, F., and Song, D. (2004). Preparation and electrochemical characterization of anatase nanorods for lithium-inserting electrode material, *J. Phys. Chem. B*, **108**, pp. 2868–2872.

52. Jiang, J., Chen, J., and Dahn, J. (2004). Comparison of the reactions between $Li_7/3Ti_5/3O_4$ or LiC_6 and nonaqueous solvents or electrolytes

using accelerating rate calorimetry, *J. Electrochem. Soc.*, **151**, pp. 2082–2087.

53. Sheppard, L. R. and Nowotny, J. (2007). Materials for photoelectrochemical energy conversion, *Adv. Appl. Ceram.*, **106**, pp. 9–20.
54. Bard, A. J. and Fox, M. A. (1995). Artificial photosynthesis: Solar splitting of water to hydrogen and oxygen, *Acc. Chem. Res.*, **28**, pp. 141–145.
55. Fujishima, A. (1972). Electrochemical photolysis of water at a semiconductor electrode, *Nature*, **238**, pp. 37–38.
56. Bak, T. (2002). Photo-electrochemical hydrogen generation from water using solar energy. Materials-related aspects, *Int. J. Hydrogen Energy*, **27**, pp. 991.
57. Ni, M., Leung, M. K., Leung, D. Y., and Sumathy, K. (2007). A review and recent developments in photocatalytic water-splitting using TiO_2 for hydrogen production, *Renew. Sustain. Energy Rev.*, **11**, pp. 401–425.
58. Grätzel, M. (2001). Photoelectrochemical cells, *Nature*, **414**, pp. 338–344.
59. Grätzel, M. (2005). Solar energy conversion by dye-sensitized photovoltaic cells, *Inorg Chem.*, **44**, pp. 6841–6851.
60. Mora-Sero, I. and Bisquert, J. (2010). Breakthroughs in the development of semiconductor-sensitized solar cells, *J. Phys. Chem. Lett.*, **1**, pp. 3046–3052.
61. O'regan, B. and Grätzel, M. (1991). A low-cost, high-efficiency solar cell based on dye-sensitized colloidal TiO_2 films, *Nature*, **353**, pp. 737–740.
62. Samuel, J. J. (2018). Nanostructured TiO_2. PhD Thesis. (Universiti Sains Malaysia).
63. Bagher, A. M., Vahid, M. M. A., and Mohsen, M. (2015). Types of solar cells and application, *Am. J. Optics Photonics*, **3**, pp. 94–113.
64. Sangster, A. J. (2014) Solar photovoltaics, in *Electromagnetic Foundations of Solar Radiation Collection*. Springer. pp. 145–172.
65. Bai, Y., Cao, Y., Zhang, J., Wang, M., Li, R., Wang, P., Zakeeruddin, S. M., and Grätzel, M. (2008). High-performance dye-sensitized solar cells based on solvent-free electrolytes produced from eutectic melts, *Nature Mater.*, **7**, pp. 626–630.
66. Nazeeruddin, M. K., De Angelis, F., Fantacci, S., Selloni, A., Viscardi, G., Liska, P., Ito, S., Takeru, B., and Grätzel, M. (2005). Combined experi-

mental and DFT-TDDFT computational study of photoelectrochemical cell ruthenium sensitizers, *J. Am. Chem. Soc.*, **127**, pp. 16835–16847.

67. Huang, S., Schlichthörl, G., Nozik, A., Grätzel, M., and Frank, A. (1997). Charge recombination in dye-sensitized nanocrystalline TiO_2 solar cells, *J. Phys. Chem. B*, **101**, pp. 2576–2582.

68. Lee, C. K., Wang, C. C., Juang, L. C., Lyu, M. D., Hung, S. H., and Liu, S. S. (2008). Effects of sodium content on the microstructures and basic dye cation exchange of titanate nanotubes, *Colloids Surf. A: Physicochem. Eng. Aspects*, **317**, pp. 164–173.

69. Zheng, J., Li, G., Ma, X., Wang, Y., Wu, G., and Cheng, Y. (2008). Polyaniline–TiO_2 nano-composite-based trimethylamine QCM sensor and its thermal behavior studies, *Sens. Actuators B: Chem.*, **133**, pp. 374–380.

70. Dong, W., Zhang, T., Epstein, J., Cooney, L., Wang, H., Li, Y., Jiang, Y. B., Cogbill, A., Varadan, V., and Tian, Z. R. (2007). Multifunctional nanowire bioscaffolds on titanium, *Chem. Mater.*, **19**, pp. 4454–4459.

71. Liu, S. and Chen, A. (2005). Coadsorption of horseradish peroxidase with thionine on TiO_2 nanotubes for biosensing, *Langmuir*, **21**, pp. 8409–8413.

72. Sunada, K., Kikuchi, Y., Hashimoto, K., and Fujishima, A. (1998). Bactericidal and detoxification effects of TiO_2 thin film photocatalysts, *Environ. Sci. Technol.*, **32**, pp. 726–728.

73. Sunada, K., Watanabe, T., and Hashimoto, K. (2003). Studies on photokilling of bacteria on TiO_2 thin film, *J. Photochem. Photobiol. A: Chem.*, **156**, pp. 227–233.

Part V
Conclusions

Chapter 16

Summary and Conclusions

16.1 Summary

TiO_2 is a new and preferred material for numerous environment friendly applications, and it has attracted much attention because of its multiple advantages, which include long-term photostability, nontoxicity, low cost, photochemical stability, and strong oxidizing power. Much research, in the past few decades, had been conducted with the specific aim of synthesizing and characterizing nanostructured TiO_2 so that it can be utilized for harvesting solar energy through photocatalysis. In this work, in situ high-temperature synchrotron radiation diffraction was used to determine the effect of air and argon, calcination time, ion implantation, and pressure on phase transformations of nanostructured TiO_2. The as-synthesized TiO_2 nanomaterials were amorphous initially, as shown by the diffraction results. Crystalline anatase and rutile phases formed at elevated temperatures after thermal annealing. The anatase-to-rutile transformation (as determined from phase analysis in different atmospheres) is more rapid in argon than in air because of oxygen vacancies. The quantity of anatase decreased slightly with an increase in isothermal heating, whereas the quantity of rutile increased by phase transformation. The implantation of TiO_2 nanostructures with different dopants affects the phase

Nanostructured Titanium Dioxide in Photocatalysis
It-Meng Low, Hani Manssor Albetran, Victor Manuel de la Prida Pidal, and Fong Kwong Yam

ISBN 978-981-4877-07-7 (Hardcover), 978-1-003-14853-1 (eBook)
www.jennystanford.com

crystallization temperature. The implantation of V and Al ions reduces the crystallization temperature because smaller ions substituted for Ti in the TiO_2 crystal structure by the dopant. A decrease in structural rigidity results in a relaxation of the Ti-bonding environment, but implantation of Cr and In ions resulted in a phase transformation inhibition. Cr and In ions substitute into the Ti sublattice and inhibit the anatase-to-rutile transformation. An improved structural rigidity is provided by In^{3+} and Cr^{3+} ions, which are larger than the Ti^{4+} ion. This increased rigidity restricts relaxation in the Ti-bonding environment. An increased capillary gas pressure resulted with an increase in temperature. The transformation from the amorphous phase to anatase phase was enhanced by nanopowder heating under pressure and that of anatase-to-rutile transformation was reduced. This behavior is believed to occur in an oxygen-rich environment, and interstitial titanium is expected to form when the material is heated under pressure. This suggests that atmospheric oxygen appears to accelerate the amorphous-to-anatase transformation, whereas interstitial titanium inhibits TiO_2 structure relaxation, which is required for the anatase-to-rutile transformation.

16.2 Conclusions

Calcination time and temperature, atmosphere, pressure, and doping are critical parameters that control the crystallization kinetics of nanostructured TiO_2. The major conclusions arising from this work are as follows:

- An easy atomic rearrangement resulted for electrospun TiO_2 nanofibers that have been heated in argon for anatase and rutile lattices because of oxygen vacancies.
- The amount of rutile increased and the anatase decreased with an increase in calcination time.
- V, Al, Cr, and In ions were implanted to determine their effect on the dynamic crystallization behavior of anodized TiO_2 nanotubes or/and electrospun TiO_2 nanofibers.
- The V- and Al-implanted material resulted in an acceleration of the TiO_2 phase transformation compared with the non-implanted sample. An accelerated titania phase transformation and reduction in activation energies resulted

when small V and Al ions were substituted for Ti in the TiO_2 crystal structure.

- In- and Cr-ion implantation inhibited the anatase-to-rutile transformation. In and Cr ions occupy the Ti sublattice substitutionally and inhibit the transformation. The relatively large In^{3+} and Cr^{3+} radius (0.080 nm and 0.0615 nm, respectively) partly replaces the Ti^{4+} (0.061 nm), provides a greater structural rigidity, and allows for no relaxation in the Ti-bonding environment, which yields the difficult anatase-to-rutile transformation in the implanted material.
- The amorphous-to-anatase transformation is accelerated by increasing the oxygen partial pressure in the sealed capillary because of the oxygen-rich environment. The anatase-to-rutile transformation is inhibited by interstitial titanium formation, which limits TiO_2 structure relaxation.

Index